An Epsilon of Room, II

pages from year three of a mathematical blog

Terence Tao

Providence, Rhode Island

2000 *Mathematics Subject Classification.* Primary 00A99.

For additional information and updates on this book, visit
www.ams.org/bookpages/mbk-77

Library of Congress Cataloging-in-Publication Data

Tao, Terence, 1975–
An epsilon of room, II: pages from year three of a mathematical blog / Terence Tao.
p. cm.
Includes bibliographical references and index.
ISBN 978-0-8218-5280-4 (alk. paper)
1. Mathematics—Miscellanea. 2. General—General and miscellaneous specific topics—Miscellaneous topics. I. Title.

QA99.T36 2010
510—22

2010036471

Printed in the United States of America.
Reprinted by the American Mathematical Society, 2023

∞ The paper used in this book is acid-free and falls within the guidelines established to ensure permanence and durability.
Visit the AMS home page at http://www.ams.org/

11 10 9 8 7 6 5 4 3 2 28 27 26 25 24 23

To Garth Gaudry, who set me on the road;

To my family, for their constant support;

And to the readers of my blog, for their feedback and contributions.

Contents

Preface

In February of 2007, I converted my "What's new" web page of research updates into a blog at `terrytao.wordpress.com`. This blog has since grown and evolved to cover a wide variety of mathematical topics, ranging from my own research updates, to lectures and guest posts by other mathematicians, to open problems, to class lecture notes, to expository articles at both basic and advanced levels.

With the encouragement of my blog readers, and also of the American Mathematical Society, I published many of the mathematical articles from the first two years of the blog as [**Ta2008**] and [**Ta2009**], which will henceforth be referred to as *Structure and Randomness* and *Poincaré's Legacies Vols. I, II* throughout this book. This gave me the opportunity to improve and update these articles to a publishable (and citeable) standard, and also to record some of the substantive feedback I had received on these articles by the readers of the blog.

The current text contains many (though not all) of the posts for the third year (2009) of the blog, focusing primarily on those posts of a mathematical nature which were not contributed primarily by other authors, and which are not published elsewhere. It has been split into two volumes.

The first volume (referred to henceforth as *Volume 1*) consisted primarily of lecture notes from my graduate courses on real analysis that I taught at UCLA. The current volume consists instead of sundry articles on a variety of mathematical topics, which I have divided (somewhat arbitrarily) into expository articles (Chapter 1) which are introductory articles on topics of relatively broad interest, and more technical articles (Chapter 2) which are narrower in scope and often related to one of my current research interests.

These can be read in any order, although they often reference each other as well as articles from previous volumes in this series.

A remark on notation

For reasons of space, we will not be able to define every single mathematical term that we use in this book. If a term is italicised for reasons other than emphasis or for definition, then it denotes a standard mathematical object, result, or concept, which can be easily looked up in any number of references. (In the blog version of the book, many of these terms were linked to their Wikipedia pages, or other on-line reference pages.)

I will however mention a few notational conventions that I will use throughout. The cardinality of a finite set E will be denoted $|E|$. We will use the asymptotic notation $X = O(Y)$, $X \ll Y$, or $Y \gg X$ to denote the estimate $|X| \leq CY$ for some absolute constant $C > 0$. In some cases we will need this constant C to depend on a parameter (e.g., d), in which case we shall indicate this dependence by subscripts, e.g., $X = O_d(Y)$ or $X \ll_d Y$. We also sometimes use $X \sim Y$ as a synonym for $X \ll Y \ll X$.

In many situations there will be a large parameter n that goes off to infinity. When that occurs, we also use the notation $o_{n\to\infty}(X)$ or simply $o(X)$ to denote any quantity bounded in magnitude by $c(n)X$, where $c(n)$ is a function depending only on n that goes to zero as n goes to infinity. If we need $c(n)$ to depend on another parameter, e.g., d, we indicate this by further subscripts, e.g., $o_{n\to\infty;d}(X)$.

We will occasionally use the averaging notation $\mathbf{E}_{x\in X} f(x) := \frac{1}{|X|}\sum_{x\in X} f(x)$ to denote the average value of a function $f : X \to \mathbf{C}$ on a nonempty finite set X.

Acknowledgments

The author is supported by a grant from the MacArthur Foundation, by NSF grant DMS-0649473, and by the NSF Waterman Award.

Thanks to Konrad Swanepoel, Blake Stacey, and anonymous commenters for global corrections to the text, and to Edward Dunne at the AMS for encouragement and editing.

Chapter 1

Expository articles

1.1. An explicitly solvable nonlinear wave equation

As is well known, the linear one-dimensional wave equation

$$-\phi_{tt} + \phi_{xx} = 0, \tag{1.1}$$

where $\phi : \mathbf{R} \times \mathbf{R} \to \mathbf{R}$ is the unknown field (which, for simplicity, we assume to be smooth), can be solved explicitly; indeed, the general solution to (1.1) takes the form

$$\phi(t, x) = f(t + x) + g(t - x) \tag{1.2}$$

for some arbitrary (smooth) functions $f, g : \mathbf{R} \to \mathbf{R}$. (One can of course determine f and g once one specifies enough initial data or other boundary conditions, but this will not be the focus of this article.)

When one moves from linear wave equations to nonlinear wave equations, then in general one does not expect to have a closed-form solution such as (1.2). So I was pleasantly surprised recently while playing with the nonlinear wave equation

$$-\phi_{tt} + \phi_{xx} = e^{\phi} \tag{1.3}$$

to discover that this equation can also be explicitly solved in closed form. (For the reason why I was interested in this equation, see [**Ta2010**].)

A posteriori, I now know the reason for this explicit solvability: (1.3) is the limiting case $a = 0, b \to -\infty$ of the more general equation

$$-\phi_{tt} + \phi_{xx} = e^{\phi+a} - e^{-\phi+b}$$

which (after applying the simple transformation

$$\phi = \frac{b-a}{2} + \psi(\sqrt{2}e^{\frac{a+b}{4}}t, \sqrt{2}e^{\frac{a+b}{4}}x))$$

becomes the *sinh-Gordon equation*

$$-\psi_{tt} + \psi_{xx} = \sinh(\psi)$$

(a close cousin of the more famous *sine-Gordon equation* $-\phi_{tt} + \phi_{xx} = \sin(\phi)$), which is known to be completely integrable and exactly solvable. However, I only realised this after the fact and stumbled upon the explicit solution to (1.3) by much more classical and elementary means. I thought I might share the computations here, as I found them somewhat cute, and they seem to serve as an example of how one might go about finding explicit solutions to PDE in general; accordingly, I will take a rather pedestrian approach to describing the hunt for the solution, rather than presenting the shortest or slickest route to the answer.

After the initial publishing of this post, Patrick Dorey pointed out to me that (1.3) is extremely classical; it is known as *Liouville's equation* and was solved by Liouville [**Li1853**], with essentially the same solution as presented here.

1.1.1. Symmetries. To simplify the discussion, let us ignore all issues of regularity, division by zero, taking square roots and logarithms of negative numbers, etc., and proceed for now in a purely formal fashion, pretending that all functions are smooth and lie in the domain of whatever algebraic operations are being performed. (It is not too difficult to go back after the fact and justify these formal computations, but I do not wish to focus on that aspect of the problem here.)

Although not strictly necessary for solving the equation (1.3), I find it convenient to bear in mind the various symmetries that (1.3) enjoys, as this provides a useful "reality check" to guard against errors (e.g., arriving at a class of solutions which is not invariant under the symmetries of the original equation). These symmetries are also useful to normalise various special families of solutions.

One easily sees that solutions to (1.3) are invariant under space-time translations

$$\phi(t,x) \mapsto \phi(t-t_0, x-x_0) \tag{1.4}$$

and also space-time reflections

$$\phi(t,x) \mapsto \phi(\pm t, \pm x). \tag{1.5}$$

Being relativistic, the equation is also invariant under Lorentz transformations

$$\phi(t,x) \mapsto \phi(\frac{t-vx}{\sqrt{1-v^2}}, \frac{x-vt}{\sqrt{1-v^2}}). \tag{1.6}$$

Finally, one has the scaling symmetry

$$\phi(t,x) \mapsto \phi(\lambda t, \lambda x) + 2\log\lambda. \tag{1.7}$$

1.1.2. Solution. Henceforth, ϕ will be a solution to (1.3). In view of the linear explicit solution (1.2), it is natural to move to null coordinates

$$u = t+x, v = t-x,$$

thus

$$\partial_u = \frac{1}{2}(\partial_t + \partial_x); \partial_v = \frac{1}{2}(\partial_t - \partial_x)$$

and (1.3) becomes

$$\phi_{uv} = -\frac{1}{4}e^{\phi}. \tag{1.8}$$

The various symmetries (1.4)–(1.7) can of course be rephrased in terms of null coordinates in a straightforward manner. The Lorentz symmetry (1.6) simplifies particularly nicely in null coordinates, to

$$\phi(u,v) \mapsto \phi(\lambda u, \lambda^{-1} v). \tag{1.9}$$

Motivated by the general theory of stress-energy tensors of relativistic wave equations (of which (1.3) is a very simple example), we now look at the null energy densities ϕ_u^2, ϕ_v^2. For the linear wave equation (1.1) (or equivalently $\phi_{uv} = 0$), these null energy densities are transported in null directions:

$$\partial_v \phi_u^2 = 0; \partial_u \phi_v^2 = 0. \tag{1.10}$$

(One can also see this from the explicit solution (1.2).)

The above transport law is not quite true for the nonlinear wave equation, of course, but we can hope to get some usable substitute. Let us just look at the first null energy ϕ_u^2 for now. By two applications of (1.10), this density obeys the transport equation

$$\begin{aligned}\partial_v \phi_u^2 &= 2\phi_u \phi_{uv} \\ &= -\frac{1}{2}\phi_u e^{\phi} \\ &= -\frac{1}{2}\partial_u(e^{\phi}) \\ &= 2\partial_u \phi_{uv} \\ &= \partial_v(2\phi_{uu}),\end{aligned}$$

and thus we have the pointwise conservation law

$$\partial_v(\phi_u^2 - 2\phi_{uu}) = 0,$$

which implies that

$$-\frac{1}{2}\phi_{uu} + \frac{1}{4}\phi_u^2 = U(u) \tag{1.11}$$

for some function $U : \mathbf{R} \to \mathbf{R}$ depending only on u. Similarly we have

$$-\frac{1}{2}\phi_{vv} + \frac{1}{4}\phi_v^2 = V(v)$$

for some function $V : \mathbf{R} \to \mathbf{R}$ depending only on v.

For any fixed v, (1.11) is a nonlinear ODE in u. To solve it, we can first look at the homogeneous ODE

$$-\frac{1}{2}\phi_{uu} + \frac{1}{4}\phi_u^2 = 0. \tag{1.12}$$

Undergraduate ODE methods (e.g., separation of variables after substituting $\psi := \phi_u$) soon reveal that the general solution to this ODE is given by $\phi(u) = -2\log(u + C) + D$ for arbitrary constants C, D (ignoring the issue of singularities or degeneracies for now). Equivalently, (1.12) is obeyed if and only if $e^{-\phi/2}$ is linear in u. Motivated by this, we become tempted to rewrite (1.11) in terms of $\Phi := e^{-\phi/2}$. One soon realises that

$$\partial_{uu}\Phi = (-\frac{1}{2}\phi_{uu} + \frac{1}{4}\phi_u^2)\Phi,$$

and hence (1.11) becomes

$$(-\partial_{uu} + U(u))\Phi = 0, \tag{1.13}$$

thus Φ is a null (generalised) eigenfunction of the Schrodinger operator (or Hill operator) $-\partial_{uu} + U(u)$. If we let $a(u)$ and $b(u)$ be two linearly independent solutions to the ODE

$$-f_{uu} + Uf = 0, \tag{1.14}$$

we thus have

$$\Phi = a(u)c(v) + b(u)d(v) \tag{1.15}$$

for some functions c, d (which one easily verifies to be smooth, since ϕ, a, b are smooth and a, b are linearly independent). Meanwhile, by playing around with the second null energy density, we have the counterpart to (1.14),

$$(-\partial_{vv} + V(v))\Phi = 0,$$

and hence (by linear independence of a, b) c, d must be solutions to the ODE

$$-g_{vv} + Vg = 0.$$

This would be a good time to pause and see whether our implications are reversible, i.e., whether any ϕ that obeys the relation (1.15) will solve (1.3) or (1.10). It is of course natural to first write (1.10) in terms of Φ. Since

$$\Phi_u = -\frac{1}{2}\phi_u\Phi; \Phi_v = -\frac{1}{2}\phi_v\Phi; \Phi_{uv} = (\frac{1}{4}\phi_u\phi_v - \frac{1}{2}\phi_{uv})\Phi,$$

one soon sees that (1.10) is equivalent to

$$\Phi\Phi_{uv} = \Phi_u\Phi_v + \frac{1}{8}. \tag{1.16}$$

If we then insert the ansatz (1.15), we soon reformulate the above equation as

$$(a(u)b'(u) - b(u)a'(u))(c(v)d'(v) - d(v)c'(v)) = \frac{1}{8}.$$

It is at this time that one should remember the classical fact that if a, u are two solutions to the ODE (1.11), then the *Wronskian* $ab' - ba'$ is constant; similarly $cd' - dc'$ is constant. Putting this all together, we see that

Theorem 1.1.1. *A smooth function ϕ solves* (1.3) *if and only if we have the relation* (1.13) *for some functions a, b, c, d obeying the Wronskian conditions $ab' - ba' = \alpha$, $cd' - dc' = \beta$ for some constants α, β multiplying to $\frac{1}{8}$.*

Note that one can generate solutions to the Wronskian equation $ab' - ba' = \alpha$ by a variety of means, for instance by first choosing a arbitrarily and then rewriting the equation as $(b/a)' = \alpha/a^2$ to recover b. (This does not quite work at the locations when a vanishes, but there are a variety of ways to resolve that; as I said above, we are ignoring this issue for the purposes of this discussion.)

This is not the only way to express solutions. Factoring $a(u)d(v)$ (say) from (1.13), we see that Φ is the product of a solution $c(v)/d(v)+b(u)/a(u)$ to the linear wave equation, plus the exponential of a solution $\log a(u)+\log d(u)$ to the linear wave equation. Thus we may write $\phi = F - 2\log G$, where F and G solve the linear wave equation. Inserting this back ansatz into (1.1), we obtain

$$2(-G_t^2 + G_x^2)/G^2 = e^F/G^2$$

and so we see that

$$\phi = \log \frac{2(-G_t^2 + G_x^2)}{G^2} = \log \frac{-8G_u G_v}{G^2}, \tag{1.17}$$

for some solution G to the free wave equation, and conversely every expression of the form (1.17) can be verified to solve (1.1) (since $\log 2(-G_t^2 + G_x^2)$ does indeed solve the free wave equation, thanks to (1.2)). Inserting (1.2) into (1.17), we thus obtain the explicit solution

$$\phi = \log \frac{-8f'(t+x)g'(t-x)}{(f(t+x)+g(t-x))^2} \tag{1.18}$$

to (1.1), where f and g are arbitrary functions (recall that we are neglecting issues such as whether the quotient and the logarithm are well defined).

I, for one, would not have expected the solution to take this form. But it is instructive to check that (1.18) does at least respect all the symmetries (1.4)–(1.7).

1.1.3. Some special solutions. If we set $U = V = 0$, then a, b, c, d are linear functions, and so Φ is affine linear in u, v. One also checks that the uv term in Φ cannot vanish. After translating in u and v, we end up with the ansatz $\Phi(u, v) = c_1 + c_2 uv$ for some constants c_1, c_2; applying (1.16), we see that $c_1 c_2 = 1/8$, and by using the scaling symmetry (1.7), we may normalise e.g., $c_1 = 8, c_2 = 1$, and so we arrive at the (singular) solution

$$\phi = -2\log(8 + uv) = \log \frac{1}{(8 + t^2 - x^2)^2}. \tag{1.19}$$

To express this solution in the form (1.18), one can take $f(u) = \frac{8}{u}$ and $g(v) = v$; some other choices of f, g are also possible. (Determining the extent to which f, g are uniquely determined by ϕ in general can be established from a closer inspection of the previous arguments; this is left as an exercise.)

We can also look at what happens when ϕ is constant in space, i.e., it solves the ODE $-\phi_{tt} = e^{\phi}$. It is not hard to see that U and V must be constant in this case, leading to a, b, c, d which are either trigonometric or exponential functions. This soon leads to the ansatz $\Phi = c_1 e^{\alpha t} + c_2 e^{-\alpha t}$ for some (possibly complex) constants c_1, c_2, α, thus $\phi = -2\log(c_1 e^{\alpha t} + c_2 e^{-\alpha t})$. By using the symmetries (1.4), (1.7) we can make $c_1 = c_2$ and specify α to be whatever we please, thus leading to the solutions $\phi = -2\log\cosh \alpha t + c_3$. Applying (1.1) we see that this is a solution as long as $e^{c_3} = 2\alpha^2$. For instance, we may fix $c_3 = 0$ and $\alpha = 1/\sqrt{2}$, leading to the solution

$$\phi = -2\log\cosh\frac{t}{\sqrt{2}}. \tag{1.20}$$

To express this solution in the form (1.18), one can take for instance $f(u) = e^{u/\sqrt{2}}$ and $g(v) = e^{-v/\sqrt{2}}$.

One can of course push around (1.19), (1.20) by the symmetries (1.4)–(1.7) to generate a few more special solutions.

Notes. This article first appeared at

`terrytao.wordpress.com/2009/01/22.`

Thanks to Jake K. for corrections.

There was some interesting discussion online regarding whether the heat equation had a natural relativistic counterpart, and more generally whether it was profitable to study nonrelativistic equations via relativistic approximations.

1.2. Infinite fields, finite fields, and the Ax-Grothendieck theorem

Jean-Pierre Serre (whose papers are, of course, always worth reading) recently wrote a lovely article [**Se2009**] in which he describes several ways in which algebraic statements over fields of zero characteristic, such as $\mathbf{C}$, can be deduced from their positive characteristic counterparts such as F_{p^m}, despite the fact that there is no nontrivial field homomorphism between the two types of fields. In particular, finitary tools, including such basic concepts as cardinality, can now be deployed to establish infinitary results. This leads to some simple and elegant proofs of nontrivial algebraic results which are not easy to establish by other means.

One deduction of this type is based on the idea that positive characteristic fields can partially *model* zero characteristic fields, and it proceeds like this: If a certain algebraic statement failed over (say) $\mathbf{C}$, then there should be a "finitary algebraic" obstruction that "witnesses" this failure over $\mathbf{C}$.

Because this obstruction is both finitary and algebraic, it must also be definable in some (large) finite characteristic, thus leading to a comparable failure over a finite characteristic field. Taking contrapositives, one obtains the claim.

Algebra is definitely not my field of expertise, but it is interesting to note that similar themes have also come up in my own area of additive combinatorics (and more generally arithmetic combinatorics), because the combinatorics of addition and multiplication on finite sets is definitely of a "finitary algebraic" nature. For instance, a recent paper of Vu, Wood, and Wood [**VuWoWo2010**] establishes a finitary "Freiman-type" homomorphism from (finite subsets of) the complex numbers to large finite fields that allows them to pull back many results in arithmetic combinatorics in finite fields (e.g., the sum-product theorem) to the complex plane; Van Vu and I also used a similar trick in [**TaVu2007**] to control the singularity property of random sign matrices by first mapping them into finite fields in which cardinality arguments became available). And I have a particular fondness for correspondences between finitary and infinitary mathematics; the correspondence Serre discusses is slightly different from the one I discuss, for instance in Section 1.3 of *Structure and Randomness*, although there seems to be a common theme of "compactness" (or of model theory) tying these correspondences together.

As one of his examples, Serre cites one of my favourite results in algebra, discovered independently by Ax [**Ax1968**] and by Grothendieck [**Gr1966**] (and then rediscovered many times since). Here is a special case of that theorem:

Theorem 1.2.1 (Ax-Grothendieck theorem, special case). *Let $P : \mathbf{C}^n \to \mathbf{C}^n$ be a polynomial map from a complex vector space to itself. If P is injective, then P is bijective.*

The full version of the theorem allows one to replace $\mathbf{C}^n$ by an algebraic variety X over any algebraically closed field, and it allows for P to be an morphism from the algebraic variety X to itself. But for simplicity I will just discuss the above special case. This theorem is not at all obvious; it is not too difficult (see Lemma 1.2.6 below) to show that the *Jacobian* of P is nondegenerate, but this does not come close to solving the problem since one would then be faced with the notorious *Jacobian conjecture*. Also, the claim fails if "polynomial" is replaced by "holomorphic", due to the existence of *Fatou-Bieberbach domains.*

In this post I would like to give the proof of Theorem 1.2.1 based on finite fields as mentioned by Serre, as well as another elegant proof of Rudin [**Ru1995**] that combines algebra with some elementary complex variable methods. (There are several other proofs of this theorem and its generalisations, for instance a topological proof by Borel [**Bo1969**], which I will not discuss here.)

1.2.1. Proof via finite fields. The first observation is that the theorem is utterly trivial in the finite field case:

Theorem 1.2.2 (Ax-Grothendieck theorem in F)**.** *Let F be a finite field, and let $P : F^n \to F^n$ be a polynomial. If P is injective, then P is bijective.*

Proof. Any injection from a finite set to itself is necessarily bijective. (The hypothesis that P is a polynomial is not needed at this stage, but becomes crucial later on.) □

Next, we pass from a finite field F to its algebraic closure $\overline{F}$.

Theorem 1.2.3 (Ax-Grothendieck theorem in $\overline{F}$)**.** *Let F be a finite field, let $\overline{F}$ be its algebraic closure, and let $P : \overline{F}^n \to \overline{F}^n$ be a polynomial. If P is injective, then P is bijective.*

Proof. Our main tool here is *Hilbert's nullstellensatz*, which we interpret here as an assertion that if an algebraic problem is insoluble, then there exists a finitary algebraic obstruction that witnesses this lack of solution (see also Section 1.15 of *Structure and Randomness*). Specifically, suppose for contradiction that we can find a polynomial $P : \overline{F}^n \to \overline{F}^n$ which is injective but not surjective. Injectivity of P means that the algebraic system

$$P(x) = P(y), \quad x \neq y,$$

has no solution over the algebraically closed field $\overline{F}$; by the nullstellensatz, this implies that there must exist an algebraic identity of the form

$$(P(x) - P(y)) \cdot Q(x, y) = (x - y)^r \tag{1.21}$$

for some $r \geq 1$ and some polynomial $Q : \overline{F}^n \times \overline{F}^n \to \overline{F}^n$ that specifically witnesses this lack of solvability. Similarly, lack of surjectivity means the existence of an $z_0 \in \overline{F}^n$ such that the algebraic system

$$P(x) = z_0$$

has no solution over the algebraically closed field $\overline{F}$. By another application of the nullstellensatz, there must exist an algebraic identity of the form

$$(P(x) - z_0) \cdot R(x) = 1 \tag{1.22}$$

for some polynomial $R : \overline{F}^n \to \overline{F}^n$ that specifically witnesses this lack of solvability.

Fix Q, z_0, R as above, and let k be the subfield of $\overline{F}$ generated by F and the coefficients of P, Q, z_0, R. Then we observe (thanks to our explicit witnesses (1.21), (1.22)) that the counterexample P descends from $\overline{F}$ to k; P is a polynomial from k^n to k^n which is injective but not surjective.

But k is finitely generated, and every element of k is algebraic over the finite field F, thus k is finite. But this contradicts Theorem 1.2.2. □

Remark 1.2.4. As pointed out to me by L. Spice, there is a simpler proof of Theorem 1.2.3 that avoids the nullstellensatz: one observes from Theorem 1.2.2 that P is bijective over any finite extension of F that contains all of the coefficients of P, and the claim then follows by taking limits.

The complex case $\mathbf{C}$ follows by a slight extension of the argument used to prove Theorem 1.2.3. Indeed, suppose for contradiction that there is a polynomial $P : \mathbf{C}^n \to \mathbf{C}^n$ which is injective but not surjective. As $\mathbf{C}$ is algebraically closed (the *fundamental theorem of algebra*), we may invoke the nullstellensatz as before and find witnesses (1.21), (1.22) for some Q, z_0, R.

Now let $k = Q[\mathcal{C}]$ be the subfield of $\mathbf{C}$ generated by the rationals $\mathbf{Q}$ and the coefficients $\mathcal{C}$ of P, Q, z_0, R. Then we can descend the counterexample to k. This time, k is not finite, but we can descend it to a finite field (and obtain the desired contradiction) by a number of methods. One approach, which is the one taken by Serre, is to quotient the ring $\mathbf{Z}[\mathcal{C}]$ generated by the above coefficients by a maximal ideal, observing that this quotient is necessarily a finite field. Another is to use a general mapping theorem of Vu, Wood, and Wood [**VuWoWo2010**]. We sketch the latter approach as follows. Being finitely generated, we know that k has a finite *transcendence basis* $\alpha_1, \ldots, \alpha_m$ over $\mathbf{Q}$. Applying the *primitive element theorem*, we can then express k as the finite extension of $\mathbf{Q}[\alpha_1, \ldots, \alpha_m]$ by an element β which is algebraic over $\mathbf{Q}[\alpha_1, \ldots, \alpha_m]$. All the coefficients $\mathcal{C}$ are thus rational combinations of $\alpha_1, \ldots, \alpha_m, \beta$. By rationalising, we can ensure that the denominators of the expressions of these coefficients are integers in $\mathbf{Z}[\alpha_1, \ldots, \alpha_m]$; dividing β by an appropriate power of the product of these denominators, we may assume that the coefficients in $\mathcal{C}$ all lie in the commutative ring $\mathbf{Z}[\alpha_1, \ldots, \alpha_m, \beta]$, which can be identified with the commutative ring $\mathbf{Z}[a_1, \ldots, a_m, b]$ generated by formal indeterminates $a_1, \ldots, a_m, b$, quotiented by the ideal generated by the minimal polynomial $f \in \mathbf{Z}[a_1, \ldots, a_m, b]$ of β; the algebraic identities (1.21), (1.22) then transfer to this ring. Now pick a large prime p, and map $a_1, \ldots, a_m$ to random elements of F_p. With high probability, the image of f (which is now in $F_p[b]$) is nondegenerate; we can then map b to a root of this image in a finite extension of F_p. (In fact, by using the *Chebotarev density theorem* (or Frobenius density theorem), we can place b back in F_p for infinitely many primes p.) This descends the identities (1.21), (1.22) to this finite extension, as desired.

Remark 1.2.5. This argument can be generalised substantially; it can be used to show that any first-order sentence in the language of fields is true in all algebraically closed fields of characteristic zero if and only if it is true for all algebraically closed fields of sufficiently large characteristic. This result can be deduced from the famous result (proved by Tarski [**Ta1951**], and independently, in an equivalent formulation, by Chevalley) that the theory of algebraically closed fields (in the language of rings) admits elimination of quantifiers. See for instance [**PCM**, Section IV.23.4]. There are also analogues for real closed fields, starting with the paper of Bialynicki-Birula and Rosenlicht [**BiRo1962**], with a general result established by Kurdyka [**Ku1999**]. Ax-Grothendieck type properties in other categories have been studied by Gromov [**Gr1999**], who calls this property "surjunctivity".

1.2.2. Rudin's proof. Now we give Rudin's proof, which does not use the nullstellensatz, instead relying on some Galois theory and the topological structure of $\mathbf{C}$. We first need a basic fact:

Lemma 1.2.6. *Let $\Omega \subset \mathbf{C}^n$ be an open set, and let $f : \Omega \to \mathbf{C}^n$ be an injective holomorphic map. Then the Jacobian of f is nondegenerate, i.e.,* $\det Df(z) \neq 0$ *for all $z \in \Omega$.*

Actually, we only need the special case of this lemma when f is a polynomial.

Proof. We use an argument of Rosay [**Ro1982**]. For $n = 1$ the claim follows from Taylor expansion. Now suppose $n > 1$, and the claim is proven for $n - 1$. Suppose for contradiction that $\det Df(z_0) = 0$ for some $z_0 \in \Omega$. We claim that $Df(z_0)$ in fact vanishes entirely. If not, then we can find $1 \leq i, j \leq n$ such that $\frac{\partial}{\partial z_j} f_i(z_0) \neq 0$; by permuting we may take $i = j = 1$. We can also normalise $z_0 = f(z_0) = 0$. Then the map $h : z \mapsto (f_1(z), z_2, \ldots, z_n)$ is holomorphic with nondegenerate Jacobian at 0 and is thus locally invertible at 0. The map $f \circ h^{-1}$ is then holomorphic at 0 and preserves the z_1 coordinate, and thus descends to an injective holomorphic map on a neighbourhood of the origin $\mathbf{C}^{n-1}$, and so its Jacobian is nondegenerate by the induction hypothesis, a contradiction.

We have just shown that the gradient of f vanishes on the zero set $\{\det Df = 0\}$, which is an analytic variety of codimension 1 (if f is polynomial, it is of course an algebraic variety). Thus f is locally constant on this variety, which contradicts injectivity, and we are done. □

From this lemma and the inverse function theorem we have

Corollary 1.2.7. *Injective holomorphic maps from $\mathbf{C}^n$ to $\mathbf{C}^n$ are* open *(i.e., they map open sets to open sets).*

Now we can give Rudin's proof. Let $P : \mathbf{C}^n \to \mathbf{C}^n$ be an injective polynomial. We let k be the field generated by $\mathbf{Q}$ and the coefficients of P; thus P is definable over k. Let $k[z] = k[z_1, \ldots, z_n]$ be the extension of k by n indeterminates $z_1, \ldots, z_n$. Inside $k[z]$ we have the subfield $k[P(z)]$ generated by k and the components of $P(z)$.

We claim that $k[P(z)]$ is all of $k[z]$. For if this were not the case, we see from Galois theory that there is a nontrivial automorphism $\phi : k[z] \to k[z]$ that fixes $k[P(z)]$; in particular, there exists a nontrivial rational (over k) combination $Q(z)/R(z)$ of z such that $P(Q(z)/R(z)) = P(z)$. Now map z to a random complex number in $\mathbf{C}$, which will almost surely be transcendental over the countable field k; this explicitly demonstrates noninjectivity of P, a contradiction.

Since $k[P(z)] = k[z]$, there exists a rational function $Q_j(z)/R_j(z)$ over k for each $j = 1, \ldots, n$ such that $z_j = Q_j(P(z))/R_j(P(z))$. We may of course assume that Q_j, R_j have no common factors.

We have the polynomial identity $Q_j(P(z)) = z_j R_j(P(z))$. In particular, this implies that on the domain $P(\mathbf{C}^n) \subset \mathbf{C}^n$ (which is open by Corollary 1.2.7) the zero set of R_j is contained in the zero set of Q_j. But as Q_j and R_j have no common factors, this is impossible by elementary algebraic geometry; thus R_j is nonvanishing on $P(\mathbf{C}^n)$. Thus the polynomial $R_j \cdot P$ has no zeroes and is thus constant; we may then normalise so that $R_j \cdot P = 1$. Thus we now have $z = Q(P(z))$ for some polynomial Q, which implies that $w = P(Q(w))$ for all w in the open set $P(\mathbf{C}^n)$. But w and $P(Q(w))$ are both polynomials and thus must agree on all of $\mathbf{C}^n$. Thus P is bijective as required.

Remark 1.2.8. Note that Rudin's proof gives the stronger statement that if a polynomial map from $\mathbf{C}^n$ to $\mathbf{C}^n$ is injective, then it is bijective and its inverse is also a polynomial.

Notes. This article first appeared at

`terrytao.wordpress.com/2009/03/07.`

Thanks to fdreher and Ricardo Menares for corrections.

Ricardo Menares and Terry Hughes also mentioned some alternate proofs and generalisations of the Ax-Grothendieck theorem.

1.3. Sailing into the wind or faster than the wind

One of the more unintuitive facts about sailing is that it is possible to harness the power of the wind to sail in a direction against that of the wind or to sail with a speed faster than the wind itself, even when the water itself is calm. It is somewhat less known, but nevertheless true, that one can (in principle) do both at the same time—sail against the wind (even directly against the wind!) at speeds faster than the wind. This does not contradict any laws of physics, such as conservation of momentum or energy (basically because the reservoir of momentum and energy in the wind far outweighs the portion that will be transmitted to the sailboat), but it is certainly not obvious at first sight how it is to be done.

The key is to exploit all three dimensions of space when sailing. The most obvious dimension to exploit is the *windward/leeward* dimension—the direction that the wind velocity v_0 is oriented in. But if this is the only dimension one exploits, one can only sail up to the wind speed $|v_0|$ and no faster, and it is not possible to sail in the direction opposite to the wind.

Things get more interesting when one also exploits the *crosswind* dimension perpendicular to the wind velocity, in particular by trimming the "sails". If one does this, then (in principle) it becomes possible to travel up to double the speed $|v_0|$ of wind, as we shall see below.

However, one still cannot sail against the wind purely by trimming the sails. To do this, one needs to not just harness the power of the wind but also that of the water *beneath* the sailboat, thus exploiting (barely) the third available dimension. By combining the use of a sail in the air with the use of sails *in the water*—better known as *keels*, *rudders*, and *hydrofoils*—one can now sail in certain directions against the wind and at certain speeds. In most sailboats, one relies primarily on the keel, which lets one sail against the wind but not directly opposite it. But if one trims the rudder or other hydrofoils as well as the sail, then in fact one can (in principle) sail in arbitrary directions (including those directly opposite to v_0) and at arbitrary speeds (even those much larger than $|v_0|$), although it is quite difficult to actually achieve this in practice. It may seem odd that the water, which we are assuming to be calm (i.e., traveling at zero velocity), can be used to increase the range of available velocities and speeds for the sailboat, but we shall see shortly why this is the case.

If one makes several simplifying and idealised (and, admittedly, rather unrealistic in practice) assumptions in the underlying physics, then sailing can in fact be analysed by a simple two-dimensional geometric model which explains all of the above statements. In this article, I would like to describe this mathematical model and how it gives the conclusions stated above.

1.3.1. One-dimensional sailing. Let us first begin with the simplest case of one-dimensional sailing, in which the sailboat lies in a one-dimensional universe (which we describe mathematically by the real line $\mathbf{R}$). To begin with, we will ignore the friction effects of the water (one might imagine sailing on an iceboat rather than a sailboat). We assume that the air is

Figure 1. The effect of a sail in one dimension.

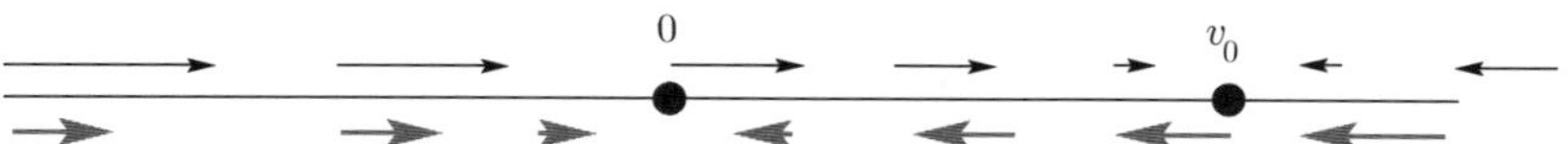

Figure 2. The effects of a sail and an anchor in one dimension.

blowing at a constant velocity $v_0 \in \mathbf{R}$, which for sake of discussion we shall take to be positive. We also assume that one can do precisely two things with a sailboat: one can either *furl* the sail, in which case the wind does not propel the sailboat at all, or one can *unfurl* the sail, in order to exploit the force of the wind.

When the sail is furled, then (ignoring friction), the velocity v of the boat stays constant, as per *Newton's first law*. When instead the sail is unfurled, the motion is instead governed by *Newton's second law*, which among other things asserts that the velocity v of the boat will be altered in the direction of the net force exerted by the sail. This net force (which, in one dimension, is purely a *drag force*) is determined not by the true wind speed v_0 as measured by an observer at rest, but by the *apparent* wind speed $v_0 - v$ as experienced by the boat, as per the (Galilean) *principle of relativity*. (Indeed, Galileo himself supported this principle with a famous thought experiment on a ship.) Thus, the sail can increase the velocity v when $v_0 - v$ is positive, and decrease it when $v_0 - v$ is negative. We can illustrate the effect of an unfurled sail by a vector field in velocity space (Figure 1).

The line here represents the space of all possible velocities v of a boat in this one-dimensional universe, including the rest velocity 0 and the wind velocity v_0. The vector field at any given velocity v represents the direction the velocity will move in if the sail is unfurled. We thus see that the effect of unfurling the sail will be to move the velocity of the sail towards v. Once one is at that speed, one is stuck there; neither furling nor unfurling the sail will affect one's velocity again in this frictionless model.

Now let us reinstate the role of the water. We will use the crudest example of a water sail, namely an *anchor*. When the anchor is raised, we assume that we are back in the frictionless situation above; but when the anchor is dropped (so that it is dragging in the water), it exerts a force on the boat which is in the direction of the apparent velocity $0 - v$ of the water with respect to the boat, and which (ideally) has a magnitude proportional to the square of the apparent speed $|0 - v|$, thanks to the drag equation. This gives a second vector field in velocity space that one is able to effect on the boat (displayed here as thick blue arrows); see Figure 2.

It is now apparent that by using either the sail or the anchor, one can reach any given velocity between 0 and v_0. However, once one is in this range, one cannot use the sail and anchor to move faster than v_0 or to move at a negative velocity.

1.3.2. Two-dimensional sailing. Now let us sail in a two-dimensional plane $\mathbf{R}^2$, thus the wind velocity v_0 is now a vector in that plane. To begin with, let us again ignore the friction effects of the water (e.g., imagine one is iceboating on a two-dimensional frozen lake).

With the square-rigged sails of the ancient era, which could only exploit drag, the net force exerted by an unfurled sail in two dimensions followed essentially the same law as in the one-dimensional case, i.e., the force was always proportional to the relative velocity $v_0 - v$ of the wind and the ship, thus leading to the black vector field in Figure 3.

We thus see that, starting from rest $v = 0$, the only thing one can do with such a sail is move the velocity v along the line segment from 0 to v_0, at which point one is stuck (unless one can exploit water friction, e.g., via an anchor, to move back down that line segment to 0). No crosswind velocity is possible at all with this type of sail.

With the invention of the curved sail, which redirects the (apparent) wind velocity $v_0 - v$ to another direction rather than stalling it to zero, it

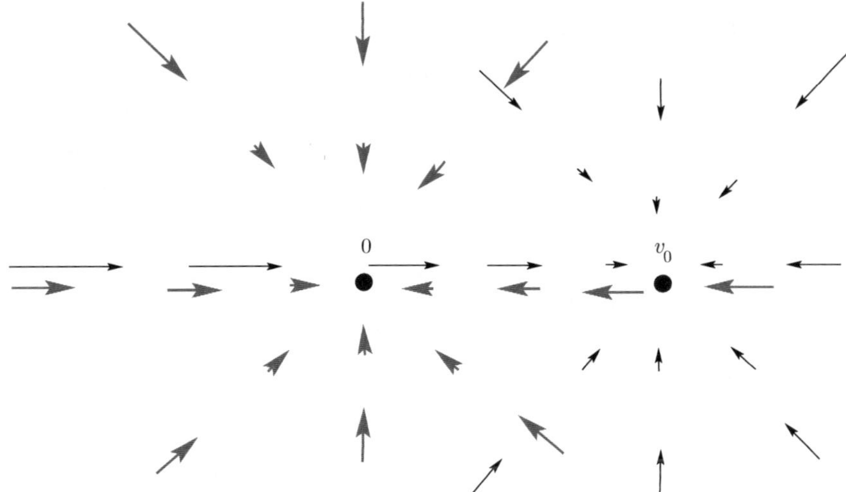

Figure 3. The effects of a pure-drag sail (black) and an anchor (blue) in two dimensions.

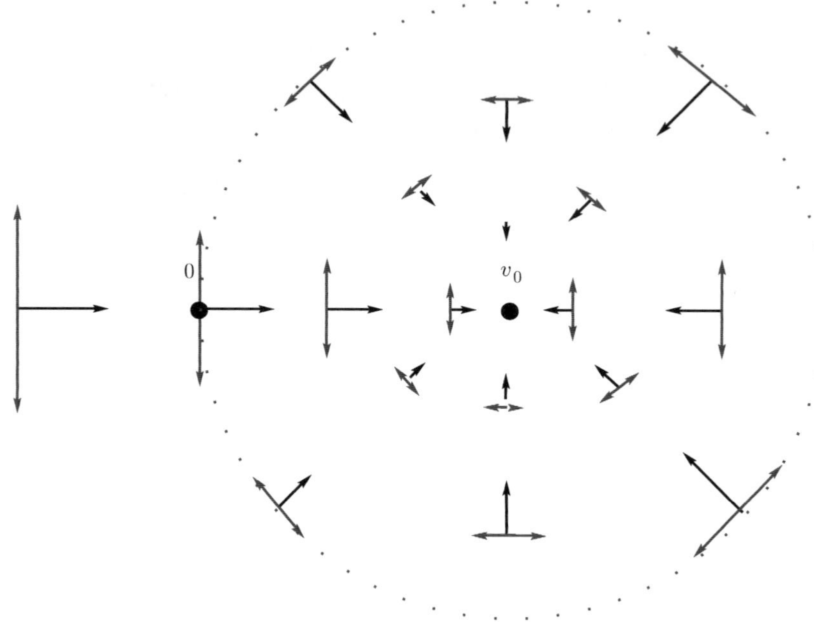

Figure 4. The effect of a pure-drag sail (black) and a pure-lift sail (red) in two dimensions. The disk enclosed by the dotted circle represents the velocities one can reach from these sails starting from the rest velocity $v = 0$.

became possible for sails to provide a *lift force*[1] which is essentially perpendicular to the (apparent) wind velocity, in contrast to the drag force that is parallel to that velocity. (Not coincidentally, such a sail has essentially the same aerofoil shape as an airplane wing, and one can also explain the lift force via *Bernoulli's principle.*)

By setting the sail in an appropriate direction, one can now use the lift force to adjust the velocity v of a sailboat in directions perpendicular to the apparent wind velocity $v_0 - v$, while using the drag force to adjust v in directions parallel to this apparent velocity; of course, one can also adjust the velocity in all intermediate directions by combining both drag and lift. This leads to the vector fields displayed in red in Figure 4.

Note that no matter how one orients the sail, the apparent wind speed $|v_0 - v|$ will decrease (or at best stay constant); this can also be seen from the law of conservation of energy in the reference frame of the wind. Thus, starting from rest and using only the sail, one can only reach speeds in the circle centred at v_0 with radius $|v_0|$ (i.e., the circle in Figure 4); thus one

[1]Despite the name, the lift force is not a vertical force in this context, but instead a horizontal one; in general, lift forces are basically perpendicular to the orientation of the aerofoil providing the lift. Unlike airplane wings, sails are vertically oriented, so the lift will be horizontal in this case.

cannot sail against the wind, but one can at least reach speeds of twice the wind speed, at least in principle.[2]

Remark 1.3.1. If all one has to work with is the air sail(s), then one cannot do any better than what is depicted in Figure 4, no matter how complicated the rigging. This can be seen by looking at the law of conservation of energy in the reference frame of the wind. In that frame, the air is at rest and thus has zero kinetic energy, while the sailboat has kinetic energy $\frac{1}{2}m|v_0|^2$. The water in this frame has an enormous reservoir of kinetic energy, but if one is not allowed to interact with this water, then the kinetic energy of the boat cannot exceed $\frac{1}{2}m|v_0|^2$ in this frame, and so the boat velocity is limited to the region inside the dotted circle. In particular, no arrangement of sails can give a negative drag force.

1.3.3. Three-dimensional sailing. Now we can turn to three-dimensional sailing, in which the sailboat is still largely confined to $\mathbf{R}^2$ but one can use both air sails and water sails as necessary to control the velocity v of the boat.[3]

As mentioned earlier, the crudest example of a water sail is an anchor, which, when dropped, exerts a pure drag force in the direction of $0-v$ on the boat; this is displayed as the blue vector field in Figure 3. Comparing this with Figure 4 (which describes all the forces available from using the air sail), we see that such a device does not increase the range of velocities attainable from a boat starting at rest (although it does allow a boat moving with the wind to return to rest, as in the one-dimensional setting). Unsurprisingly, anchors are not used all that much for sailing in practice.

However, we can do better by using other water sails. For instance, the *keel* of a boat is essentially a water sail oriented in the direction of the boat (which in practice is kept close to parallel to v, e.g., by use of the rudder, else one would encounter substantial (and presumably unwanted) water drag and torque effects). The effect of the keel is to introduce significant resistance to any lateral movement of the boat. Ideally, the effect this has on the net force acting on the boat is that it should orthogonally project that force to be parallel to the direction of the boat (which, as stated before, is usually parallel to v). Applying this projection to the vector fields arising from the air sail, we obtain some new vector fields along which we can modify the boat's velocity; see Figure 5.

---

[2]In practice, friction effects of air and water, such as wave making resistance, and the difficulty in forcing the sail to provide purely lift and no drag, mean that one cannot quite reach this limit, but it has still been possible to exceed the wind speed with this type of technique.

[3]Some boats do in fact exploit the third dimension more substantially than this, e.g., using sails to vertically lift the boat to reduce water drag, but we will not discuss these more advanced hull designs here.

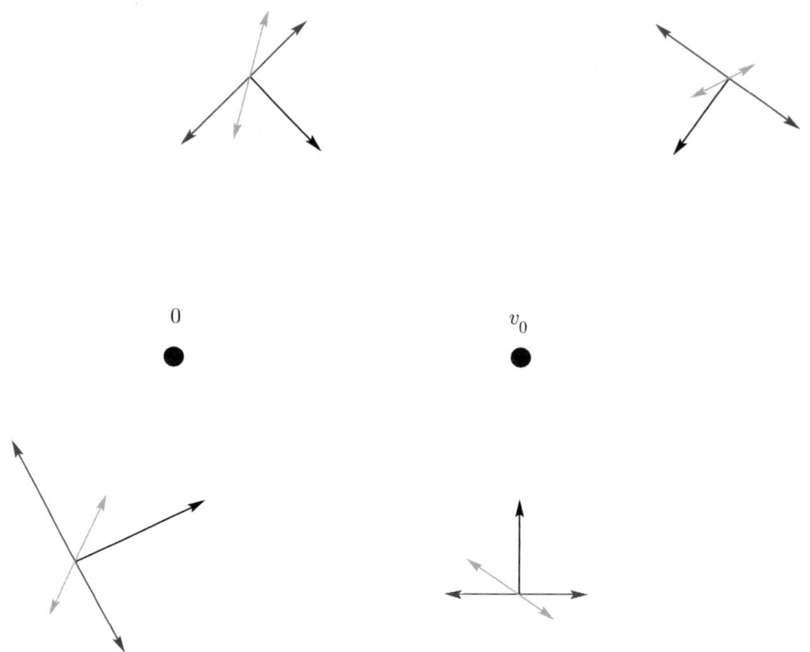

Figure 5. The effect of a pure-drag sail (black), a pure-lift sail (red), and a pure-lift sail combined with a keel (green). Note that one now has the ability to shift the velocity v away from both 0 and v_0 no matter how fast one is already traveling, so long as v is not collinear with 0 and v_0.

In particular, it becomes possible to sail against the wind, or faster than the wind, so long as one is moving at a nontrivial angle to the wind (i.e., v is not parallel to v_0 or $-v_0$).

What is going on here is as follows: By using lift instead of drag, and trimming the sail appropriately, one can make the force exerted by the sail be at any angle of up to 90° from the actual direction of apparent wind. By then using the keel, one can make the net force on the boat be at any angle up to 90° from the force exerted by the sail. Putting the two together, one can create a force on the boat at any angle up to 180° from the apparent wind speed—i.e., in any direction other than directly against the wind. (In practice, because it is impossible have a pure lift force free of drag and because the keel does not perfectly eliminate all lateral forces, most sailboats can only move at angles up to about 135° or so from the apparent wind direction, though one can then create a net movement at larger angles by tacking and beating. For similar reasons, water drag prevents one from using these methods to move too much faster than the wind speed.)

In theory, one can also sail at any desired speed and direction by combining the use of an air sail (or aerofoil) with the use of a water sail (or

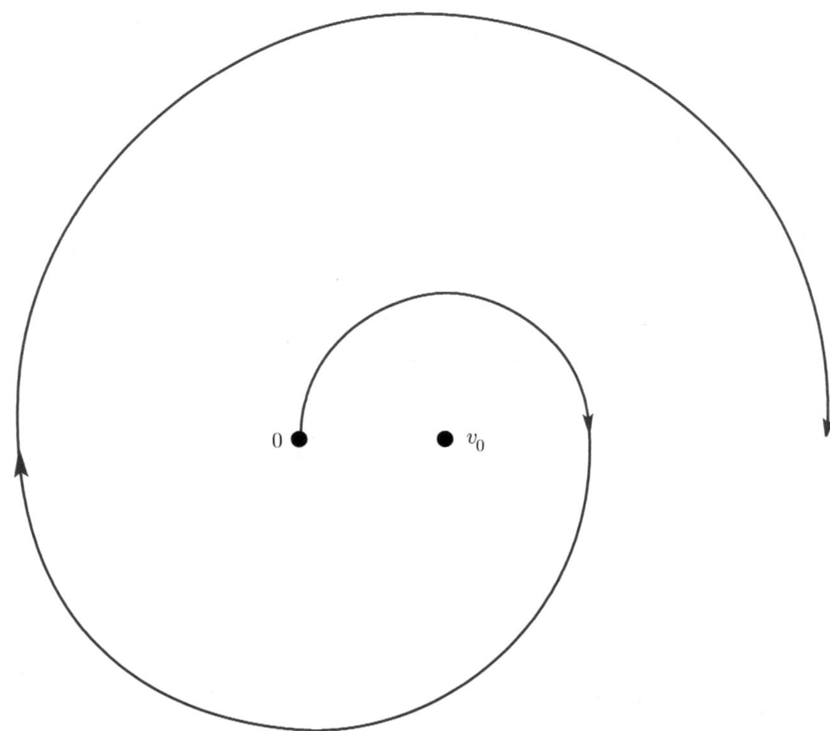

Figure 6. By alternating between a pure-lift aerofoil (red) and a pure-lift hydrofoil (purple), one can in principle reach arbitrarily large speeds in any direction.

hydrofoil). While water is a rather different fluid from air in many respects (it is far denser, and virtually incompressible), one could in principle deploy hydrofoils to exert lift forces on a boat perpendicular to the apparent water velocity $0 - v$, much as an aerofoil can be used to exert lift forces on the boat perpendicular to the apparent wind velocity $v_0 - v$. We saw in the previous section that if the effects of air resistance could somehow be ignored, then one could use lift to alter the velocity v along a circle centred at the true wind speed v_0; similarly, if the effects of water resistance could also be ignored (e.g., by *planing*, which reduces, but does not completely eliminate, these effects), then one could alter the velocity v along a circle centred at the true water speed 0. By alternately using the aerofoil and hydrofoil, one could in principle reach arbitrarily large speeds and directions, as illustrated in Figure 6.

I do not know however if one could actually implement such a strategy with a physical sailing vessel. (Iceboats, however, have been known to reach speeds of up to six times the wind speed or more, though not exactly by the technique indicated in Figure 6. Thanks to kanyonman for this fact.)

It is reasonable (in light of results such as the Kutta-Joukowski theorem) to assume that the amount of lift provided by an aerofoil or hydrofoil is linearly proportional to the apparent wind speed or water speed. If so, then some basic trigonometry reveals that (assuming negligible drag) one can use either of the above techniques to increase one's speed at what is essentially a constant rate; in particular, one can reach speeds of $n|v_0|$ for any $n > 0$ in time $O(n)$. On the other hand, as drag forces are quadratically proportional to apparent wind or water speed, one can decrease one's speed at a very rapid rate simply by dropping anchor; in fact one can drop speed from $n|v_0|$ to $|v_0|$ in bounded time $O(1)$ no matter how large n is! (This fact is the time reversal of the well-known fact that the Riccati ODE $u' = u^2$ blows up in finite time.) These appear to be the best possible rates for acceleration or deceleration using only air and water sails, though I do not have a formal proof of this fact.

Notes. This article first appeared at

`terrytao.wordpress.com/2009/03/23.`

Izabella Łaba pointed out several real-world sailing features not covered by the above simplified model, notably the interaction between multiple sails, and noted that the model was closer in many ways to windsurfing (or iceboating) than to traditional sailing.

Meichenl pointed out the relevance of the drag equation.

1.4. The completeness and compactness theorems of first-order logic

The famous *Gödel completeness theorem* in logic (not to be confused with the even more famous *Gödel incompleteness theorem*) roughly states the following:

Theorem 1.4.1 (Gödel completeness theorem, informal statement). *Let Γ be a* theory *(a* formal language *$\mathcal{L}$, together with a set of axioms, i.e.,* sentences *assumed to be true), and let ϕ be a sentence in the formal language. Assume also that the language $\mathcal{L}$ has at most countably many symbols. Then the following are equivalent:*

(i) Syntactic consequence. *ϕ can be deduced from the axioms in Γ by a finite number of applications of the* laws of deduction *in first-order logic. (This property is abbreviated as $\Gamma \vdash \phi$.)*

(ii) Semantic consequence. *Every* structure *$\mathfrak{U}$ which* satisfies *or* models *Γ, also satisfies ϕ. (This property is abbreviated as $\Gamma \models \phi$.)*

(iii) Semantic consequence for at most countable models. *Every structure $\mathfrak{U}$ which is at most countable and which models Γ, also satisfies ϕ.*

One can also formulate versions of the completeness theorem for languages with uncountably many symbols, but I will not do so here. One can also force other cardinalities on the model $\mathfrak{U}$ by using the *Löwenheim-Skolem theorem*.

To state this theorem even more informally, any (first-order) result which is true in *all* models of a theory, must be logically deducible from that theory, and vice versa. (For instance, any result which is true for all groups, must be deducible from the group axioms; any result which is true for all systems obeying *Peano arithmetic*, must be deducible from the Peano axioms; and so forth.) In fact, it suffices to check countable and finite models only. For instance, any first-order statement which is true for all finite or countable groups, is in fact true for all groups! Informally, a first-order language with only countably many symbols cannot "detect" whether a given structure is countably or uncountably infinite. Thus for instance even the *Zermelo-Frankel-Choice (ZFC)* axioms of set theory must have some at most countable model, even though one can use ZFC to prove the existence of uncountable sets; this is known as *Skolem's paradox*. (To resolve the paradox, one needs to carefully distinguish between an object in a set theory being "externally" countable in the structure that models that theory and being "internally" countable within that theory.)

Of course, a theory Γ may contain *undecidable* statements ϕ—sentences which are neither provable nor disprovable in the theory. By the completeness theorem, this is equivalent to saying that ϕ is satisfied by some models of Γ but not by other models. Thus the completeness theorem is compatible with the incompleteness theorem: *recursively enumerable* theories such as Peano arithmetic are modeled by the natural numbers $\mathbf{N}$, but are also modeled by other structures, and there are sentences satisfied by $\mathbf{N}$ that are not satisfied by other models of Peano arithmetic which are thus undecidable within that arithmetic.

An important corollary of the completeness theorem is the *compactness theorem*:

Corollary 1.4.2 (Compactness theorem, informal statement). *Let Γ be a first-order theory whose language has at most countably many symbols. Then the following are equivalent:*

(i) *Γ is* consistent, *i.e., it is not possible to logically deduce a contradiction from the axioms in Γ.*

(ii) *Γ is* satisfiable, *i.e., there exists a structure $\mathfrak{U}$ that models Γ.*

(iii) *There exists a structure* $\mathfrak{U}$ *which is at most countable, that models* Γ.

(iv) *Every finite subset* Γ' *of* Γ *is consistent.*

(v) *Every finite subset* Γ' *of* Γ *is satisfiable.*

(vi) *Every finite subset* Γ' *of* Γ *is satisfiable with an at most countable model.*

Indeed, the equivalence of (i)–(iii), or (iv)–(vi), follows directly from the completeness theorem, while the equivalence of (i) and (iv) follows from the fact that any logical deduction has finite length and so can involve at most finitely many of the axioms in Γ. (Again, the theorem can be generalised to uncountable languages, but the models become uncountable also.)

There is a consequence of the compactness theorem which more closely resembles the sequential concept of compactness. Let a sequence $\mathfrak{U}_1, \mathfrak{U}_2, \ldots$ be a sequence of structures for $\mathcal{L}$, and given another structure $\mathfrak{U}$ for $\mathcal{L}$, let us say that $\mathfrak{U}_n$ *converges elementarily* to $\mathfrak{U}$ if every sentence ϕ which is satisfied by $\mathfrak{U}$ is also satisfied by $\mathfrak{U}_n$ for sufficiently large n. (Replacing ϕ by its negation $\neg\phi$, we also see that every sentence that is not satisfied by $\mathfrak{U}$, is not satisfied by $\mathfrak{U}_n$ for sufficiently large n.) Note that the limit $\mathfrak{U}$ is only unique up to *elementary equivalence*. Clearly, if each of the $\mathfrak{U}_n$ models some theory Γ, then the limit $\mathfrak{U}$ will also; thus for instance the elementary limit of a sequence of groups is still a group, the elementary limit of a sequence of rings is still a ring, etc.

Corollary 1.4.3 (Sequential compactness theorem). *Let* $\mathcal{L}$ *be a language with at most countably many symbols, and let* $\mathfrak{U}_1, \mathfrak{U}_2, \ldots$ *be a sequence of structures for* $\mathcal{L}$. *Then there exists a subsequence* $\mathfrak{U}_{n_j}$ *which converges elementarily to a limit* $\mathfrak{U}$ *which is at most countable.*

Proof. For each structure $\mathfrak{U}_n$, let $\mathrm{Th}(\mathfrak{U}_n)$ be the theory of that structure, i.e., the set of all sentences that are satisfied by that structure. One can view that theory as a point in $\{0,1\}^{\mathcal{S}}$, where $\mathcal{S}$ is the set of all sentences in the language $\mathcal{L}$. Since $\mathcal{L}$ has at most countably many symbols, $\mathcal{S}$ is at most countable, and so (by the sequential *Tychonoff theorem*) $\{0,1\}^{\mathcal{S}}$ is sequentially compact in the product topology. (This can also be seen directly by the usual *Arzelá-Ascoli* diagonalisation argument.) Thus we can find a subsequence $\mathrm{Th}(\mathfrak{U}_{n_j})$ which converges in the product topology to a limit theory $\Gamma \in \{0,1\}^{\mathcal{S}}$, thus every sentence in Γ is satisfied by $\mathfrak{U}_{n_j}$ for sufficiently large j (and every sentence not in Γ is not satisfied by $\mathfrak{U}_{n_j}$ for sufficiently large j). In particular, any finite subset of Γ is satisfiable, hence consistent; by the compactness theorem, Γ itself is therefore consistent, and has an at most countable model $\mathfrak{U}$. Also, each of the theories $\mathrm{Th}(\mathfrak{U}_{n_j})$ is clearly complete (given any sentence ϕ, either ϕ or $\neg\phi$ is in the theory), and so Γ is complete as well. One concludes that Γ is the theory of $\mathfrak{U}$, and hence $\mathfrak{U}$ is the elementary limit of the $\mathfrak{U}_{n_j}$ as claimed. □

Remark 1.4.4. It is also possible to state the compactness theorem using the topological notion of compactness, as follows: Let X be the space of all structures of a given language $\mathcal{L}$, quotiented by elementary equivalence. One can define a topology on X by taking the sets $\{\mathfrak{U} \in X : \mathfrak{U} \models \phi\}$ as a subbase, where ϕ ranges over all sentences. Then the compactness theorem is equivalent to the assertion that X is topologically compact.

One can use the sequential compactness theorem to build a number of interesting "nonstandard" models to various theories. For instance, consider the language $\mathcal{L}$ used by Peano arithmetic (which contains the operations $+, \times$ and the successor operation S, the relation $=$, and the constant 0), and adjoint a new constant N to create an expanded language $\mathcal{L} \cup \{N\}$. For each natural number $n \in \mathbf{N}$, let $\mathbf{N}_n$ be a structure for $\mathcal{L} \cup \{N\}$ which consists of the natural numbers $\mathbf{N}$ (with the usual interpretations of $+$, $\times$, etc.) and interprets the symbol N as the natural number n. By the compactness theorem, some subsequence of $\mathbf{N}_n$ must converge elementarily to a new structure $*\mathbf{N}$ of $\mathcal{L} \cup \{N\}$, which still models Peano arithmetic, but now has the additional property that $N > n$ for every (standard) natural number n. Thus we have managed to create a nonstandard model of Peano arithmetic which contains a nonstandardly large number (one which is larger than every standard natural number).

The sequential compactness theorem also lets us construct infinitary limits of various sequences of finitary objects; for instance, one can construct infinite pseudo-finite fields as the elementary limits of sequences of finite fields. It also appears to be related to a number of *correspondence principles* between finitary and infinitary objects, such as the Furstenberg correspondence principle between sets of integers and dynamical systems, or the more recent correspondence principles concerning graph limits.

In this article, I will review the proof of the completeness (and hence compactness) theorem. The material here is quite standard (I basically follow the usual proof of Henkin and take advantage of *Skolemisation*), but I wish to popularise the notion of an *elementary limit*, which is not particularly well known.[4]

[4]The closely related concept of an *ultraproduct* is better known and can be used to prove most of the compactness theorem already, thanks to *Los's theorem*, but I do not know how to use ultraproducts to ensure that the limiting model is countable. However, one can think (intuitively, at least), of the limit model $\mathfrak{U}$ in the above theorem as being the set of "constructible" elements of an ultraproduct of the $\mathfrak{U}_n$.

In order to emphasise the main ideas in the proof, I will gloss over some of the more technical details in the proofs, relying instead on informal arguments and examples at various points.

1.4.1. Completeness and compactness in propositional logic. The completeness and compactness theorems are results in first-order logic. But to motivate some of the ideas in proving these theorems, let us first consider the simpler case of *propositional logic*. The language $\mathcal{L}$ of a propositional logic consists of the following:

- A finite or infinite collection $A_1, A_2, A_3, \ldots$ of *propositional variables*—*atomic formulae* which could be true or false, depending on the interpretation.
- A collection of *logical connectives*, such as *conjunction* $\wedge$, *disjunction* $\vee$, *negation* $\neg$, or *implication* $\implies$. (The exact choice of which logical connectives to include in the language is to some extent a matter of taste.)
- Parentheses (in order to indicate the order of operations).

Of course, we assume that the symbols used for atomic formulae are distinct from those used for logical connectives, or for parentheses; we will implicitly make similar assumptions of this type in later sections without further comment.

Using this language, one can form *sentences* (or *formulae*) by some standard formal rules which I will not write down here. Typical examples of sentences in propositional logic are $A_1 \implies (A_2 \vee A_3)$, $(A_1 \wedge \neg A_1) \implies A_2$, and $(A_1 \wedge A_2) \vee (A_1 \wedge A_3)$. Each sentence is of finite length, and thus involves at most finitely many of the propositional variables. Observe that if $\mathcal{L}$ is at most countable, then there are at most countably many sentences.

The analogue of a structure in propositional logic is a *truth assignment*. A truth assignment $\mathfrak{U}$ for a propositional language $\mathcal{L}$ consists of a truth value $A_n^{\mathfrak{U}} \in \{\text{true}, \text{false}\}$ assigned to each propositional variable A_n. (Thus, for instance, if there are N propositional variables in the language, then there are 2^N possible truth assignments.) Once a truth assignment $\mathfrak{U}$ has assigned a truth value $A_n^{\mathfrak{U}}$ to each propositional variable A_n, it can then assign a truth value $\phi^{\mathfrak{U}}$ to any other sentence ϕ in the language $\mathcal{L}$ by using the usual truth tables for conjunction, negation, etc.; we write $\mathfrak{U} \models \phi$ if $\mathfrak{U}$ assigns a true value to ϕ (and say that ϕ is *satisfied* by $\mathfrak{U}$), and $\mathfrak{U} \not\models \phi$ otherwise. Thus, for instance, if $A_1^{\mathfrak{U}} = \text{false}$ and $A_2^{\mathfrak{U}} = \text{true}$, then $\mathfrak{U} \models A_1 \vee A_2$ and $\mathfrak{U} \models A_1 \implies A_2$, but $\mathfrak{U} \not\models A_2 \implies A_1$. Some sentences, e.g., $A_1 \vee \neg A_1$, are true in every truth assignment; these are the (semantic) *tautologies*. At the other extreme, the negation of a tautology will of course be false in every truth assignment.

A *theory* Γ is a language $\mathcal{L}$, together with a (finite or infinite) collection of sentences (also called Γ) in that language. A truth assignment $\mathfrak{U}$ *satisfies* (or *models*) the theory Γ, and we write $\mathfrak{U} \models \Gamma$, if we have $\mathfrak{U} \models \phi$ for all $\phi \in \Gamma$. Thus, for instance, if $\mathfrak{U}$ is as in the preceding example and $\Gamma := \{A_1, A_1 \implies A_2\}$, then $\mathfrak{U} \models \Gamma$.

The analogue of the Gödel completeness theorem is then

Theorem 1.4.5 (Completeness theorem for propositional logic). *Let Γ be a theory for a propositional language $\mathcal{L}$, and let ϕ be a sentence in $\mathcal{L}$. Then the following are equivalent:*

(i) Syntactic consequence. *ϕ can be deduced from the axioms in Γ by a finite number of applications of the* laws of propositional logic.

(ii) Semantic consequence. *Every truth assignment $\mathfrak{U}$ which satisfies (or models) Γ, also satisfies ϕ.*

One can list a complete set of laws of propositional logic used in (i), but we will not do so here.

To prove the completeness theorem, it suffices to show the following equivalent version.

Theorem 1.4.6 (Completeness theorem for propositional logic, again). *Let Γ be a theory for a propositional language $\mathcal{L}$. Then the following are equivalent:*

(i) *Γ is* consistent, *i.e., it is not possible to logically deduce a contradiction from the axioms in Γ.*

(ii) *Γ is* satisfiable, *i.e., there exists a truth assignment $\mathfrak{U}$ that models Γ.*

Indeed, Theorem 1.4.5 follows from Theorem 1.4.6 by applying Theorem 1.4.6 to the theory $\Gamma \cup \{\neg\phi\}$ and taking contrapositives.

It remains to prove Theorem 1.4.6. It is easy to deduce (i) from (ii), because the laws of propositional logic are *sound*: given any truth assignment, it is easy to verify that these laws can only produce true conclusions given true hypotheses. The more interesting implication is to obtain (ii) from (i)—given a consistent theory Γ, one needs to produce a truth assignment that models that theory.

Let us first consider the case when the propositional language $\mathcal{L}$ is finite, so that there are only finitely many propositional variables $A_1, \ldots, A_N$. Then we can argue using the following "greedy algorithm".

- We begin with a consistent theory Γ.

- Observe that at least one of $\Gamma \cup \{A_1\}$ or $\Gamma \cup \{\neg A_1\}$ must be consistent. For if both $\Gamma \cup \{A_1\}$ and $\Gamma \cup \{\neg A_1\}$ led to a logical contradiction, then by the laws of logic one can show that Γ must also lead to a logical contradiction.
- If $\Gamma \cup \{A_1\}$ is consistent, we set $A_1^{\mathfrak{U}} :=$ true and $\Gamma_1 := \Gamma \cup \{A_1\}$; otherwise, we set $A_1^{\mathfrak{U}} :=$ false and $\Gamma_1 := \Gamma \cup \{\neg A_1\}$.
- Γ_1 is consistent, so arguing as before, we know that at least one of $\Gamma_1 \cup \{A_2\}$ or $\Gamma_1 \cup \{\neg A_2\}$ must be consistent. If the former is consistent, we set $A_2^{\mathfrak{U}} :=$ true and $\Gamma_2 := \Gamma_1 \cup \{A_2\}$; otherwise set $A_2^{\mathfrak{U}} :=$ false and $\Gamma_2 := \Gamma_1 \cup \{\neg A_2\}$.
- We continue in this fashion, eventually ending up with a consistent theory Γ_N containing Γ, and a complete truth assignment $\mathfrak{U}$ such that $A_n \in \Gamma_N$ whenever $1 \leq n \leq N$ is such that $A_n^{\mathfrak{U}} =$ true, and such that $\neg A_n \in \Gamma_N$ whenever $1 \leq n \leq N$ is such that $A_n^{\mathfrak{U}} =$ false.
- From the laws of logic and an induction argument, one then sees that if ϕ is any sentence with $\phi^{\mathfrak{U}} =$ true, then ϕ is a logical consequence of Γ_N, and hence (since Γ_N is consistent) $\neg\phi$ is not a consequence of Γ_N. Taking contrapositives, we see that $\phi^{\mathfrak{U}} =$ false whenever $\neg\phi$ is a consequence of Γ_N; replacing ϕ by $\neg\phi$, we conclude that $\mathfrak{U}$ satisfies every sentence in Γ_N, and the claim follows.

Remark 1.4.7. The above argument shows in particular that any finite theory either has a model or a proof of a contradictory statement (such as $A \wedge \neg A$). Actually producing a model if it exists, though, is essentially the infamous *satisfiability problem*, which is known to be NP-complete, and thus (if $P \neq NP$) would require superpolynomial time to execute.

The case of an infinite language can be obtained by combining the above argument with *Zorn's lemma* (or *transfinite induction* and the axiom of choice, if the set of propositional variables happens to be well ordered). Alternatively, one can proceed by establishing

Theorem 1.4.8 (Compactness theorem for propositional logic)**.** *Let Γ be a theory for a propositional language $\mathcal{L}$. Then the following are equivalent:*

(i) *Γ is satisfiable.*

(ii) *Every finite subset Γ' of Γ is satisfiable.*

It is easy to see that Theorem 1.4.8 will allow us to use the finite case of Theorem 1.4.6 to deduce the infinite case, so it remains to prove Theorem 1.4.8. The implication of (ii) from (i) is trivial; the interesting implication is the converse.

Observe that there is a one-to-one correspondence between truth assignments $\mathfrak{U}$ and elements of the product space $\{0,1\}^{\mathcal{A}}$, where $\mathcal{A}$ is the set of propositional variables. For every sentence ϕ, let $F_\phi \subset \{0,1\}^{\mathcal{A}}$ be the collection of all truth assignments that satisfy ϕ; observe that this is a closed (and open) subset of $\{0,1\}^{\mathcal{A}}$ in the product topology (basically because ϕ only involves finitely many of the propositional variables). If every finite subset Γ' of Γ is satisfiable, then $\bigcup_{\phi\in\Gamma'} F_\phi$ is nonempty; thus the family $(F_\phi)_{\phi\in\Gamma}$ of closed sets enjoys the *finite intersection property*. On the other hand, from *Tychonoff's theorem*, $\{0,1\}^{\mathcal{A}}$ is compact. We conclude that $\bigcap_{\phi\in\Gamma} F_\phi$ is nonempty, and the claim follows.

Remark 1.4.9. While Tychonoff's theorem in full generality is equivalent to the axiom of choice, it is possible to prove the compactness theorem using a weaker version of this axiom, namely the *ultrafilter lemma*. In fact, the compactness theorem is logically equivalent to this lemma.

1.4.2. Zeroth-order logic. Propositional logic is far too limited a language to do much mathematics. Let us make the language a bit more expressive, by adding constants, operations, relations, and (optionally) the equals sign; however, we refrain at this stage from adding variables or quantifiers, making this a *zeroth-order logic* rather than a first-order one.

A zeroth-order language $\mathcal{L}$ consists of the following objects:

- A (finite or infinite) collection $A_1, A_2, A_3, \ldots$ of propositional variables;
- A collection $R_1, R_2, R_3, \ldots$ of *relations* (or *predicates*), with each R_i having an *arity* (or *valence*) $a[R_i]$ (e.g., unary relation, binary relation, etc.);
- A collection $c_1, c_2, c_3, \ldots$ of constants;
- A collection $f_1, f_2, f_3, \ldots$ of *operators* (or functions), with each operator f_i having an arity $a[f_i]$ (e.g., unary operator, binary operator, etc.);
- Logical connectives;
- Parentheses;
- Optionally, the equals sign $=$.

For instance, a zeroth-order language for arithmetic on the natural numbers might include the constants $0, 1, 2, \ldots$, the binary relations $<, \leq, >, \geq$, the binary operations $+, \times$, the unary successor operation S, and the equals sign $=$. A zeroth-order language for studying all groups generated by six elements might include six generators $a_1, \ldots, a_6$ and the identity element e as constants, as well as the binary operation $\cdot$ of group multiplication and the unary operation $()^{-1}$ of group inversion, together with the equals sign $=$. And so forth.

Note that one could shorten the description of such languages by viewing propositional variables as relations of arity zero, and similarly viewing constants as operators of arity zero, but I find it conceptually clearer to leave these two operations separate, at least initially. As we shall see shortly, one can also essentially eliminate the privileged role of the equals sign $=$ by treating it as just another binary relation, which happens to have some standard axioms[5] attached to it.

By combining constants and operators together in the usual fashion, one can create *terms*; for instance, in the zeroth-order language for arithmetic, $3 + (4 \times 5)$ is a term. By inserting terms into a predicate or relation (or the equals sign $=$) or using a propositional variable, one obtains an *atomic formula*; thus for instance $3 + (4 \times 5) > 25$ is an atomic formula. By combining atomic formulae using logical connectives, one obtains a sentence (or *formula*); thus for instance $((4 \times 5) > 22) \implies (3 + (4 \times 5) > 25)$ is a sentence.

In order to assign meaning to sentences, we need the notion of a *structure* $\mathfrak{U}$ for a zeroth-order language $\mathcal{L}$. A structure consists of the following objects:

- A *domain of discourse* (or *universe of discourse*) $\mathrm{Dom}(\mathfrak{U})$;
- An assignment of a value $c_n^{\mathfrak{U}} \in \mathrm{Dom}(\mathfrak{U})$ to every constant c_n;
- An assignment of a function $f_n^{\mathfrak{U}} : \mathrm{Dom}(\mathfrak{U})^{a[f_n]} \to \mathrm{Dom}(\mathfrak{U})$ to every operation f_n;
- An assignment of a truth value $A_n^{\mathfrak{U}} \in \{\mathrm{true}, \mathrm{false}\}$ to every propositional variable A_n;
- An assignment of a function $R_n^{\mathfrak{U}} : \mathrm{Dom}(\mathfrak{U})^{a[R_n]}\{\mathrm{true}, \mathrm{false}\}$ to every relation R_n.

For instance, if $\mathcal{L}$ is the language of groups with six generators discussed above, then a structure $\mathfrak{U}$ would consist of a set $G = \mathrm{Dom}(\mathfrak{U})$, seven elements $a_1^{\mathfrak{U}}, \ldots, a_6^{\mathfrak{U}}, e^{\mathfrak{U}} \in G$ in that set, a binary operation $\cdot^{\mathfrak{U}} : G \times G \to G$, and a unary operation $(()^{-1})^{\mathfrak{U}} : G \to G$. At present, no group-type properties are assumed on these operations; the structure here is little more than a *magma* at this point.

Every sentence ϕ in a zeroth-order language $\mathcal{L}$ can be interpreted in a structure $\mathfrak{U}$ for that language to give a truth value $\phi^{\mathfrak{U}} \in \{\mathrm{true}, \mathrm{false}\}$, simply by substituting all symbols α in the language with their interpreted counterparts $\alpha^{\mathfrak{U}}$ (note that the equals sign $=$ does not need any additional data in order to be interpreted). For instance, the sentence $a_1 \cdot a_2 = a_3$ is true in $\mathfrak{U}$ if $a_1^{\mathfrak{U}} \cdot^{\mathfrak{U}} a_2^{\mathfrak{U}} = a_3^{\mathfrak{U}}$. Similarly, every term t in the language can be interpreted to give a value $t^{\mathfrak{U}}$ in the domain of discourse $\mathrm{Dom}(\mathfrak{U})$.

[5]Namely, that equality is reflexive, transitive, and symmetric, and it can be substituted in any expression to create an equal expression or in any formula to create an equivalent formula.

As before, a theory is a collection of sentences; we can define satisfiability $\mathfrak{U} \models \phi$, $\mathfrak{U} \models \Gamma$ of a sentence ϕ or a theory Γ by a structure $\mathfrak{U}$ just as in the previous section. For instance, to describe groups with at most six generators in the language $\mathcal{L}$, one might use the theory Γ which consists of all the group axioms, specialised to terms, e.g., Γ would contain the associativity axioms $t_1 \cdot (t_2 \cdot t_3) = (t_1 \cdot t_2) \cdot t_3$ for all choices of terms t_1, t_2, t_3. (Note that this theory is not quite strong enough to capture the concept of a structure $\mathfrak{U}$ being a group generated by six elements, because the domain of $\mathfrak{U}$ may contain some "inaccessible" elements which are not the interpretation of any term in $\mathcal{L}$, but without the universal quantifier, there is not much we can do in zeroth-order logic to say anything about those elements, and so this is pretty much the best we can do with this limited logic.)

Now we can state the completeness theorem:

Theorem 1.4.10 (Completeness theorem for zeroth-order logic). *Let Γ be a theory for a zeroth-order language $\mathcal{L}$, and let ϕ be a sentence in $\mathcal{L}$. Then the following are equivalent:*

(i) Syntactic consequence. *ϕ can be deduced from the axioms in Γ by a finite number of applications of the laws of zeroth-order logic (i.e., all the* laws of first-order logic *that do not involve variables or quantifiers).*

(ii) Semantic consequence. *Every truth assignment $\mathfrak{U}$ which satisfies (or models) Γ also satisfies ϕ.*

To prove this theorem, it suffices as before to show that every consistent theory Γ in a zeroth-order logic is satisfiable, and conversely. The converse implication is again straightforward (the laws of zeroth-order logic are easily seen to be sound); the main task is to show the forward direction, i.e.,

Proposition 1.4.11. *Let Γ be a consistent zeroth-order theory. Then Γ has at least one model.*

Proof. It is convenient to begin by eliminating the equality symbol from the language. Suppose we have already proven Proposition 1.4.11 has already been shown for languages without the equality symbol. Then we claim that the proposition also holds for languages with the equality symbol. Indeed, given a consistent theory Γ in a language $\mathcal{L}$ with equality, we can form a companion theory Γ' in the language $\mathcal{L}'$ formed by removing the equality symbol from $\mathcal{L}$ and replacing it with a new binary relation $='$, by taking all the sentences in Γ and replacing $=$ by $='$, and then adding in all the axioms of equality (with $=$ replaced by $='$) to Γ'. Thus, for instance, one would add the transitivity axioms $(x =' y) \wedge (y =' z) \implies (x =' z)$ to Γ for each triple of terms x, y, z, as well as substitution axioms such as

$(x =' y) \implies (B(x,z) =' B(y,z))$ for any terms x, y, z and binary functions B. It is straightforward to verify that if Γ is consistent, then Γ' is also consistent, because any contradiction derived in Γ' can be translated to a contradiction derived in Γ simply by replacing $='$ with $=$ throughout and using the axioms of equality. By hypothesis, we conclude that Γ' has some model $\mathfrak{U}'$. By the axioms of equality, the interpretation $(=')^{\mathfrak{U}'}$ of $='$ in this model is then an *equivalence relation* on the domain $\mathrm{Dom}(\mathfrak{U}')$ of $\mathfrak{U}'$. One can also remove from the domain of $\mathfrak{U}'$ any element which is not of the form $t^{\mathfrak{U}'}$ for some term t, as such "inaccessible" elements will not influence the satisfiability of Γ'. We can then define a structure $\mathfrak{U}$ for the original language $\mathfrak{L}$ by *quotienting* the domain of $\mathfrak{U}'$ by the equivalence relation $='$, and also quotienting all the interpretations of the relations and operations of $\mathcal{L}$. The axioms of equality ensure that this quotienting is possible, and that the quotiented structure $\mathfrak{U}$ satisfies $\mathcal{L}$; we omit the details.

Henceforth we assume that $\mathcal{L}$ does not contain the equality sign. We will then choose a "tautological" domain of discourse $\mathrm{Dom}(\mathfrak{U})$, by setting this domain to be nothing more than the collection of all terms in the language $\mathcal{L}$. For instance, in the language of groups on six generators, the domain $\mathrm{Dom}(\mathfrak{U})$ is basically the free *magma* (with "inverse") on six generators plus an "identity", consisting of terms such as $(a_1 \cdot a_2)^{-1} \cdot a_1$, $(e \cdot a_3) \cdot ((a_4)^{-1})^{-1}$, etc. With this choice of domain, there is an obvious tautological interpretation of constants ($c^{\mathfrak{U}} := c$) and operations (e.g., $B^{\mathfrak{U}}(t_1, t_2) := B(t_1, t_2)^{\mathfrak{U}}$ for binary operations B and terms t_1, t_2), which leads to every term t being interpreted as itself: $t^{\mathfrak{U}} = t$.

It remains to figure out how to interpret the propositional variables $A_1, A_2, \ldots$ and relations $R_1, R_2, \ldots$. Actually, one can replace each relation with an equivalent collection of new propositional variables by substituting in all possible terms in the relation. For instance, if one has a binary relation $R(,)$, one can replace this single relation symbol in the language by a (possibly infinite) collection of propositional variables $R(t_1, t_2)$, one for each pair of terms t_1, t_2, leading to a new (and probably much larger) language $\tilde{L}$ without any relation symbols. It is not hard to see that if theory Γ is consistent in $\mathcal{L}$, then the theory $\tilde{\Gamma}$ in $\tilde{L}$ formed by interpreting all atomic formulae such as $R(t_1, t_2)$ as propositional variables is also consistent. If $\tilde{\Gamma}$ has a model $\tilde{\mathfrak{U}}$ with the tautological domain of discourse, it is not hard to see that this can be converted to a model $\mathfrak{U}$ of Γ with the same domain by defining the interpretation $R^{\mathfrak{U}}$ of relations R in the obvious manner.

So now we may assume that there are no relation symbols, so that Γ now consists entirely of propositional sentences involving the propositional variables. But the claim now follows from the completeness theorem in propositional logic. □

Remark 1.4.12. The above proof can be viewed as a combination of the completeness theorem in propositional logic and the familiar procedure in algebra of constructing an algebraic object (e.g., a group) that obeys various relations, by starting with the free version of that object (e.g., a free group) and then quotienting out by the equivalence relation generated by those relations.

Remark 1.4.13. Observe that if $\mathfrak{L}$ is at most countable, then the structures $\mathfrak{U}$ constructed by the above procedure are at most countable (because the set of terms is at most countable, and quotienting by an equivalence relation cannot increase the cardinality). Thus we see (as in Theorem 1.4.1 or Corollary 1.4.2) that if a zeroth-order theory in an at most countable language is satisfiable, then it is in fact satisfiable with an at most countable model.

From the completeness theorem for zeroth-order logic and the above remark, we obtain the compactness theorem for zeroth-order logic, which is formulated exactly as in Corollary 1.4.2.

1.4.3. First-order logic. We are now ready to study languages which are expressive enough to do some serious mathematics, namely the languages of first-order logic, which are formed from zeroth-order logics by adding variables and quantifiers. (There are *higher-order logics* as well, but unfortunately the completeness and compactness theorems typically fail for these, and they will not be discussed here.)

A language $\mathcal{L}$ for a first-order logic consists of the following:

- A (finite or infinite) collection $A_1, A_2, A_3, \ldots$ of propositional variables;
- A collection $R_1, R_2, R_3, \ldots$ of relations, with each R_i having an arity $a[R_i]$;
- A collection $c_1, c_2, c_3, \ldots$ of constants;
- A collection $f_1, f_2, f_3, \ldots$ of operators, with each f_i having an arity $a[f_i]$;
- A collection $x_1, x_2, x_3, \ldots$ of variables;
- Logical connectives;
- The *quantifiers* $\forall, \exists$;
- Parentheses;
- Optionally, the equals sign $=$.

For instance, the language for Peano arithmetic includes a constant 0, a unary operator S, binary operators $+, \times$, the equals sign $=$, and a countably infinite number of variables $x_1, x_2, \ldots$.

By combining constants, variables, and operators, one creates terms; by inserting terms into predicates or relations or using propositional variables, one obtains atomic formulae. These atomic formulae can contain a number of free variables. Using logical connectives as well as quantifiers to bind any or all of these variables, one obtains *well-formed formulae*; a formula with no free variables is a *sentence.* Thus, for instance, $\forall x_2 : x_1 + x_2 = x_2 + x_1$ is a well-formed formula, and $\forall x_1 \forall x_2 : x_1 + x_2 = x_2 + x_1$ is a sentence.

A structure $\mathfrak{U}$ for a first-order language $\mathfrak{L}$ is exactly the same concept as for a zeroth-order language: a domain of discourse, together with an interpretation of all the constants, operators, propositional variables, and relations in the language. Given a structure $\mathfrak{U}$, one can interpret terms t with no free variables as elements $t^{\mathfrak{U}}$ of $\mathrm{Dom}(\mathfrak{U})$, and interpret sentences ϕ as truth values $\phi^{\mathfrak{U}} \in \{\text{true}, \text{false}\}$, in the standard fashion.

A theory is, once again, a collection of sentences in the first-order language $\mathcal{L}$; one can define what it means for a structure to satisfy a sentence or a theory just as before.

Remark 1.4.14. In most fields of mathematics, one wishes to discuss several types of objects (e.g., numbers, sets, points, group elements, etc.) at once. For this, one would prefer to use a *typed* language, in which variables, constants, and functions take values in one type of object, and relations and functions take only certain types of objects as input. However, one can easily model a typed theory using a typeless theory by the trick of adding some additional unary predicates to capture type (e.g., $N(x)$ to indicate the assertion "x is a natural number", $S(x)$ to indicate the assertion "x is a set", etc.) and modifying the axioms of the theory being considered accordingly. (For instance, in a language involving both natural numbers and other mathematical objects, one might impose a new closure axiom $\forall x \forall y : N(x) \wedge N(y) \implies N(x+y)$, and axioms such as the commutativity axiom $\forall x \forall y : x + y = y + x$ would need to be modified to $\forall x \forall y : N(x) \wedge N(y) \implies x + y = y + x$.) It is a tedious but routine matter to show that the completeness and compactness theorems for typeless first-order logic imply analogous results for typed first-order logic; we omit the details.

To prove the completeness (and hence compactness) theorem, it suffices as before to show that

Proposition 1.4.15. *Let Γ be a consistent first-order theory, with an at most countable language $\mathcal{L}$. Then Γ has at least one model $\mathcal{U}$, which is also at most countable.*

We shall prove this result using a series of reductions. Firstly, we can mimic the arguments in the zeroth-order case and reduce to the case when

$\mathcal{L}$ does not contain the equality symbol. (We no longer need to restrict the domain of discourse to those elements which can be interpreted by terms, because the universal quantifier $\forall$ is now available for use when stating the axioms of equality.) Henceforth we shall assume that the equality symbol is not present in the language.

Next, by using the laws of first-order logic to push all quantifiers in the sentences in Γ to the beginning (e.g., replacing $(\forall x : P(x)) \wedge (\forall y : Q(y))$ with $\forall x \forall y : P(x) \wedge Q(y))$ one may assume that all sentences in Γ are in *prenex normal form*, i.e., they consist of a "matrix" of quantifiers, followed by an *quantifier-free formula*—a well-formed formula with no quantifiers. For instance, $\forall x \exists y \forall z \exists w : P(x, y, z, w)$ is in prenex normal form, where $P(x, y, z, w)$ is a quantifier-free formula with four free variables x, y, z, w.

Now we will start removing the existential quantifiers $\exists$ from the sentences in Γ. Let us begin with a simple case, when Γ contains a sentence of the form $\exists x : P(x)$ for some quantifier-free formula of one free variable x. Then one can eliminate the existential quantifier by introducing a *witness*, or more precisely adjoining a new constant c to the language $\mathcal{L}$ and replacing the statement $\exists x : P(x)$ with the statement $P(c)$, giving rise to a new theory Γ' in a new language Λ'. The consistency of Γ easily implies the consistency of Γ', while any at most countable model for Γ' can be easily converted to an at most countable model for Γ (by "forgetting" the symbol c). (In fact, Γ' is a *conservative extension* of Γ.) We can apply this reduction simultaneously to all sentences of the form $\exists x : P(x)$ in Γ (thus potentially expanding the collection of constants in the language by a countable amount).

The same argument works for any sentence in prenex normal form in which all the existential quantifiers are to the left of the universal quantifiers, e.g., $\exists x \exists y \forall z \forall w : P(x, y, z, w)$; this statement requires two constants to separately witness x and y, but otherwise one proceeds much as in the previous paragraph. But what about if one or more of the existential quantifiers is buried behind a universal quantifier? The trick is then to use *Skolemisation.* We illustrate this with the simplest case of this type, namely that of a sentence $\forall x \exists y : P(x, y)$. Here, one cannot use a constant witness for y. But this is no problem: one simply introduces a witness that depends on x. More precisely, one adjoins a new unary operator c to the language $\mathcal{L}$ and replaces the statement $\forall x \exists y : P(x, y)$ by $\forall x : P(x, c(x))$, creating a new theory Γ' in a new language Λ'. One can again show (though this is not entirely trivial) that the consistency of Γ implies the consistency of Γ', and that every countable model for Γ' can be converted to a countable model for Γ (again by "forgetting" c). So one can eliminate the existential quantifier from this sentence also. Similar methods work for any other prenex normal form. For instance, with the sentence

$$\forall x \exists y \forall z \exists w : P(x, y, z, w),$$

one can obtain a conservative extension of that theory by introducing a unary operator c and a binary operator d and replacing the above sentence with

$$\forall x \forall z : P(x, c(x), z, d(x, z)).$$

One can show that one can perform Skolemisation on all the sentences in Γ simultaneously, which has the effect of eliminating all existential quantifiers from Γ while still keeping the language $\mathcal{L}$ at most countable (since Γ is at most countable). (Intuitively, what is going on here is that we are interpreting all existential axioms in the theory as implicitly defining functions, which we then explicitly formalise as a new symbol in the language. For instance, if we had some theory of sets which contained the *axiom of choice* (every family of nonempty sets $(X_\alpha)_{\alpha \in A}$ admits a *choice function* $f : A \to \bigcup_{\alpha \in A} X_\alpha$), then we can Skolemise this by introducing a "choice function function" $\mathcal{F} : (X_\alpha)_{\alpha \in A} \mapsto \mathcal{F}((X_\alpha)_{\alpha \in A})$ that witnessed this axiom to the language. Note that we do not need uniqueness in the existential claim in order to be able to perform Skolemisation.)

After performing Skolemisation and adding all the witnesses to the language, we are reduced to the case in which all the sentences in Γ are in fact universal statements, i.e., of the form $\forall x_1 \cdots \forall x_k : P(x_1, \ldots, x_k)$, where $P(x_1, \ldots, x_k)$ is a quantifier-free formula of k free variables. In this case one can repeat the zeroth-order arguments, selecting a structure $\mathfrak{U}$ whose domain of discourse is the tautological one, indexed by all the terms with no free variables (in particular, this structure will be countable). One can then replace each first-order statement $\forall x_1 \cdots \forall x_k : P(x_1, \ldots, x_k)$ in Γ by the family of zeroth-order statements $P(t_1, \ldots, t_k)$, where $t_1, \ldots, t_k$ ranges over all terms with no free variables, thus creating a zeroth-order theory Γ_0. As Γ is consistent, Γ_0 is also, so by the zeroth-order theory, we can find a model $\mathfrak{U}$ for Γ_0 with the tautological domain of discourse, and it is clear that this structure will also be a model for the original theory Γ. The proof of the completeness theorem (and thus the compactness theorem) is now complete.

In summary: To create a countable model from a consistent first-order theory, one first replaces the equals sign $=$ (if any) by a binary relation $='$, then one uses Skolemisation to make all implied functions and operations explicit elements of the language. Next, one makes the zeroth-order terms of the new language the domain of discourse, applies a greedy algorithm to decide the truth assignment of all zeroth-order sentences, and then finally quotients out by the equivalence relation given by $='$ to recover the countable model.

Remark 1.4.16. I find the use of Skolemisation to greatly clarify, at a conceptual level, the proof of the completeness theorem. However, at a technical level it does make things more complicated: in particular, showing that the Skolemisation of a consistent theory is still consistent does require some nontrivial effort (one has to take all arguments involving the Skolem function $c()$, and replace every occurence of $c()$ by a "virtual" function, defined implicitly using existential quantifiers). On the other hand, this fact is easy to prove once one already *has* the completeness theorem, though we of course cannot formally take advantage of this while trying to *prove* that theorem!

The more traditional Henkin approach is based instead on adding a large number of constant witnesses, one for every existential statement: roughly speaking, for each existential sentence $\exists x : P(x)$ in the language, one adds a new constant c to the language and inserts an axiom $(\exists x : P(x)) \implies P(c)$ to the theory; it is easier to show that this preserves consistency than it is with a more general Skolemisation. Unfortunately, every time one adds a constant to the language, one increases the number of existential sentences for which one needs to perform this type of witnessing. But it turns out that after applying this procedure a countable number of times, one can get to the point where every existential sentence is automatically witnessed by some constant. This has the same ultimate effect as Skolemisation, namely one can convert sentences containing existential quantifiers to ones which are purely universal, and so the rest of the proof is much the same as the proof described above. On the other hand, the Henkin approach avoids the axiom of choice (though one still must use the ultrafilter lemma, of course).

Notes. This article first appeared at

`terrytao.wordpress.com/2009/04/10.`

Thanks to Carson Chow, Ernie Cohen, Eric, John Goodrick, and anonymous commenters for corrections.

1.5. Talagrand's concentration inequality

In the theory of discrete random matrices (e.g., matrices whose entries are random signs ± 1), one often encounters the problem of understanding the distribution of the random variable $\mathrm{dist}(X, V)$, where $X = (x_1, \ldots, x_n) \in \{-1, +1\}^n$ is an n-dimensional random sign vector (so X is uniformly distributed in the discrete cube $\{-1, +1\}^n$), and V is some d-dimensional subspace of $\mathbf{R}^n$ for some $0 \leq d \leq n$.

It is not hard to compute the second moment of this random variable. Indeed, if $P = (p_{ij})_{1 \leq i,j \leq n}$ denotes the orthogonal projection matrix from

$\mathbf{R}^n$ to the orthogonal complement $V^\perp$ of V, then one observes that

$$\operatorname{dist}(X,V)^2 = X \cdot PX = \sum_{i=1}^n \sum_{j=1}^n x_i x_j p_{ij},$$

and so upon taking expectations we see that

$$\mathbf{E}\operatorname{dist}(X,V)^2 = \sum_{i=1}^n p_{ii} = \operatorname{tr} P = n - d, \tag{1.23}$$

since P is a rank $n-d$ orthogonal projection. So we expect $\operatorname{dist}(X,V)$ to be about $\sqrt{n-d}$ on the average.

In fact, one has sharp concentration around this value, in the sense that $\operatorname{dist}(X,V) = \sqrt{n-d} + O(1)$ with high probability. More precisely, we have

Proposition 1.5.1 (Large deviation inequality). *For any $t > 0$, one has*

$$\mathbf{P}(|\operatorname{dist}(X,V) - \sqrt{n-d}| \geq t) \leq C\exp(-ct^2)$$

for some absolute constants $C, c > 0$.

In fact the constants C, c are very civilised; for large t one can basically take $C = 4$ and $c = 1/16$, for instance. This type of concentration, particularly for subspaces V of moderately large codimension[6] $n-d$, is fundamental to much of my work on random matrices with Van Vu, starting with our first paper [**TaVu2006**] (in which this proposition first appears).

Proposition 1.5.1 is an easy consequence of the second moment computation and *Talagrand's inequality* [**Ta1996**], which among other things provides a sharp concentration result for convex Lipschitz functions on the cube $\{-1,+1\}^n$; since $\operatorname{dist}(x,V)$ is indeed a convex Lipschitz function, this inequality can be applied immediately. The proof of Talagrand's inequality is short and can be found in several textbooks (e.g., [**AlSp2008**]), but I thought I would reproduce the argument here (specialised to the convex case), mostly to force myself to learn the proof properly. Note the concentration of $O(1)$ obtained by Talagrand's inequality is much stronger than what one would get from more elementary tools such as *Azuma's inequality* or *McDiarmid's inequality*, which would only give concentration of about $O(\sqrt{n})$ or so (which is in fact trivial, since the cube $\{-1,+1\}^n$ has diameter $2\sqrt{n}$). The point is that Talagrand's inequality is very effective at exploiting the convexity of the problem, as well as the Lipschitz nature of the function in all directions, whereas Azuma's inequality can only easily take advantage of the Lipschitz nature of the function in coordinate directions. On the other hand, Azuma's inequality works just as well if the ℓ^2 metric is replaced with

[6]For subspaces of small codimension (such as hyperplanes) one has to use other tools to get good results, such as *inverse Littlewood-Offord theory* or the *Berry-Esséen central limit theorem*, but that is another story.

the larger ℓ^1 metric, and one can conclude that the ℓ^1 distance between X and V concentrates around its median to a width $O(\sqrt{n})$, which is a more nontrivial fact than the ℓ^2 concentration bound given by that inequality. (The computation of the median of the ℓ^1 distance is more complicated than for the ℓ^2 distance, though, and depends on the orientation of V.)

Remark 1.5.2. If one makes the coordinates of X independent and identically distributed (iid) Gaussian variables $x_i \equiv N(0,1)$ rather than random signs, then Proposition 1.5.1 is much easier to prove; the probability distribution of a Gaussian vector is rotation-invariant, so one can rotate V to be, say, $\mathbf{R}^d$, at which point $\operatorname{dist}(X,V)^2$ is clearly the sum of $n-d$ independent squares of Gaussians (i.e., a *chi-square distribution*), and the claim follows from direct computation (or one can use the *Chernoff inequality*). The Gaussian counterpart of Talagrand's inequality is more classical, being essentially due to Lévy, and will also be discussed later in this article.

1.5.1. Concentration on the cube.

Proposition 1.5.1 follows easily from the following statement, which asserts that if a convex set $A \subset \mathbf{R}^n$ occupies a nontrivial fraction of the cube $\{-1,+1\}^n$, then the neighbourhood $A_t := \{x \in \mathbf{R}^n : \operatorname{dist}(x,A) \leq t\}$ will occupy almost all of the cube for $t \gg 1$:

Proposition 1.5.3 (Talagrand's concentration inequality). *Let A be a convex set in $\mathbf{R}^d$. Then*

$$\mathbf{P}(X \in A)\mathbf{P}(X \notin A_t) \leq \exp(-ct^2)$$

for all $t > 0$ and some absolute constant $c > 0$, where $X \in \{-1,+1\}^n$ is chosen uniformly from $\{-1,+1\}^n$.

Remark 1.5.4. It is crucial that A is convex here. If instead A is, say, the set of all points in $\{-1,+1\}^n$ with fewer than $n/2-\sqrt{n}$ +1's, then $\mathbf{P}(X \in A)$ is comparable to 1, but $\mathbf{P}(X \notin A_t)$ only starts decaying once $t \gg \sqrt{n}$, rather than $t \gg 1$. Indeed, it is not hard to show that Proposition 1.5.3 implies the variant

$$\mathbf{P}(X \in A)\mathbf{P}(X \notin A_t) \leq \exp(-ct^2/n)$$

for nonconvex A (by restricting A to $\{-1,+1\}^n$ and then passing from A to the convex hull, noting that distances to A on $\{-1,+1\}^n$ may be contracted by a factor of $O(\sqrt{n})$ by this latter process). This inequality can also be easily deduced from *Azuma's inequality*.

To apply this proposition to the situation at hand, observe that if A is the cylindrical region $\{x \in \mathbf{R}^n : \operatorname{dist}(x,V) \leq r\}$ for some r, then A is convex and A_t is contained in $\{x \in \mathbf{R}^n : \operatorname{dist}(x,V) \leq r+t\}$. Thus

$$\mathbf{P}(\operatorname{dist}(X,V) \leq r)\mathbf{P}(\operatorname{dist}(X,V) > r+t) \leq \exp(-ct^2).$$

Applying this with $r := M$ or $r := M - t$, where M is the median value of $\operatorname{dist}(X, V)$, one soon obtains concentration around the median:

$$\mathbf{P}(|\operatorname{dist}(X, V) - M| > t) \leq 4\exp(-ct^2).$$

This is only compatible with (1.23) if $M = \sqrt{n-d} + O(1)$, and the claim follows.

To prove Proposition 1.5.3, we use the exponential moment method. Indeed, it suffices by Markov's inequality to show that

$$\mathbf{P}(X \in A)\mathbf{E}\exp(c\operatorname{dist}(X, A)^2) \leq 1 \tag{1.24}$$

for a sufficiently small absolute constant $c > 0$ (in fact one can take $c = 1/16$).

We prove (1.24) by an induction on the dimension n. The claim is trivial for $n = 0$, so suppose $n \geq 1$ and the claim has already been proven for $n-1$.

Let us write $X = (X', x_n)$ for $x_n = \pm 1$. For each $t \in \mathbf{R}$, we introduce the slice $A_t := \{x' \in \mathbf{R}^{n-1} : (x', t) \in A\}$, then A_t is convex. We now try to bound the left-hand side of (1.24) in terms of X', A_t rather than X, A. Clearly,

$$\mathbf{P}(X \in A) = \frac{1}{2}[\mathbf{P}(X' \in A_{-1}) + \mathbf{P}(X' \in A_{+1})].$$

By symmetry we may assume that $\mathbf{P}(X' \in A_{+1}) \geq \mathbf{P}(X' \in A_{-1})$, thus we may write

$$\mathbf{P}(X' \in A_{\pm 1}) = p(1 \pm q), \tag{1.25}$$

where $p := \mathbf{P}(X \in A)$ and $0 \leq q \leq 1$.

Now we look at $\operatorname{dist}(X, A)^2$. For $t = \pm 1$, let $Y_t \in \mathbf{R}^{n-1}$ be the closest point of (the closure of) A_t to X', thus

$$|X' - Y_t| = \operatorname{dist}(X', A_t). \tag{1.26}$$

Let $0 \leq \lambda \leq 1$ be chosen later; then the point $(1-\lambda)(Y_{x_n}, x_n) + \lambda(Y_{-x_n}, -x_n)$ lies in A by convexity, and so

$$\operatorname{dist}(X, A) \leq |(1-\lambda)(Y_{x_n}, x_n) + \lambda(Y_{-x_n}, -x_n) - (X', x_n)|.$$

Squaring this and using Pythagoras, one obtains

$$\operatorname{dist}(X, A)^2 \leq 4\lambda^2 + |(1-\lambda)(X' - Y_{x_n}) + \lambda(X' - Y_{-x_n})|^2.$$

As we will shortly be exponentiating the left-hand side, we need to linearise the right-hand side. Accordingly, we will exploit the convexity of the function $x \mapsto |x|^2$ to bound

$$\begin{aligned}
&|(1-\lambda)(X - Y_{x_n}) + \lambda(X - Y_{-x_n})|^2 \\
&\quad \leq (1-\lambda)|X' - Y_{x_n}|^2 + \lambda|X' - Y_{-x_n}|^2,
\end{aligned}$$

and thus by (1.26)

$$\operatorname{dist}(X, A)^2 \leq 4\lambda^2 + (1-\lambda)\operatorname{dist}(X', A_{x_n})^2 + \lambda \operatorname{dist}(X', A_{-x_n})^2.$$

We exponentiate this and take expectations in X' (holding x_n fixed for now) to get

$$\mathbf{E}_{X'} e^{c \operatorname{dist}(X,A)^2} \leq e^{4c\lambda^2} \mathbf{E}_{X'} (e^{c \operatorname{dist}(X', A_{x_n})^2})^{1-\lambda} (e^{c \operatorname{dist}(X', A_{-x_n})^2})^{\lambda}.$$

Meanwhile, from the induction hypothesis and (1.25) we have

$$\mathbf{E}_{X'} e^{c \operatorname{dist}(X', A_{x_n})^2} \leq \frac{1}{p(1+x_n q)},$$

and similarly for A_{-x_n}. By Hölder's inequality, we conclude

$$\mathbf{E}_{X'} e^{c \operatorname{dist}(X,A)^2} \leq e^{4c\lambda^2} \frac{1}{p(1+x_n q)^{1-\lambda}(1-x_n q)^{\lambda}}.$$

For $x_n = +1$, the optimal choice of λ here is 0, obtaining

$$\mathbf{E}_{X'} e^{c \operatorname{dist}(X,A)^2} = \frac{1}{p(1+q)};$$

for $x_n = -1$, the optimal choice of λ is to be determined. Averaging, we obtain

$$\mathbf{E}_{X} e^{c \operatorname{dist}(X,A)^2} = \frac{1}{2}[\frac{1}{p(1+q)} + e^{4c\lambda^2} \frac{1}{p(1-q)^{1-\lambda}(1+q)^{\lambda}}],$$

so to establish (1.24), it suffices to pick $0 \leq \lambda \leq 1$ such that

$$\frac{1}{1+q} + e^{4c\lambda^2} \frac{1}{(1-q)^{1-\lambda}(1+q)^{\lambda}} \leq 2.$$

If q is bounded away from zero, then by choosing $\lambda = 1$ we would obtain the claim if c is small enough, so we may take q to be small. But then a Taylor expansion allows us to conclude if we take λ to be a constant multiple of q, and again pick c to be small enough. The point is that $\lambda = 0$ already almost works up to errors of $O(q^2)$, and increasing λ from zero to a small nonzero quantity will decrease the left-hand side by about $O(\lambda q) - O(c\lambda^2)$.

By optimising everything using first-year calculus, one eventually gets the constant $c = 1/16$ claimed earlier.

Remark 1.5.5. Talagrand's inequality is in fact far more general than this; it applies to arbitrary products of probability spaces, rather than just to $\{-1,+1\}^n$, and to nonconvex A, but the notion of distance needed to define A_t becomes more complicated; the proof of the inequality, though, is essentially the same. Besides its applicability to convex Lipschitz functions, Talagrand's inequality is also very useful for controlling combinatorial Lipschitz functions F which are "locally certifiable" in the sense that whenever $F(x)$ is larger than some threshold t, then there exist some bounded number $f(t)$ of coefficients of x which "certify" this fact (in the sense that $F(y) \geq t$ for any other y which agrees with x on these coefficients). See e.g., [**AlSp2008**] for a more precise statement and some applications.

1.5.2. Gaussian concentration. As mentioned earlier, there are analogous results when the uniform distribution on the cube $\{-1,+1\}^n$ are replaced by other distributions, such as the n-dimensional Gaussian distribution. In fact, in this case convexity is not needed:

Proposition 1.5.6 (Gaussian concentration inequality). *Let A be a measurable set in $\mathbf{R}^d$. Then*

$$\mathbf{P}(X \in A)\mathbf{P}(X \notin A_t) \leq \exp(-ct^2)$$

for all $t > 0$ and some absolute constant $c > 0$, where $X \equiv N(0,1)^n$ is a random Gaussian vector.

This inequality can be deduced from *Lévy's classical concentration of measure inequality for the sphere* (with the optimal constant), but we will give an alternate proof due to Maurey and Pisier. It suffices to prove the following variant of Proposition 1.5.6:

Proposition 1.5.7 (Gaussian concentration inequality for Lipschitz functions). *Let $f : \mathbf{R}^d \to \mathbf{R}$ be a function which is Lipschitz with constant* 1 *(i.e., $|f(x) - f(y)| \leq |x - y|$ for all $x, y \in \mathbf{R}^d$). Then for any t we have*

$$\mathbf{P}(|f(X) - \mathbf{E}f(X)| \geq t) \leq \exp(-ct^2)$$

for all $t > 0$ and some absolute constant $c > 0$, where $X \equiv N(0,1)^n$ is a random variable.

Indeed, if one sets $f(x) := \operatorname{dist}(x, A)$, one can soon deduce Proposition 1.5.6 from Proposition 1.5.7.

Informally, Proposition 1.5.7 asserts that Lipschitz functions of Gaussian variables concentrate as if they were Gaussian themselves; for comparison, Talagrand's inequality implies that *convex* Lipschitz functions of *Bernoulli* variables concentrate as if they were Gaussian.

Now we prove Proposition 1.5.7. By the *epsilon regularisation argument* (Section 2.7 of *Volume I*) we may take f to be smooth, and so by the Lipschitz property we have

$$|\nabla f(x)| \leq 1 \tag{1.27}$$

for all x. By subtracting off the mean, we may assume $\mathbf{E}f = 0$. By replacing f with $-f$ if necessary, it suffices to control the upper tail probability $\mathbf{P}(f(X) \geq t)$ for $t > 0$.

We again use the exponential moment method. It suffices to show that

$$\mathbf{E}\exp(tf(X)) \leq \exp(Ct^2)$$

for some absolute constant C.

Now we use a variant of the *square and rearrange* trick. Let Y be an independent copy of X. Since $\mathbf{E} f(Y) = 0$, we see from *Jensen's inequality* that $\mathbf{E}\exp(-tf(Y)) \geq 1$, and so

$$\mathbf{E}\exp(tf(X)) \leq \mathbf{E}\exp(t(f(X) - f(Y))).$$

With an eye to exploiting (1.27), one might seek to use the fundamental theorem of calculus to write

$$f(X) - f(Y) = \int_0^1 \frac{d}{d\lambda} f((1-\lambda)Y + \lambda X)\ d\lambda.$$

But actually it turns out to be smarter to use a circular arc of integration, rather than a line segment:

$$f(X) - f(Y) = \int_0^{\pi/2} \frac{d}{d\theta} f(Y\cos\theta + X\sin\theta)\ d\theta.$$

The reason for this is that $X_\theta := Y\cos\theta + X\sin\theta$ is another Gaussian random variable equivalent to $N(0,1)^n$, as is its derivative $X'_\theta := -Y\sin\theta + X\cos\theta$; furthermore, and crucially, these two random variables are *independent*.

To exploit this, we first use Jensen's inequality to bound

$$\exp(t(f(X) - f(Y))) \leq \frac{\pi}{2}\int_0^{\pi/2} \exp(\frac{2t}{\pi}\frac{d}{d\theta} f(X_\theta))\ d\theta.$$

Applying the chain rule and taking expectations, we have

$$\mathbf{E}\exp(t(f(X) - f(Y))) \leq \frac{\pi}{2}\int_0^{\pi/2} \mathbf{E}\exp(\frac{2t}{\pi}\nabla f(X_\theta)\cdot X'_\theta)\ d\theta.$$

Let us condition X_θ to be fixed, then $X'_\theta \equiv N(0,1)^n$; applying (1.27), we conclude that $\frac{2t}{\pi}\nabla f(X_\theta)\cdot X'_\theta$ is normally distributed with standard deviation at most $\frac{2t}{\pi}$. As such we have

$$\mathbf{E}\exp(\frac{2t}{\pi}\nabla f(X_\theta)\cdot X'_\theta) \leq \exp(Ct)$$

for some absolute constant C; integrating out the conditioning on X_θ we obtain the claim.

Notes. This article first appeared at

`terrytao.wordpress.com/2009/06/09.`

Thanks to Oded and vedadi for corrections.

1.6. The Szemerédi-Trotter theorem and the cell decomposition

The celebrated *Szemerédi-Trotter theorem* gives a bound for the set of incidences $I(P, L) := \{(p, \ell) \in P \times L : p \in \ell\}$ between a finite set of points P and a finite set of lines L in the Euclidean plane $\mathbf{R}^2$. Specifically, the bound is

$$|I(P, L)| \ll |P|^{2/3}|L|^{2/3} + |P| + |L|, \tag{1.28}$$

where we use the asymptotic notation $X \ll Y$ or $X = O(Y)$ to denote the statement that $X \leq CY$ for some absolute constant C. In particular, the number of incidences between n points and n lines is $O(n^{4/3})$. This bound is sharp; consider for instance the discrete box $P := \{(a, b) \in \mathbf{Z}^2 : 1 \leq a \leq N; 1 \leq b \leq 2N^2\}$ with L being the collection of lines $\{(x, mx + b) : m, b \in \mathbf{Z}, 1 \leq m \leq N, 1 \leq b \leq N^2\}$. One easily verifies that $|P| = 2N^3$, $|L| = N^3$, and $|I(P, L)| = N^4$, showing that (1.28) is essentially sharp in the case $|P| \sim |L|$; one can concoct similar examples for other regimes of $|P|$ and $|L|$.

On the other hand, if one replaces the Euclidean plane $\mathbf{R}^2$ by a finite field geometry F^2, where F is a finite field, then the estimate (1.28) is false. For instance, if P is the entire plane F^2, and L is the set of all lines in F^2, then $|P|, |L|$ are both comparable to $|F|^2$, but $|I(P, L)|$ is comparable to $|F|^3$, thus violating (1.28) when $|F|$ is large. Thus any proof of the Szemerédi-Trotter theorem must use a special property of the Euclidean plane which is not enjoyed by finite field geometries. In particular, this strongly suggests that one cannot rely purely on algebra and combinatorics to prove (1.28); one must also use some Euclidean geometry or topology as well.

Nowadays, the slickest proof of the Szemerédi-Trotter theorem is via the crossing number inequality (as discussed in Section 1.10 of *Structure and Randomness*), which ultimately relies on *Euler's formula* $|V| - |E| + |F| = 2$; thus in this argument it is *topology* that is the feature of Euclidean space which one is exploiting, and which is not present in the finite field setting. In this article, though, I would like to mention a different proof (closer in spirit to the original proof of Szemerédi and Trotter [**SzTr1983**], and closer still to the later paper [**ClEdGuShWe1990**]), based on the method of *cell decomposition*, which has proven to be a very flexible method in combinatorial incidence geometry. Here, the distinctive feature of Euclidean geometry one is exploiting is *convexity*, which again has no finite field analogue.

Roughly speaking, the idea is this. Using nothing more than the axiom that two points determine at most one line, one can obtain the bound

$$|I(P, L)| \ll |P||L|^{1/2} + |L|, \tag{1.29}$$

which is inferior to (1.28). (On the other hand, this estimate works in both Euclidean and finite field geometries, and is sharp in the latter case, as shown by the example given earlier.) Dually, the axiom that two lines determine at most one point gives the bound

$$|I(P, L)| \ll |L||P|^{1/2} + |P| \tag{1.30}$$

(or alternatively, one can use projective duality to interchange points and lines and deduce (1.30) from (1.29)).

An inspection of the proof of (1.29) shows that it is only expected to be sharp when the *bushes* $L_p := \{\ell \in L : \ell \ni p\}$ associated to each point $p \in P$ behave like "independent" subsets of L, so that there is no significant correlation between the bush L_p of one point and the bush of another point L_q.

However, in Euclidean space, we have the phenomenon that the bush of a point L_p is influenced by the region of space that p lies in. Clearly, if p lies in a set Ω (e.g., a convex polygon), then the only lines $\ell \in L$ that can contribute to L_p are those lines which pass through Ω. If Ω is a small convex region of space, one expects only a fraction of the lines in L to actually pass through Ω. As such, if p and q both lie in Ω, then L_p and L_q are compressed inside a smaller subset of L, namely the set of lines passing through Ω, and so should be more likely to intersect than if they were independent. This should lead to an improvement to (1.29) (and indeed, as we shall see below, ultimately leads to (1.28)).

More formally, the argument proceeds by applying the following lemma:

Lemma 1.6.1 (Cell decomposition). *Let L be a finite collection of lines in $\mathbf{R}^2$, and let $r \geq 1$. Then it is possible to find a set R of $O(r)$ lines in the plane (which may or may not be in L), which subdivide $\mathbf{R}^2$ into $O(r^2)$ convex regions (or* cells*), such that the interior of each such cell is incident to at most $O(|L|/r)$ lines.*

The deduction of (1.28) from (1.29), (1.30) and Lemma 1.6.1 is very quick. Firstly we may assume we are in the range

$$|L|^{1/2} \ll |P| \ll |L|^2; \tag{1.31}$$

otherwise, the bound (1.28) follows already from either (1.29) or (1.30) and some high-school algebra.

Let $r \geq 1$ be a parameter to be optimised later. We apply the cell decomposition to subdivide $\mathbf{R}^2$ into $O(r^2)$ open convex regions, plus a family R of $O(r)$ lines. Each of the $O(r^2)$ convex regions Ω has only $O(|L|/r)$ lines through it, and so by (1.29) contributes $O(|P \cap \Omega||L|^{1/2}/r^{1/2} + |L|/r)$ incidences. Meanwhile, on each of the lines ℓ in R used to perform this decomposition, there are at most $|L|$ transverse incidences (because each line in L distinct from ℓ can intersect ℓ at most once), plus all the incidences along ℓ itself. Putting all this together, one obtains

$$|I(P,L)| \leq |I(P, L \cap R)| + O(|P||L|^{1/2}/r^{1/2} + |L|r).$$

We optimise this by selecting $r \sim |P|^{2/3}/|L|^{1/3}$; from (1.31) we can ensure that $r \leq |L|/2$, so that $|L \cap R| \leq |L|/2$. One then obtains

$$|I(P,L)| \leq |I(P, L \cap R)| + O(|P|^{2/3}|L|^{2/3}).$$

We can iterate away the $L \cap R$ error (halving the number of lines each time) and sum the resulting geometric series to obtain (1.28).

It remains to prove (1.6.1). If one subdivides $\mathbf{R}^2$ using r arbitrary lines, one creates at most $O(r^2)$ cells (because each new line intersects the existing lines at most once and so can create at most $O(r)$ distinct cells), and for a similar reason, every line in L visits at most r of these regions, and so by double counting one expects $O(|L|/r)$ lines per cell "on the average". The key difficulty is then to get $O(|L|/r)$ lines through *every* cell, not just on the average. It turns out that a probabilistic argument will almost work, but with a logarithmic loss (thus having $O(|L|\log|L|/r)$ lines per cell rather than $O(|L|/r)$); but with a little more work one can then iterate away this loss also. The arguments here are loosely based on those of [**ClEdGuShWe1990**]; a related (deterministic) decomposition also appears in [**SzTr1983**]. But I wish to focus here on the probabilistic approach.

It is also worth noting that the original (somewhat complicated) argument of Szemerédi and Trotter has been adapted to establish the analogue of (1.28) in the complex plane $\mathbf{C}^2$ by Toth [**To2005**], while the other known proofs of Szemerédi and Trotter, so far, have not been able to be extended to this setting (the Euler characteristic argument clearly breaks down, as does any proof based on using lines to divide planes into half-spaces). So all three proofs have their advantages and disadvantages.

1.6.1. The trivial incidence estimate. We first give a quick proof of the trivial incidence bound (1.29). We have

$$|I(P,L)| = \sum_{\ell \in L} |P \cap \ell|,$$

and thus by Cauchy-Schwarz

$$\sum_{\ell \in L} |P \cap \ell|^2 \geq \frac{|I(P,L)|^2}{|L|}.$$

On the other hand, observe that

$$\sum_{\ell \in L} |P \cap \ell|^2 - |P \cap \ell| = |\{(p,q,\ell) \in P \times P \times L : p \neq q; p, q \in \ell\}|.$$

Because two distinct points p, q are incident to at most one line, the right-hand side is at most $|P|^2$, thus

$$\sum_{\ell \in L} |P \cap \ell|^2 \leq |I(P,L)| + |P|^2.$$

Comparing this with the Cauchy-Schwarz bound and using a little high-school algebra, we obtain (1.29). A dual argument (swapping the role of lines and points) gives (1.30).

A more informal proof of (1.29) can be given as follows. Suppose for contradiction that $|I(P,L)|$ was much larger than $|P||L|^{1/2} + |L|$. Since $|I(P,L)| = \sum_{p \in P} |L_p|$, this implies that that the $|L_p|$ are much larger than $|L|^{1/2}$ on the average. By the *birthday paradox*, one then expects two randomly chosen L_p, L_q to intersect in at least two places ℓ, ℓ'; but this would mean that two lines intersect in two points, a contradiction. The use of Cauchy-Schwarz in the rigorous argument given above can thus be viewed as an assertion that the average intersection of L_p and L_q is at least as large as what random chance predicts.

As mentioned in the introduction, we now see (intuitively, at least) that if nearby p, q are such that L_p, L_q are drawn from a smaller pool of lines than L, then their intersection is likely to be higher, and so one should be able to improve upon (1.29).

1.6.2. The probabilistic bound. Now we start proving Lemma 1.6.1. We can assume that $r < |L|$, since the claim is trivial otherwise (we just use all the lines in L to subdivide the plane, and there are no lines left in L to intersect any of the cells). Similarly we may assume that $r > 1$, and that $|L|$ is large. We can also perturb all the lines slightly and assume that the lines are in general position (no three are concurrent), as the general claim then follows from a limiting argument (note that this may send some of the cells to become empty). (Of course, the Szemerédi-Trotter theorem is quite easy under the assumption of general position, but this theorem is not our current objective right now.)

We use the *probabilistic method*, i.e., we construct R by some random recipe and aim to show that the conclusion of the lemma holds with positive probaility.

The most obvious approach would be to choose the r lines R at random from L, thus each line $\ell \in L$ has a probability of $r/|L|$ of lying in R. Actually, for technical reasons it is slightly better to use a Bernoulli process to select R, thus each line $\ell \in L$ lies in R with an *independent* probability of $r/|L|$. This can cause R to occasionally have size much larger than r, but this probability can be easily controlled (e.g., using the *Chernoff inequality*). So with high probability, R consists of $O(r)$ lines, which therefore carve out

$O(r^2)$ cells. The remaining task is to show that each cell is incident to at most $O(|L|/r)$ lines from L.

Observe that each cell is a (possibly unbounded) polygon, whose edges come from lines in R. Note that (except in the degenerate case when R consists of at most one line, which we can ignore) any line ℓ which meets a cell in R, must intersect at least one of the edges of R. If we pretend for the moment that all cells have a bounded number of edges, it would then suffice to show that each edge of each cell was incident to $O(|L|/r)$ lines.

Let us see how this would go. Suppose that one line $\ell \in L$ was picked for the set R, and consider all the other lines in L that intersect ℓ; there are $O(|L|)$ of these lines ℓ', which (by the general position hypothesis) intersect ℓ at distinct points $\ell \cap \ell'$ on the line. If one of these lines ℓ' intersecting ℓ is also selected for R, then the corresponding point $\ell \cap \ell'$ will become a vertex of one of the cells (indeed, it will be the vertex of four cells). Thus each of these points on ℓ has an independent probability of $r/|L|$ of becoming a vertex for a cell.

Now consider m consecutive such points on ℓ. The probability that they all fail to be chosen as cell vertices is $(1 - r/|L|)^m$; if $m = k|L|/r$, then this probability is $O(\exp(-k))$. Thus, runs of much more than $|L|/r$ points without vertices are unlikely. In particular, setting $k = 100 \log |L|$, we see that the probability that any given $100|L| \log |L|/r$ consecutive points on any given line ℓ are skipped is $O(|L|^{-100})$. By the union bound, we thus see that with probability $1 - O(|L|^{-98})$, that *every* line ℓ has at most $O(|L| \log |L|/r)$ points between any two adjacent vertices. Or in other words, the edge of every cell is incident to at most $O(|L| \log |L|/r)$ lines from L. This yields Lemma 1.6.1 except for two things: first, the logarithmic loss of $O(\log |L|)$, and second, the assumption that each cell had only a bounded number of edges.

To fix the latter problem, we will have to modify the construction of R, allowing the use of some lines outside of L. First, we randomly rotate the plane so that none of the lines in L are vertical. Then we do the following modified construction: We select $O(r)$ lines from L as before, creating $O(r^2)$ cells, some of which may have a very large number of edges. But then for each cell, and each vertex in that cell, we draw a vertical line segment from that vertex (in either the up or down direction) to bisect the cell into two pieces. (If the vertex is on the far left or far right of the cell, we do nothing.) Doing this once for each vertex, we see that we have subdivided each of the old cells into a number of new cells, each of which have at most four sides (two vertical sides, and two nonvertical sides). So we have now achieved a bounded number of sides per cell. But what about the number of such cells? Well, each vertex of each cell is responsible for at most two subdivisions of one cell into two, and the number of such vertices is at most $O(r^2)$ (as they are formed by intersecting two lines from the original selection of $O(r)$ lines together), so the total number of cells is still $O(r^2)$.

Is it still true that each edge of each cell is incident to $O(|L| \log |L|/r)$ lines in L? We have already proven this (with high probability) for all the old edges—the ones that were formed from lines in L. But there are now some new edges, caused by dropping a vertical line segment from the intersection of two lines in L. But it is not hard to see that one can use much the same argument as before to see that with high probability, each of these line segments is incident to at most $O(|L| \log |L|/r)$ lines in L as desired.

Finally, we have to get rid of the logarithm. An inspection of the above arguments (and a use of the first moment method) reveals the following refinement: for any $k \geq 1$, there are expected to be at most $O(\exp(-k)r^2)$ cells which are incident to more than $Ck|L|/r$ lines, where C is an absolute constant. This is already enough to improve the $O(|L| \log |L|/r)$ bound slightly to $O(|L| \log r/r)$. But one can do even better by using Lemma 1.6.1 as an induction hypothesis, i.e., assume that for any smaller set L' of lines with $|L'| < |L|$, and any $r' \geq 1$, one can partition L' into at most $C_1(r')^2$ cells using at most $C_0 r'$ lines such that each cell is incident to at most $C_2|L'|/r'$ lines, where C_1, C_2, C_3 are absolute constants. (When using induction, asymptotic notation becomes quite dangerous to use, and it is much safer to start writing out the constants explicitly. To close the induction, one has to end up with the same constants C_0, C_1, C_2 as one started with.) For each k between C_2/C and $O(\log r)$ which is a power of two, one can apply the induction hypothesis to all the cells which are incident to between $Ck|L|/r$ and $2Ck|L|/r$ (with L' set equal to the lines in L incident to this cell, and r' set comparable to $2Ck$), and sum up (using the fact that $\sum_k k^2 \exp(-k)$ converges, especially if k is restricted to powers of two) to close the induction if the constants C_0, C_1, C_2 are chosen properly. We leave the details as an exercise.

Notes. This article first appeared at

`terrytao.wordpress.com/2009/06/12.`

Thanks to Oded and vedadi for corrections.

Jozsef Solymosi noted that there is still no good characterisation of the point-line configurations for which the Szemerédi-Trotter theorem is close to sharp; such a characterisation may well lead to improvements to a variety of bounds which are currently proven using this theorem.

Jordan Ellenberg raised the interesting possibility of using algebraic methods to attack the finite field analogue of this problem.

1.7. Benford's law, Zipf's law, and the Pareto distribution

A remarkable phenomenon in probability theory is that of *universality*—that many seemingly unrelated probability distributions, which ostensibly involve large numbers of unknown parameters, can end up converging to a universal law that may only depend on a small handful of parameters. One of the most famous examples of the universality phenomenon is the *central limit theorem*; another rich source of examples comes from *random matrix theory*, which is one of the areas of my research.

Analogous universality phenomena also show up in *empirical* distributions—the distributions of a statistic X from a large population of "real-world" objects. Examples include *Benford's law*, *Zipf's law*, and the *Pareto distribution* (of which the *Pareto principle* or 80–20 *law* is a special case). These laws govern the asymptotic distribution of many statistics X which

(i) take values as positive numbers;
(ii) range over many different orders of magnitude;
(iii) arise from a complicated combination of largely independent factors (with different samples of X arising from different independent factors); and
(iv) have not been artificially rounded, truncated, or otherwise constrained in size.

Examples here include the population of countries or cities, the frequency of occurrence of words in a language, the mass of astronomical objects, or the net worth of individuals or corporations. The laws are then as follows:

- **Benford's law:** For $k = 1, \ldots, 9$, the proportion of X whose first digit is k is approximately $\log_{10} \frac{k+1}{k}$. Thus, for instance, X should have a first digit of 1 about 30% of the time, but a first digit of 9 only about 5% of the time.
- **Zipf's law:** The nth largest value of X should obey an approximate power law, i.e., it should be approximately $Cn^{-\alpha}$ for the first few $n = 1, 2, 3, \ldots$ and some parameters $C, \alpha > 0$. In many cases, α is close to 1.
- **Pareto distribution:** The proportion of X with at least m digits (before the decimal point), where m is above the median number of digits, should obey an approximate exponential law, i.e., be approximately of the form $c10^{-m/\alpha}$ for some $c, \alpha > 0$. Again, in many cases α is close to 1.

Benford's law and Pareto distribution are stated here for base 10, which is what we are most familiar with, but the laws hold for any base (after replacing all the occurrences of 10 in the above laws with the new base, of course). The laws tend to break down if the hypotheses (i)–(iv) are dropped. For instance, if the statistic X concentrates around its mean (as opposed to being spread over many orders of magnitude), then the *normal distribution* tends to be a much better model (as indicated by such results as the central limit theorem). If instead the various samples of the statistics are highly correlated with each other, then other laws can arise (for instance, the eigenvalues of a random matrix, as well as many empirically observed matrices, are correlated to each other, with the behaviour of the largest eigenvalues being governed by laws such as the *Tracy-Widom law* rather than Zipf's law, and the bulk distribution being governed by laws such as the *semicircular law* rather than the normal or Pareto distributions).

To illustrate these laws, let us take as a data set the populations of 235 countries and regions of the world in 2007.[7] This is a relatively small sample (cf. Section 1.9 of *Poincaré's Legacies, Vol. I*), but is already enough to discern these laws in action. For instance, here is how the data set tracks with Benford's law (rounded to three significant figures):

[7]This data was taken from the CIA world factbook at `http://www.umsl.edu/services/govdocs/wofact2007/index.html`; I have put the raw data at `http://spreadsheets.google.com/pub?key=rj_3TkLJrrVuvOXkijCHelQ&output=html`.

k	Countries	Number	Benford
1	Angola, Anguilla, Aruba, Bangladesh, Belgium, Botswana, Brazil, Burkina Faso, Cambodia, Cameroon, Chad, Chile, China, Christmas Island, Cook Islands, Cuba, Czech Republic, Ecuador, Estonia, Gabon, (The) Gambia, Greece, Guam, Guatemala, Guinea-Bissau, India, Japan, Kazakhstan, Kiribati, Malawi, Mali, Mauritius, Mexico, (Federated States of) Micronesia, Nauru, Netherlands, Niger, Nigeria, Niue, Pakistan, Portugal, Russia, Rwanda, Saint Lucia, Saint Vincent and the Grenadines, Senegal, Serbia, Swaziland, Syria, Timor-Leste (East-Timor), Tokelau, Tonga, Trinidad and Tobago, Tunisia, Tuvalu, (U.S.) Virgin Islands, Wallis and Futuna, Zambia, Zimbabwe	59 (25.1%)	71 (30.1%)
2	Armenia, Australia, Barbados, British Virgin Islands, Cote d'Ivoire, French Polynesia, Ghana, Gibraltar, Indonesia, Iraq, Jamaica, (North) Korea, Kosovo, Kuwait, Latvia, Lesotho, Macedonia, Madagascar, Malaysia, Mayotte, Mongolia, Mozambique, Namibia, Nepal, Netherlands Antilles, New Caledonia, Norfolk Island, Palau, Peru, Romania, Saint Martin, Samoa, San Marino, Sao Tome and Principe, Saudi Arabia, Slovenia, Sri Lanka, Svalbard, Taiwan, Turks and Caicos Islands, Uzbekistan, Vanuatu, Venezuela, Yemen	44 (18.7%)	41 (17.6%)
3	Afghanistan, Albania, Algeria, (The) Bahamas, Belize, Brunei, Canada, (Rep. of the) Congo, Falkland Islands, Iceland, Kenya, Lebanon, Liberia, Liechtenstein, Lithuania, Maldives, Mauritania, Monaco, Morocco, Oman, (Occupied) Palestinian Territory, Panama, Poland, Puerto Rico, Saint Kitts and Nevis, Uganda, United States of America, Uruguay, Western Sahara	29 (12.3%)	29 (12.5%)

k	Countries	Number	Benford
4	Argentina, Bosnia and Herzegovina, Burma (Myanmar), Cape Verde, Cayman Islands, Central African Republic, Colombia, Costa Rica, Croatia, Faroe Islands, Georgia, Ireland, (South) Korea, Luxembourg, Malta, Moldova, New Zealand, Norway, Pitcairn Islands, Singapore, South Africa, Spain, Sudan, Suriname, Tanzania, Ukraine, United Arab Emirates	27 (11.4%)	22 (9.7%)
5	(Macao SAR) China, Cocos Islands, Denmark, Djibouti, Eritrea, Finland, Greenland, Italy, Kyrgyzstan, Montserrat, Nicaragua, Papua New Guinea, Slovakia, Solomon Islands, Togo, Turkmenistan	16 (6.8%)	19 (7.9%)
6	American Samoa, Bermuda, Bhutan, (Dem. Rep. of the) Congo, Equatorial Guinea, France, Guernsey, Iran, Jordan, Laos, Libya, Marshall Islands, Montenegro, Paraguay, Sierra Leone, Thailand, United Kingdom	17 (7.2%)	16 (6.7%)
7	Bahrain, Bulgaria, (Hong Kong SAR) China, Comoros, Cyprus, Dominica, El Salvador, Guyana, Honduras, Israel, (Isle of) Man, Saint Barthelemy, Saint Helena, Saint Pierre and Miquelon, Switzerland, Tajikistan, Turkey	17 (7.2%)	14 (5.8%)
8	Andorra, Antigua and Barbuda, Austria, Azerbaijan, Benin, Burundi, Egypt, Ethiopia, Germany, Haiti, Holy See (Vatican City), Northern Mariana Islands, Qatar, Seychelles, Vietnam	15 (6.4%)	12 (5.1%)
9	Belarus, Bolivia, Dominican Republic, Fiji, Grenada, Guinea, Hungary, Jersey, Philippines, Somalia, Sweden	11 (4.5%)	11 (4.6%)

Here is how the same data tracks Zipf's law for the first twenty values of n, with the parameters $C \approx 1.28 \times 10^9$ and $\alpha \approx 1.03$ (selected by log-linear regression), again rounding to three significant figures:

n	Country	Population	Zipf prediction	Error
1	China	1,330,000,000	1,280,000,000	+4.1%
2	India	1,150,000,000	626,000,000	+83.5%
3	USA	304,000,000	412,000,000	−26.3%
4	Indonesia	238,000,000	307,000,000	−22.5%
5	Brazil	196,000,000	244,000,000	−19.4%
6	Pakistan	173,000,000	202,000,000	−14.4%
7	Bangladesh	154,000,000	172,000,000	−10.9%
8	Nigeria	146,000,000	150,000,000	−2.6%
9	Russia	141,000,000	133,000,000	+5.8%
10	Japan	128,000,000	120,000,000	+6.7%
11	Mexico	110,000,000	108,000,000	+1.7%
12	Philippines	96,100,000	98,900,000	−2.9%
13	Vietnam	86,100,000	91,100,000	−5.4%
14	Ethiopia	82,600,000	84,400,000	−2.1%
15	Germany	82,400,000	78,600,000	+4.8%
16	Egypt	81,700,000	73,500,000	+11.1%
17	Turkey	71,900,000	69,100,000	+4.1%
18	Congo	66,500,000	65,100,000	+2.2%
19	Iran	65,900,000	61,600,000	+6.9%
20	Thailand	65,500,000	58,400,000	+12.1%

As one sees, Zipf's law is not particularly precise at the extreme edge of the statistics (when n is very small), but becomes reasonably accurate (given the small sample size, and given that we are fitting twenty data points using only two parameters) for moderate sizes of n.

This data set has too few scales in base 10 to illustrate the Pareto distribution effectively—over half of the country populations are either seven or eight digits in that base. But if we instead work in base 2, then country populations range in a decent number of scales (the majority of countries have population between 2^{23} and 2^{32}), and we begin to see the law emerge, where m is now the number of digits in binary, the best-fit parameters are $\alpha \approx 1.18$ and $c \approx 1.7 \times 2^{26}/235$:

m	Countries with m-binary-digit populations	Number	Pareto
31	China, India	2	1
30	(None)	2	2
29	United States of America	3	5
28	Indonesia, Brazil, Pakistan, Bangladesh, Nigeria, Russia	9	8
27	Japan, Mexico, Philippines, Vietnam, Ethiopia, Germany, Egypt, Turkey	17	15
26	(Dem. Rep. of the) Congo, Iran, Thailand, France, United Kingdom, Italy, South Africa, (South) Korea, Burma (Myanmar), Ukraine, Colombia, Spain, Argentina, Sudan, Tanzania, Poland, Kenya, Morocco, Algeria	36	27
25	Canada, Afghanistan, Uganda, Nepal, Peru, Iraq, Saudi Arabia, Uzbekistan, Venezuela, Malaysia, (North) Korea, Ghana, Yemen, Taiwan, Romania, Mozambique, Sri Lanka, Australia, Cote d'Ivoire, Madagascar, Syria, Cameroon	58	49
24	Netherlands, Chile, Kazakhstan, Burkina Faso, Cambodia, Malawi, Ecuador, Niger, Guatemala, Senegal, Angola, Mali, Zambia, Cuba, Zimbabwe, Greece, Portugal, Belgium, Tunisia, Czech Republic, Rwanda, Serbia, Chad, Hungary, Guinea, Belarus, Somalia, Dominican Republic, Bolivia, Sweden, Haiti, Burundi, Benin	91	88
23	Austria, Azerbaijan, Honduras, Switzerland, Bulgaria, Tajikistan, Israel, El Salvador, (Hong Kong SAR) China, Paraguay, Laos, Sierra Leone, Jordan, Libya, Papua New Guinea, Togo, Nicaragua, Eritrea, Denmark, Slovakia, Kyrgyzstan, Finland, Turkmenistan, Norway, Georgia, United Arab Emirates, Singapore, Bosnia and Herzegovina, Croatia, Central African Republic, Moldova, Costa Rica	123	159

Thus, with each new scale, the number of countries introduced increases by a factor of a little less than 2, on the average. This approximate doubling of countries with each new scale begins to falter at about the population 2^{23} (i.e., at around 4 million), for the simple reason that one has begun to run out of countries. (Note that the median-population country in this set, Singapore, has a population with 23 binary digits.)

These laws are not merely interesting statistical curiosities; for instance, Benford's law is often used to help detect fraudulent statistics (such as those arising from accounting fraud), as many such statistics are invented by choosing digits at random, and will therefore deviate significantly from Benford's law. (This is nicely discussed in [**Ma1999**].) In a somewhat analogous spirit, Zipf's law and the Pareto distribution can be used to mathematically test various models of real-world systems (e.g., formation of astronomical objects, accumulation of wealth, population growth of countries, etc.), without necessarily having to fit all the parameters of that model with the actual data.

Being empirically observed phenomena rather than abstract mathematical facts, Benford's law, Zipf's law, and the Pareto distribution cannot be "proved" the same way a mathematical theorem can be proved. However, one can still *support* these laws mathematically in a number of ways, for instance showing how these laws are compatible with each other, and with other plausible hypotheses on the source of the data. In this article I would like to describe a number of ways (both technical and nontechnical) in which one can do this. These arguments do not fully explain these laws (in particular, the empirical fact that the exponent α in Zipf's law or the Pareto distribution is often close to 1 is still quite a mysterious phenomenon) and do not always have the same universal range of applicability as these laws seem to have, but I hope that they do demonstrate that these laws are not completely arbitrary, and they ought to have a satisfactory basis of mathematical support.

1.7.1. Scale invariance. One consistency check that is enjoyed by all of these laws is that of *scale invariance*—they are invariant under rescalings of the data (for instance, by changing the units).

For example, suppose for sake of argument that the country populations X of the world in 2007 obey Benford's law, thus for instance about 30.7% of the countries have population with first digit 1, 17.6% have population with first digit 2, and so forth. Now, imagine that several decades in the future, say in 2067, all of the countries in the world double their population, from X to a new population $\tilde{X} := 2X$. (This makes the somewhat implausible assumption that growth rates are uniform across all countries; I will discuss what happens when one omits this hypothesis later.) To further simplify the thought experiment, suppose that no countries are created or dissolved during this time period. What happens to Benford's law when passing from X to $\tilde{X}$?

The key observation here, of course, is that the first digit of X is linked to the first digit of $\tilde{X} = 2X$. If, for instance, the first digit of X is 1, then the first digit of $\tilde{X}$ is either 2 or 3; conversely, if the first digit of $\tilde{X}$ is 2 or 3, then the first digit of X is 1. As a consequence, the proportion of X's with first digit 1 is equal to the proportion of $\tilde{X}$'s with first digit 2, plus the proportion of $\tilde{X}$'s with first digit 3. This is consistent with Benford's law holding for both X and $\tilde{X}$, since

$$\log_{10}\frac{2}{1} = \log_{10}\frac{3}{2} + \log_{10}\frac{4}{3}(= \log_{10}\frac{4}{2})$$

(or numerically, 30.7% = 17.6% + 12.5% after rounding). Indeed one can check the other digit ranges also and that conclude that Benford's law for X is compatible with Benford's law for $\tilde{X}$; to pick a contrasting example, a uniformly distributed model in which each digit from 1 to 9 is the first digit of X occurs with probability $1/9$ totally fails to be preserved under doubling.

One can be even more precise. Observe (through telescoping series) that Benford's law implies that

$$\mathbf{P}(\alpha 10^n \leq X < \beta 10^n \text{ for some integer } n) = \log_{10} \frac{\beta}{\alpha} \tag{1.32}$$

for all integers $1 \leq \alpha \leq \beta < 10$, where the left-hand side denotes the proportion of data for which X lies between $\alpha 10^n$ and $\beta 10^n$ for some integer n. Suppose now that we generalise Benford's law to the *continuous Benford's law*, which asserts that (1.32) is true for all *real numbers* $1 \leq \alpha \leq \beta < 10$. Then it is not hard to show that a statistic X obeys the continuous Benford's law if and only if its dilate $\tilde{X} = 2X$ does, and similarly with 2 replaced by any other constant growth factor. (This is most easily seen by observing that (1.32) is equivalent to asserting that the fractional part of $\log_{10} X$ is uniformly distributed.) In fact, the continuous Benford law is the *only* distribution for the quantities on the left-hand side of (1.32) with this scale-invariance property; this fact is a special case of the general fact that Haar measures are unique (see Section 1.12 of *Volume I*).

It is also easy to see that Zipf's law and the Pareto distribution also enjoy this sort of scale-invariance property, as long as one generalises the Pareto distribution

$$\mathbf{P}(X \geq 10^m) = c10^{-m/\alpha} \tag{1.33}$$

from integer m to real m, just as with Benford's law. Once one does that, one can phrase the Pareto distribution law independently of any base as

$$\mathbf{P}(X \geq x) = cx^{-1/\alpha} \tag{1.34}$$

for any x much larger than the median value of X, at which point the scale-invariance is easily seen.

One may object that the above thought experiment was too idealised, because it assumed uniform growth rates for all the statistics at once. What happens if there are nonuniform growth rates? To keep the computations simple, let us consider the following toy model, where we take the same 2007 population statistics X as before, and assume that half of the countries (the "high-growth" countries) will experience a population doubling by 2067, while the other half (the "zero-growth" countries) will keep their population constant, thus the 2067 population statistic $\tilde{X}$ is equal to $2X$ half the time and X half the time. (We will assume that our sample sizes are large enough that the *law of large numbers* kicks in, and we will therefore ignore issues such as what happens to this "half the time" if the number of samples is odd.) Furthermore, we make the plausible but crucial assumption that the event that a country is a high-growth or a zero-growth country is *independent* of the first digit of the 2007 population; thus, for instance, a country whose population begins with 3 is assumed to be just as likely to be high-growth as one whose population begins with 7.

Let us have a look again at the proportion of countries whose 2067 population $\tilde{X}$ begins with either 2 or 3. There are exactly two ways in which a country can fall into this category: either it is a zero-growth country whose 2007 population X also began with either 2 or 3, or it was a high-growth country whose population in 2007 began with 1. Since all countries have a probability 1/2 of being high-growth regardless of the first digit of their population, we conclude the identity

$$\mathbf{P}(\tilde{X} \text{ has first digit } 2,3) = \frac{1}{2}\mathbf{P}(X \text{ has first digit } 2,3) + \frac{1}{2}\mathbf{P}(X \text{ has first digit } 1), \tag{1.35}$$

which is once again compatible with Benford's law for $\tilde{X}$ since

$$\log_{10}\frac{4}{2} = \frac{1}{2}\log_{10}\frac{4}{2} + \frac{1}{2}\log\frac{2}{1}.$$

More generally, it is not hard to show that if X obeys the continuous Benford's law (1.32) and if one multiplies X by some positive multiplier Y which is independent of the first digit of X (and, *a fortiori*, is independent of the fractional part of $\log_{10} X$), one obtains another quantity $\tilde{X} = XY$ which also obeys the continuous Benford's law. (Indeed, we have already seen this to be the case when Y is a deterministic constant, and the case when Y is random then follows simply by conditioning Y to be fixed.)

In particular, we see an absorptive property of Benford's law: if X obeys Benford's law, and Y is any positive statistic independent of X, then the product $\tilde{X} = XY$ also obeys Benford's law—*even if* Y *did not obey this law.* Thus, if a statistic is the product of many independent factors, then it only requires a single factor to obey Benford's law in order for the whole product to obey the law also. For instance, the population of a country is the product of its area and its population density. Assuming that the population density of a country is independent of the area of that country (which is not a completely reasonable assumption, but let us take it for the sake of argument), then we see that Benford's law for the population would follow if just one of the area or population density obeyed this law. It is also clear that Benford's law is the only distribution with this absorptive property. (If there was another law with this property, what would happen if one multiplied a statistic with that law with an independent statistic with Benford's law?) Thus we begin to get a glimpse as to why Benford's law is universal for quantities which are the product of many separate factors, in a manner that no other law could be.

As an example: for any given number N, the uniform distribution from 1 to N does not obey Benford's law. For instance, if one picks a random number from 1 to $999,999$ then each digit from 1 to 9 appears as the first digit with an equal probability of $1/9$ each. However, if N is not fixed, but instead obeys Benford's law, then a random number selected from 1 to N also obeys Benford's law (ignoring for now the distinction between continuous and discrete distributions), as it can be viewed as the product of N with an independent random number selected from between 0 and 1.

Actually, one can say something even stronger than the absorption property. Suppose that the continuous Benford's law (1.32) for a statistic X did not hold exactly, but instead held with some accuracy $\varepsilon > 0$, thus

$$\begin{aligned} \log_{10}\frac{\beta}{\alpha} - \varepsilon \leq \mathbf{P}(\alpha 10^n \leq X < \beta 10^n \text{ for some integer } n) \\ \leq \log_{10}\frac{\beta}{\alpha} + \varepsilon \end{aligned} \tag{1.36}$$

for all $1 \leq \alpha \leq \beta < 10$. Then it is not hard to see that any dilated statistic, such as $\tilde{X} = 2X$, or more generally $\tilde{X} = XY$ for any fixed deterministic Y, also obeys (1.36) with exactly the same accuracy ε. But now suppose one uses a variable multiplier; for instance, suppose one uses the model discussed earlier in which $\tilde{X}$ is equal to $2X$ half the time and X half the time. Then the relationship between the distribution of the first digit of $\tilde{X}$ and the first digit of X is given by formulae such as (1.35). Now, in the right-hand side of (1.35), each of the two terms $\mathbf{P}(X$ has first digit $2, 3)$ and $\mathbf{P}(X$ has first digit $1)$ differs from the Benford's law predictions of $\log_{10}\frac{4}{2}$ and $\log_{10}\frac{2}{1}$, respectively, by at most ε. Since the left-hand side of (1.35) is the average of these two terms, it also differs from the Benford law prediction by at most ε. But the averaging opens up an opportunity for cancelling; for instance, an overestimate of $+\varepsilon$ for $\mathbf{P}(X$ has first digit $2, 3)$ could cancel an underestimate of $-\varepsilon$ for $\mathbf{P}(X$ has first digit $1)$ to produce a spot-on prediction for $\tilde{X}$. Thus we see that variable multipliers (or variable growth rates) not only preserve Benford's law, but in fact *stabilise* it by averaging out the errors. In fact, if one started with a distribution which did not initially obey Benford's law, and then started applying some variable (and independent) growth rates to the various samples in the distribution, then under reasonable assumptions one can show that the resulting distribution will converge to Benford's law over time. This helps explain the universality of Benford's law for statistics such as populations, for which the independent variable growth law is not so unreasonable (at least, until the population hits some maximum capacity threshold).

Note that the independence property is crucial; if for instance population growth always slowed down for some inexplicable reason to a crawl whenever the first digit of the population was 6, then there would be a noticeable deviation from Benford's law, particularly in digits 6 and 7, due to this growth bottleneck. But this is not a particularly plausible scenario (being somewhat analogous to *Maxwell's demon* in thermodynamics).

The above analysis can also be carried over to some extent to the Pareto distribution and Zipf's law. If a statistic X obeys these laws approximately, then after multiplying by an independent variable Y, the product $\tilde{X} = XY$ will obey the same laws with equal or higher accuracy, so long as Y is small compared to the number of scales that X typically ranges over. (One needs a restriction such as this because the Pareto distribution and Zipf's law must break down below the median. Also, Zipf's law loses its stability at the very extreme end of the distribution, because there are no longer enough samples for the law of large numbers to kick in; this is consistent with the empirical observation that Zipf's law tends to break down *in extremis.*) These laws are also stable under other multiplicative processes, for instance if some fraction of the samples in X spontaneously split into two smaller pieces, or conversely if two samples in X spontaneously merge into one; as before, the key is that the occurrence of these events should be independent of the actual size of the objects being split. If one considers a generalisation of the Pareto distribution or Zipf's law in which the exponent α is not fixed but varies with n or k, then the effect of these sorts of multiplicative changes is to blur and average together the various values of α, thus "flattening" the α curve over time and making the distribution approach Zipf's law and/or the Pareto distribution. This helps explain why α eventually becomes constant; however, I do not have a good explanation as to why α is often close to 1.

1.7.2. Compatibility between laws. Another mathematical line of support for Benford's law, Zipf's law, and the Pareto distribution is that the laws are highly compatible with each other. For instance, Zipf's law and the Pareto distribution are formally equivalent: if there are N samples of X, then applying (1.34) with x equal to the nth largest value X_n of X gives

$$\frac{n}{N} = \mathbf{P}(X \geq X_n) = cX_n^{-1/\alpha},$$

which implies Zipf's law $X_n = Cn^{-\alpha}$ with $C := (Nc)^\alpha$. Conversely, one can deduce the Pareto distribution from Zipf's law. These deductions are only formal in nature, because the Pareto distribution can only hold exactly for continuous distributions, whereas Zipf's law only makes sense for discrete distributions, but one can generate more rigorous variants of these deductions without much difficulty.

In some literature, Zipf's law is applied primarily near the extreme edge of the distribution (e.g., the top 0.1% of the sample space), whereas the Pareto distribution in regions closer to the bulk (e.g., between the top 0.1%

and and top 50%). But this is mostly a difference of degree rather than of kind, though in some cases (such as with the example of the 2007 country populations data set) the exponent α for the Pareto distribtion in the bulk can differ slightly from the exponent for Zipf's law at the extreme edge.

The relationship between Zipf's law or the Pareto distribution and Benford's law is more subtle. For instance Benford's law predicts that the proportion of X with initial digit 1 should equal the proportion of X with initial digit 2 or 3. But if one formally uses the Pareto distribution (1.34) to compare those X between 10^m and 2×10^m, and those X between 2×10^m and 4×10^m, it seems that the former is larger by a factor of $2^{1/\alpha}$, which upon summing by m appears inconsistent with Benford's law (unless α is extremely large). A similar inconsistency is revealed if one uses Zipf's law instead.

However, the fallacy here is that the Pareto distribution (or Zipf's law) does not apply on the entire range of X, but only on the upper tail region when X is significantly higher than the median; it is a law for the *outliers* of X only. In contrast, Benford's law concerns the behaviour of *typical* values of X; the behaviour of the top 0.1% is of negligible significance to Benford's law, though it is of prime importance for Zipf's law and the Pareto distribution. Thus the two laws describe different components of the distribution and thus complement each other. Roughly speaking, Benford's law asserts that the bulk distribution of $\log_{10} X$ is locally uniform at unit scales, while the Pareto distribution (or Zipf's law) asserts that the tail distribution of $\log_{10} X$ decays exponentially. Note that Benford's law only describes the fine-scale behaviour of the bulk distribution; the coarse-scale distribution can be a variety of distributions (e.g., log-Gaussian).

Notes. This article first appeared at

`terrytao.wordpress.com/2009/07/03.`

Thanks to Kevin O'Bryant for corrections. Several other derivations of Benford's law and the Pareto distribution, such as those relying on max-entropy principles, were also discussed in the comments.

1.8. Selberg's limit theorem for the Riemann zeta function on the critical line

The *Riemann zeta function* $\zeta(s)$, defined for $\mathrm{Re}(s) > 1$ by

$$\zeta(s) := \sum_{n=1}^{\infty} \frac{1}{n^s} \tag{1.37}$$

and then continued *meromorphically* to other values of s by *analytic continuation*, is a fundamentally important function in analytic number theory, as it is connected to the primes $p = 2, 3, 5, \ldots$ via the *Euler product formula*

$$\zeta(s) = \prod_p (1 - \frac{1}{p^s})^{-1} \tag{1.38}$$

(for $\mathrm{Re}(s) > 1$, at least), where p ranges over primes. (The equivalence between (1.37) and (1.38) is essentially the *generating function* version of the *fundamental theorem of arithmetic*.) The function ζ has a pole at 1 and a number of zeroes ρ. A formal application of the *factor theorem* gives

$$\zeta(s) = \frac{1}{s-1} \prod_\rho (s - \rho) \times \cdots, \tag{1.39}$$

where ρ ranges over zeroes of ζ, and we will be vague about what the $\cdots$ factor is, how to make sense of the infinite product, and exactly which zeroes of ζ are involved in the product. Equating (1.38) and (1.39) and taking logarithms gives the formal identity

$$-\log \zeta(s) = \sum_p \log(1 - \frac{1}{p^s}) = \log(s-1) - \sum_\rho \log(s - \rho) + \cdots; \tag{1.40}$$

using the Taylor expansion

$$\log(1 - \frac{1}{p^s}) = -\frac{1}{p^s} - \frac{1}{2p^{2s}} - \frac{1}{3p^{3s}} - \cdots \tag{1.41}$$

and differentiating the above identity in s yields the formal identity

$$-\frac{\zeta'(s)}{\zeta(s)} = \sum_n \frac{\Lambda(n)}{n^s} = \frac{1}{s-1} - \sum_\rho \frac{1}{s-\rho} + \cdots, \tag{1.42}$$

where $\Lambda(n)$ is the *von Mangoldt function*, defined to be $\log p$ when n is a power of a prime p and zero otherwise. Thus we see that the behaviour of the primes (as encoded by the von Mangoldt function) is intimately tied to the distribution of the zeroes ρ. For instance, if we knew that the zeroes were far away from the axis $\mathrm{Re}(s) = 1$, then we would heuristically have

$$\sum_n \frac{\Lambda(n)}{n^{1+it}} \approx \frac{1}{it}$$

for real t. On the other hand, the integral test suggests that

$$\sum_n \frac{1}{n^{1+it}} \approx \frac{1}{it},$$

and thus we see that $\frac{\Lambda(n)}{n}$ and $\frac{1}{n}$ have essentially the same (multiplicative) *Fourier transform*:

$$\sum_n \frac{\Lambda(n)}{n^{1+it}} \approx \sum_n \frac{1}{n^{1+it}}.$$

Inverting the Fourier transform (or performing a contour integral closely related to the inverse Fourier transform), one is led to the *prime number theorem*

$$\sum_{n \leq x} \Lambda(n) \approx \sum_{n \leq x} 1.$$

In fact, the standard proof of the prime number theorem basically proceeds by making all of the above formal arguments precise and rigorous.

Unfortunately, we do not know as much about the zeroes ρ of the zeta function (and hence, about the ζ function itself) as we would like. The *Riemann hypothesis* (RH) asserts that all the zeroes (except for the "trivial" zeroes at the negative even numbers) lie on the *critical line* $\mathrm{Re}(s) = 1/2$; this hypothesis would make the error terms in the above proof of the prime number theorem significantly more accurate. Furthermore, the stronger *GUE hypothesis* asserts in addition to the RH that the local distribution of these zeroes on the critical line should behave like the local distribution of the eigenvalues of a random matrix drawn from the *Gaussian unitary ensemble* (GUE). I will not give a precise formulation of this hypothesis here, except to say that the adjective "local" in the context of distribution of zeroes ρ means something like "at scale $O(1/\log T)$ when $\mathrm{Im}(s) = O(T)$".

Nevertheless, we do know some reasonably nontrivial facts about the zeroes ρ and the zeta function ζ, either unconditionally, or assuming RH (or GUE). Firstly, there are no zeroes for $\mathrm{Re}(s) > 1$ (as one can already see from the convergence of the Euler product (1.38) in this case) or for $\mathrm{Re}(s) = 1$ (this is trickier, relying on (1.42) and the elementary observation that

$$\mathrm{Re}(3\frac{\Lambda(n)}{n^\sigma} + 4\frac{\Lambda(n)}{n^{\sigma+it}} + \frac{\Lambda(n)}{n^{\sigma+2it}}) = 2\frac{\Lambda(n)}{n^\sigma}(1 + \cos(t \log n))^2$$

is nonnegative for $\sigma > 1$ and $t \in \mathbf{R}$); from the *functional equation*

$$\pi^{-s/2}\Gamma(s/2)\zeta(s) = \pi^{-(1-s)/2}\Gamma((1-s)/2)\zeta(1-s)$$

(which can be viewed as a consequence of the *Poisson summation formula*, see e.g., Section 1.5 of *Poincaré's Legacies, Vol. I*) we know that there are no zeroes for $\mathrm{Re}(s) \leq 0$ either (except for the trivial zeroes at negative even integers, corresponding to the poles of the Gamma function). Thus all the nontrivial zeroes lie in the *critical strip* $0 < \mathrm{Re}(s) < 1$.

We also know that there are infinitely many nontrivial zeroes, and we can approximately count how many zeroes there are in any large bounded region of the critical strip. For instance, for large T, the number of zeroes ρ in this strip with $\mathrm{Im}(\rho) = T + O(1)$ is $O(\log T)$. This can be seen by applying (1.42) to $s = 2 + iT$ (say); the trivial zeroes at the negative integers end up giving a contribution of $O(\log T)$ to this sum (this is a heavily disguised variant of *Stirling's formula*, as one can view the trivial zeroes as essentially being poles of the Gamma function), while the $\frac{1}{s-1}$ and $\cdots$ terms end up being negligible (of size $O(1)$), while each nontrivial zero ρ contributes a term which has a nonnegative real part, and furthermore has size comparable to 1 if $\mathrm{Im}(\rho) = T + O(1)$. (Here, I am glossing over a technical renormalisation needed to make the infinite series in (1.42) converge properly.) Meanwhile, the left-hand side of (1.42) is absolutely convergent for $s = 2 + iT$ and of size $O(1)$, and the claim follows. A more refined version of this argument shows that the number of nontrivial zeroes with $0 \leq \mathrm{Im}(\rho) \leq T$ is $\frac{T}{2\pi} \log \frac{T}{2\pi} - \frac{T}{2\pi} + O(\log T)$, but we will not need this more precise formula here. (A fair fraction—at least 40%, in fact—of these zeroes are known to lie on the critical line; see [**Co1989**].)

Another thing that we happen to know is how the *magnitude* $|\zeta(1/2+it)|$ of the zeta function is distributed as $t \to \infty$. It turns out to be *log-normally* distributed with log-variance about $\frac{1}{2} \log \log t$. More precisely, we have the following result of Selberg:

Theorem 1.8.1. *Let T be a large number, and let t be chosen uniformly at random from between T and $2T$ (say). Then the distribution of $\frac{1}{\sqrt{\frac{1}{2}\log\log T}} \log|\zeta(1/2 + it)|$ converges (*in distribution*) to the normal distribution $N(0,1)$.*

To put it more informally, $\log|\zeta(1/2 + it)|$ behaves like $\sqrt{\frac{1}{2}\log\log t} \times N(0,1)$ plus lower order terms for "typical" large values of t. (Zeroes ρ of ζ are, of course, certainly not typical, but one can show that one can usually stay away from these zeroes.) In fact, Selberg showed a slightly more precise result, namely that for any fixed $k \geq 1$, the kth moment of $\frac{1}{\sqrt{\frac{1}{2}\log\log T}} \log|\zeta(1/2 + it)|$ converges to the kth moment of $N(0,1)$.

Remarkably, Selberg's result does not need RH or GUE, though it is certainly consistent with such hypotheses. (For instance, the determinant of a GUE matrix asymptotically obeys a remarkably similar log-normal law to that given by Selberg's theorem.) Indeed, the net effect of these hypotheses only affects some error terms in $\log|\zeta(1/2 + it)|$ of magnitude $O(1)$, and are thus asymptotically negligible compared to the main term, which has magnitude about $O(\sqrt{\log\log T})$. So Selberg's result, while very pretty, manages to finesse the question of what the zeroes ρ of ζ are actually doing—he makes the primes do most of the work rather than the zeroes.

Selberg never actually published the above result, but it is reproduced in a number of places (e.g., in [**Jo1986**] or [**La1996**]). As with many other results in analytic number theory, the actual details of the proof can get somewhat technical; but I would like to record here (partly for my own benefit) an informal sketch of some of the main ideas in the argument.

1.8.1. Informal overview of argument. The first step is to get a usable (approximate) formula for $\log|\zeta(s)|$. On the one hand, from the second part of (1.40) one has

$$-\log|\zeta(s)| = \log|s-1| - \sum_\rho \log|s-\rho| + \cdots. \tag{1.43}$$

This formula turns out not to be directly useful because it requires much more knowledge about the distribution of the zeroes ρ than we currently possess. On the other hand, from the first part of (1.40) and (1.41) one also has the formula

$$\log|\zeta(s)| = \sum_p \operatorname{Re} \frac{1}{p^s} + \cdots. \tag{1.44}$$

This formula also turns out not to be directly useful, because it requires much more knowledge about the distribution of the primes p than we currently possess.

However, it turns out that we can "split the difference" between (1.43), (1.44), and get a formula for $\log|\zeta(s)|$ which involves some zeroes ρ and some primes p, in such a manner that one can control them both. Roughly speaking, the formula looks like this:[8]

$$\log|\zeta(s)| = \sum_{p \leq T^\varepsilon} \operatorname{Re}\frac{1}{p^s} + O(\sum_{\rho = s+O(1/\log T)} 1 + |\log \frac{|s-\rho|}{1/\log T}|) + \cdots \tag{1.45}$$

for $s = 1/2 + it$ and $t = O(T)$, where ε is a small parameter that we can choose (e.g., $\varepsilon = 0.01$); thus we have localised the prime sum to the primes p of size $O(T^{O(1)})$, and the zero sum to those zeroes at a distance $O(1/\log T)$ from s.

It turns out that all of these expressions can be controlled. The error term coming from the zeroes (as well as the $\cdots$ error term) turn out to be of size $O(1)$ for most values of t, so are a lower order term. (As mentioned before, it is this error term that would be better controlled if one had RH or GUE, but this is not necessary to establish Selberg's result.) The main term is the one coming from the primes.

We can heuristically argue as follows. The expression $X_p := \operatorname{Re}\frac{1}{p^s} = \frac{1}{\sqrt{p}}\cos(t\log p)$, for t ranging between T and $2T$, is a random variable of mean zero and variance approximately $\frac{1}{2p}$ (if $p \leq T^\varepsilon$ and ε is small). Making the heuristic assumption that the X_p behave as if they were independent, the *central limit theorem* then suggests that the sum $\sum_{p\leq T^\varepsilon} X_p$ should behave

[8] This is an oversimplification: there is a "tail" coming from those zeroes that are more distant from s than $O(1/\log T)$, and also one has to smooth out the sum in p a little bit, and allow the implied constants in the $O()$ notation to depend on ε. But let us ignore these technical issues here as well as the issue of what exactly is hiding in the $\cdots$ error.

like a normal distribution of mean zero and variance $\sum_{p \leq T^\varepsilon} \frac{1}{2p}$. But the claim now follows from the classical estimate

$$\sum_{p \leq x} \frac{1}{p} = \log \log x + O(1)$$

(which follows from the prime number theorem, but can also be deduced from the formula (1.44) for $s = 1 + O(1/\log x)$, using the fact that ζ has a simple pole at 1).

To summarise, there are three main tasks to establish Selberg's theorem:

(1) Establish a formula along the lines of (1.45);

(2) Show that the error terms arising from zeroes are $O(1)$ on the average;

(3) Justify the central limit calculation for $\sum_p X_p$.

I'll now briefly discuss (informally) each of the three steps in turn.

1.8.2. The explicit formula. To get a formula such as (1.45), the basic strategy is to take a suitable average of the formula (1.43) and the formula (1.44). Traditionally, this is done by contour integration; however, I prefer (perhaps idiosyncratically) to take a more Fourier-analytic perspective, using convolutions rather than contour integrals. (The two approaches are largely equivalent, though.) The basic point is that the imaginary part $\mathrm{Im}(\rho)$ of the zeroes inhabits the same space as the imaginary part $t = \mathrm{Im}(s)$ of the s variable, which in turn is the Fourier analytic dual of the variable that the logarithm $\log p$ of the primes p live in; this can be seen by writing (1.43), (1.44) in a more Fourier-like manner[9] as

$$\sum_\rho \log |1/2 + it - \rho| + \cdots = \mathrm{Re} \sum_p \frac{1}{\sqrt{p}} e^{-it \log p} + \cdots .$$

The uncertainty principle then predicts that localising $\log p$ to the scale $O(\log T^\varepsilon)$ should result in blurring out the zeroes ρ at scale $O(1/\log T^\varepsilon)$, which is where (1.45) is going to come from.

Let us see how this idea works in practice. We consider a convolution of the form

$$\int_{\mathbf{R}} \log |\zeta(s + \frac{iy}{\log T^\varepsilon})| \psi(y)\, dy, \tag{1.46}$$

where ψ is some bump function with total mass 1; informally, this is $\log |\zeta(s)|$ averaged out in the vertical direction at scale $O(1/\log T^\varepsilon) = O(1/\log T)$ (we allow implied constants to depend on ε).

[9]These sorts of Fourier-analytic connections are often summarised by the slogan "the zeroes of the zeta function are the music of the primes".

We can express (1.46) in two different ways, one using (1.43), and one using (1.44). Let us look at (1.43) first. If one modifies s by $O(1/\log T)$, then the quantity $\log|s-\rho|$ does not fluctuate very much, unless ρ is within $O(1/\log T)$ of s, in which case it can move by about $O(1+\log\frac{|s-\rho|}{1/\log T})$. As a consequence, we see that

$$\int_{\mathbf{R}} \log|s+\frac{iy}{\log T^\varepsilon}-\rho|\psi(y)\,dy \approx \log|s-\rho|$$

when $|\rho-s| \gg 1/\log T$, and

$$\int_{\mathbf{R}} \log|s+\frac{iy}{\log T^\varepsilon}-\rho|\psi(y)\,dy = \log|s-\rho| + O(1+\log\frac{|s-\rho|}{1/\log T}).$$

The quantity $\log|s-1|$ also does not move very much by this shift (we are assuming the imaginary part of s to be large). Inserting these facts into (1.43), we thus see that (1.46) is (heuristically) equal to

$$\log|\zeta(s)| + \sum_{\rho=s+O(1/\log T)} O(1+\log\frac{|s-\rho|}{1/\log T}) + \cdots . \tag{1.47}$$

Now let us compute (1.46) using (1.44) instead. Writing $s=1/2+it$, we express (1.46) as

$$\sum_p \mathrm{Re}\,\frac{1}{p^s}\int_{\mathbf{R}} e^{-iy\log p/\log T^\varepsilon}\psi(y)\,dy + \cdots .$$

Introducing the Fourier transform $\hat\psi(\xi) := \int_{\mathbf{R}} e^{-iy\xi}\psi(y)\,dy$ of ψ, one can write this as

$$\sum_p \mathrm{Re}\,\frac{1}{p^s}\hat\psi(\log p/\log T^\varepsilon) + \cdots .$$

Now we took ψ to be a bump function, so its Fourier transform should also be like a bump function (or perhaps a Schwartz function). As a first approximation, one can thus think of $\hat\psi$ as a smoothed truncation to the region $\{\xi : \xi = O(1)\}$, thus the $\hat\psi(\log p/\log T^\varepsilon)$ weight is morally restricting p to the region $p \le T^\varepsilon$. Thus we (morally) can express (1.46) as

$$\sum_{p\le T^\varepsilon} \mathrm{Re}\,\frac{1}{p^s} + \cdots .$$

Comparing this with the other formula (1.47) we have for (1.46), we obtain (1.45) as required (formally, at least).

1.8.3. Controlling the zeroes. Next, we want to show that the quantity

$$\sum_{\rho=s+O(1/\log T)} 1+|\log \frac{|s-\rho|}{1/\log T}|$$

is $O(1)$ on the average, when $s = 1/2 + it$ and t is chosen uniformly at random from T to $2T$.

For this, we can use the *first moment method*. For each zero ρ, let I_ρ be the random variable which equals $1+|\log \frac{|s-\rho|}{1/\log T}|$ when $\rho = s + O(1/\log T)$ and zero otherwise, thus we are trying to control the expectation of $\sum_\rho I_\rho$. The only zeroes which are relevant are those which are of size $O(T)$, and we know that there are $O(T \log T)$ of these (indeed, we have an even more precise formula, as remarked earlier). On the other hand, a randomly chosen s has a probability of $O(1/T \log T)$ of falling within $O(1/\log T)$ of ρ, and so we expect each I_ρ to have an expected value of $O(1/T \log T)$. (The logarithmic factor in the definition of I_ρ turns out not to be an issue, basically because $\log x$ is locally integrable.) By linearity of expectation, we conclude that $\sum_\rho I_\rho$ has expectation $O(T \log T) \times O(1/T \log T) = O(1)$, and the claim follows.

Remark 1.8.2. One can actually do a bit better than this, showing that higher order moments of $\sum_\rho I_\rho$ are also $O(1)$, by using a variant of (1.45) together with the moment bounds in the next section; but we will not need that refinement here.

1.8.4. The central limit theorem. Finally, we have to show that $\sum_{p \leq T^\varepsilon} X_p$ behaves like a normal distribution, as predicted by the central limit theorem heuristic. The key is to show that the X_p behave "as if" they were jointly independent. In particular, as the X_p all have mean zero, one would like to show that products such as

$$X_{p_1} \cdots X_{p_k} \tag{1.48}$$

have a negligible expectation as long as at least one of the primes in $p_1, \ldots, p_k$ occurs at most once. Once one has this (as well as a similar formula for the case when all primes appear at least twice), one can then do a standard moment computation of the kth moment $(\sum_{p \leq T^\varepsilon} X_p)^k$ and verify that this moment then matches the answer predicted by the central limit theorem, which by standard arguments (involving the *Weierstrass approximation theorem*) is enough to establish the distributional law. Note that to get close to the normal distribution by a fixed amount of accuracy, it suffices to control a bounded number of moments, which ultimately means that we can treat k as being bounded, $k = O(1)$.

If we expand out the product (1.48), we get

$$\frac{1}{\sqrt{p_1}\cdots\sqrt{p_k}}\cos(t\log p_1)\cdots\cos(t\log p_k).$$

Using the product formula for cosines (or *Euler's formula*), the product of cosines here can be expressed as a linear combination of cosines $\cos(t\xi)$, where the frequency ξ takes the form

$$\xi = \pm\log p_1 \pm \log p_2 \cdots \pm \log p_k.$$

Thus, ξ is the logarithm of a rational number, whose numerator and denominator are the product of some of the $p_1, \ldots, p_k$. Since all the p_j are at most T^ε, we see that the numerator and denominator here are at most $T^{k\varepsilon}$.

Now for the punchline. If there is a prime in $p_1, \ldots, p_k$ that appears only once, then the numerator and denominator cannot fully cancel, by the *fundamental theorem of arithmetic*. Thus ξ cannot be 0. Furthermore, since the denominator is at most $T^{k\varepsilon}$, we see that ξ must stay away from 0 by a distance of about $1/T^{k\varepsilon}$ or more, and so $\cos(t\xi)$ has a wavelength of at most $O(T^{k\varepsilon})$. On the other hand, t ranges between T and $2T$. If k is fixed and ε is small enough (much smaller than $1/k$), we thus see that the average value of $\cos(t\xi)$ between T and $2T$ is close to zero, and so (1.48) does indeed have negligible expectation as claimed. (A similar argument lets one compute the expectation of (1.48) when all primes appear at least twice.)

Remark 1.8.3. A famous theorem of Erdős and Kac [**ErKa1940**] gives a normal distribution for the number of prime factors of a large number n, with mean $\log\log n$ and variance $\log\log n$. One can view Selberg's theorem as a sort of Fourier-analytic variant of the Erdős-Kac theorem.

Remark 1.8.4. The Fourier-like correspondence between zeroes of the zeta function and primes can be used to convert statements about zeroes, such as the Riemann hypothesis and the GUE hypothesis, into equivalent statements about primes. For instance, the Riemann hypothesis is equivalent to having the square root error term

$$\sum_{x\le n\le x+y}\Lambda(n) = y + O_\varepsilon(y^{1/2+\varepsilon})$$

in the prime number theorem holding asymptotically as $x\to\infty$ for all $\varepsilon > 0$ and all intervals $[x, x+y]$ which are *large* in the sense that y is comparable to x. Meanwhile, the pair correlation conjecture (the simplest component of the GUE hypothesis) is equivalent (on RH) to the square root error term holding (with the expected variance) for all $\varepsilon > 0$ and *almost* all intervals $[x, x+y]$ which are *short* in the sense that $y = x^\theta$ for some small (fixed) $\theta > 0$. (This is a rough statement; a more precise formulation can be found in [**GoMo1987**].) It seems to me that reformulation of the full GUE hypothesis in terms of primes should be similar, but would assert that the error term in the prime number theorem (as well as variants of this theorem for almost primes) in short intervals enjoys the expected normal distribution; I do not know of a precise formulation of this assertion, but calculations in this direction lie in [**BoKe1996**].

Notes. This article first appeared at

`terrytao.wordpress.com/2009/07/12.`

Thanks to anonymous commenters for corrections.

Emmanuel Kowalski discusses the relationship between Selberg's limit theorem and the Erdős-Kac theorem further at

`http://blogs.ethz.ch/kowalski/2009/02/28/a-beautiful-analogy-2/.`

1.9. $P = NP$, relativisation, and multiple-choice exams

The most fundamental unsolved problem in complexity theory is undoubtedly the $P = NP$ *problem*, which asks (roughly speaking) whether a problem which can be solved by a *nondeterministic polynomial-time (NP)* algorithm, can also be solved by a *deterministic polynomial-time (P)* algorithm. The general belief is that $P \neq NP$, i.e., there exist problems which can be solved by nondeterministic polynomial-time algorithms but not by deterministic polynomial-time algorithms.

One reason why the $P \neq NP$ question is so difficult to resolve is that a certain generalisation of this question has an affirmative answer in some cases, and a negative answer in other cases. More precisely, if we give all the algorithms access to an *oracle*, then for one choice A of this oracle, all the problems that are solvable by nondeterministic polynomial-time algorithms that calls A (NP^A), can also be solved by a deterministic polynomial-time algorithm algorithm that calls A (P^A), thus $P^A = NP^A$. But for another choice B of this oracle, there exist problems solvable by nondeterministic polynomial-time algorithms that call B, which *cannot* be solved by a deterministic polynomial-time algorithm that calls B, thus $P^B \neq NP^B$. One particular consequence of this result (which is due to Baker, Gill, and Solovay [**BaGiSo1975**]) is that there cannot be any *relativisable* proof of either $P = NP$ or $P \neq NP$, where "relativisable" means that the proof would also work without any changes in the presence of an oracle.

The Baker-Gill-Solovay result was quite surprising, but the idea of the proof turns out to be rather simple. To get an oracle A such that $P^A = NP^A$, one basically sets A to be a powerful simulator that can simulate nondeterministic machines (and, furthermore, can also simulate *itself*); it turns out that any *PSPACE-complete* oracle would suffice for this task. To get an oracle B for which $P^B \neq NP^B$, one has to be a bit sneakier, setting B to be a query device for a sparse set of random (or high-complexity) strings, which are too complex to be guessed at by any deterministic polynomial-time algorithm.

Unfortunately, the simple idea of the proof can be obscured by various technical details (e.g., using *Turing machines* to define P and NP precisely), which require a certain amount of time to properly absorb. To help myself try to understand this result better, I have decided to give a sort of "allegory" of the proof, based around a (rather contrived) story about various students trying to pass a multiple-choice test, which avoids all the technical details but still conveys the basic ideas of the argument.

1.9.1. P and NP students.

In this story, two students, named P and NP (and which for sake of grammar, I will arbitrarily assume to be male), are preparing for their final exam in a mathematics course, which will consist of a long, tedious sequence of multiple-choice questions, or more precisely true-false questions. The exam has a reasonable but fixed time limit (e.g., three hours), and unlimited scratch paper is available during the exam. Students are allowed to bring one small index card into the exam. Other than scratch paper, an index card, and a pencil, no other materials are allowed. Students cannot leave questions blank; they must answer each question true or false. The professor for this course is dull and predictable; everyone knows in advance the type of questions that will be on the final, the only issue being the precise numerical values that will be used in the actual questions.

For each student response to a question, there are three possible outcomes:

- **Correct answer.** The student answers the question correctly.
- **False negative.** The student answers "false", but the actual answer is "true".
- **False positive.** The student answers "true", but the actual answer is "false".

We will assume a certain asymmetry in the grading: a few points are deducted for false negatives, but a large number of points are deducted for false positives. (There are many real-life situations in which one type of error is considered less desirable than another; for instance, when deciding on guilt in a capital crime, a false positive is generally considered a much worse mistake than a false negative.) So, while students would naturally like to ace the exam by answering all questions correctly, they would tend to err on the side of caution and put down "false" when in doubt.

Student P is hard working and careful, but unimaginative and with a poor memory. His exam strategy is to put all the techniques needed to solve the exam problems on the index card, so that they can be applied by rote during the exam. If the nature of the exam is such that P can be guaranteed to ace it by this method, we say that the exam *is in class* P. For instance, if the exam will consist of verifying various multiplication problems (e.g., "Is $231 * 136 = 31516$?"), then this exam is in class P, since P can put the algorithm for long multiplication, together with a multiplication table, on the index card, and perform these computations during the exam. A more nontrivial example of an exam in class P would be an exam consisting solely of determining whether various large numbers are prime; here P could be guaranteed to ace the test by writing down on his index card the details of the *AKS primality test*.

Student NP is similar to P, but is substantially less scrupulous; he has bribed the proctor of the exam to supply him with a full solution key, containing not only the answers, but also the worked computations that lead to that answer (when the answer is "true"). The reason he has asked (and paid) for the latter is that he does not fully trust the proctor to give reliable answers, and is terrified of the impact to his grades if he makes a false positive. Thus, if the answer key asserts that the answer to a question is "true", he plans to check the computations given to the proctor himself before putting down "true"; if he cannot follow these computations, and cannot work out the problem himself, he will play it safe and put down "false" instead.

We will say that the exam *is in class* NP if

- NP is guaranteed to ace the exam if the information given to him by the proctor is reliable;
- NP is guaranteed not to make a false positive, even if the proctor has given him unreliable information.

For instance, imagine an exam consisting of questions such as "Is Fermat's Last Theorem provable in ten pages or less?". Such an exam is in the class NP, as the student can bribe the proctor to ask for a ten-page proof of FLT, if such exists, and then would check that proof carefully before putting down "True". This way, the student is guaranteed not to make a false positive (which, in this context, would be a severe embarrassment to any reputable mathematician), and will ace the exam if the proctor actually does happen to have all the relevant proofs available.

It is clear that NP is always going to do at least as well as P, since NP always has the option of ignoring whatever the proctor gives him, and copying P's strategy instead. But how much of an advantage does NP have over P? In particular, if we give P a little bit more time (and a somewhat larger index card), could every exam that is in class NP, also be in class P? This, roughly speaking, is the $P = NP$ problem. It is believed that $P \neq NP$, thus there are exams which NP will ace (with reliable information) and will at least not make a false positive (even with unreliable information), but for which P is not guaranteed to ace, even with a little extra time and space.

1.9.2. Oracles. Now let us modify the exams a bit by allowing a limited amount of computer equipment in the exam. In addition to the scratch paper, pencil, and index card, every student in the exam is now also given access to a computer A which can perform a carefully limited set of tasks that are intended to assist the student. Examples of tasks permitted by A could include a scientific calculator, a mathematics package such as Matlab or SAGE, or access to Wikipedia or Google. We say that an exam is *in class* P^A if it can be guaranteed to be aced by P if he has access to A, and similarly the exam *is in class* NP^A if it can be guaranteed to be aced by NP if he has access to A and the information obtained from the proctor was reliable, and if he is at least guaranteed not to make a false positive with access to A if the information from the proctor turned out to be unreliable. Again, it is clear that NP will have the advantage over P, in the sense that every exam in class P^A will also be in class NP^A. (In other words, the proof that $P \subset NP$ *relativises*.) But what about the converse—is every exam in class NP^A also in class P^A (if we give P a little more time and space, and perhaps also a slightly larger and faster version of A)?

We now give an example of a computer A with the property that $P^A = NP^A$, i.e., that every exam in class NP^A, is also in class P^A. Here, A is an extremely fast computer with reasonable amount of memory and a compiler for a general-purpose programming language, but with no additional capabilities. (More precisely, A should be a *PSPACE-complete* language, but let me gloss over the precise definition of this term here.)

Suppose that an exam is in class NP^A, thus NP will ace the exam if he can access A and has reliable information, and will not give any false positives if he can access A and has unreliable information. We now claim that P can also ace this exam, if he is given a little bit more time and a slightly larger version of A. The way he does it is to program his version of A to simulate NP's strategy, by looping through all possible values of the solution key that NP might be given, and also simulating NP's copy of A as well. (The latter task is possible as long as P's version of A is slightly larger and faster than NP's version.) There are of course an extremely large number of combinations in the solution key to loop over (for instance, consider how many possible proofs of Fermat's Last Theorem under ten pages there could be), but we assume that the computer is so fast that it can handle all these combinations without difficulty. If at least one of the possible choices for a solution key causes the simulation of NP to answer "true", then P will answer "true" also; if instead none of the solution keys cause NP to answer "true", then P will answer "false" instead. If the exam is in class NP^A, it is then clear that P will ace the exam.

Now we give an example of a computer B with the property that $P^B \neq NP^B$, i.e., there exists an exam which is in class NP^B, but that P is not guaranteed to ace even with the assistance of B. The only software loaded on B is a web browser, which can fetch any web page desired after typing in the correct URL. However, rather than being connected to the Internet, the browser can only access a local file system of pages. Furthermore, there is no directory or search feature in this file system—the only way to find a page is to type in its URL, and if you cannot guess the URL correctly, there is no way to access that page. (In particular, there are no links between pages.)

Furthermore, to make matters worse, the URLs are not designed according to any simple scheme, but have in fact been generated randomly, by the following procedure. For each positive integer n, flip a coin. If the coin is heads, then create a URL of n random characters and place a web page at that URL. Otherwise, if the coin is tails, do nothing. Thus, for each n, there will either be one web page with a URL of length n, or there will be no web pages of this length; but in the former case, the web page will have an address consisting of complete gibberish, and there will be no means to obtain this address other than by guessing.

The exam will consist of a long series of questions such as "Is there a web page on B with a URL of 1254 characters in length?"

It is clear that this exam is in class NP^B. Indeed, for NP to ace this exam, he just needs to bribe the proctor for the URLs of all the relevant web pages (if they exist). He can then confirm their existence by typing them into B, and then answer "true" if he finds the page, and "false" otherwise. It is clear that NP will ace the exam if the proctor information is reliable, and will avoid false positives otherwise.

On the other hand, poor P will have no chance to ace this exam if the length of the URLs are long enough, for two reasons. Firstly, the browser B is useless to him: any URL he can guess will have almost no chance of being the correct one, and so the only thing he can generate on the browser is an endless stream of "404 Not Found" messages. (Indeed, these URLs are very likely to have a high *Kolmogorov complexity*, and thus cannot be guessed by P. Admittedly, P does have B available, but one can show by induction on the number of queries that B is useless to P. We also make the idealised assumption that *side-channel attacks* are not available.) As B is useless, the only hope P has is to guess the sequence of coin flips that were used to determine the set of n for which URLs exist of that length. But the random sequence of coin flips is also likely to have high Kolmogorov complexity, and thus cannot be guaranteed to be guessed by P either. Thus $P^B \neq NP^B$.

Remark 1.9.1. Note how the existence of long random strings could be used to make an oracle that separates P from NP. In the absence of oracles, it appears that separation of P from NP is closely connected to the existence of long *pseudo-random* strings—strings of numbers which can be deterministically generated (perhaps from a given seed) in a reasonable amount of time, but are difficult to distinguish from genuinely random strings by any quick tests.

Notes. This article first appeared at

`terrytao.wordpress.com/2009/08/01.`

Thanks to Tom for corrections.

There was some discussion on the relationship between $P = NP$ and $P = BPP$. Greg Kuperberg gave some further examples of oracles that shed some light on this:

- Consider as an oracle an extremely large book of randomly generated numbers. This oracle could be used to simulate any probabilistic algorithm, so $P = BPP$ relative to this oracle. On the other hand, if one assigns the task to determine whether a given string of numbers exists in some range in the book, this question is in NP but not in P.
- Another example of an oracle would be an extremely large book, in which most of the pages contained the answer to the problem at hand, but for which the nth page was blank for every natural number n that could be quickly created by any short deterministic algorithm. This type of oracle could be used to create a scenario in which $P \neq BPP$ and $P \neq NP$.
- A third example, this time of an *advice function* rather than an oracle, would be if the proctor wrote a long random string on the board before starting the exam (with the length of the string depending on the length of the exam). This can be used to show the inclusion $BPP \subset P/poly$.

By using written oracles instead of computer oracles, it also became more obvious that the oracles were noninteractive (i.e., subsequent responses by the oracle did not depend on earlier queries).

1.10. Moser's entropy compression argument

There are many situations in combinatorics in which one is running some sort of iteration algorithm to continually "improve" some object A; each loop of the algorithm replaces A with some better version A' of itself, until some desired property of A is attained and the algorithm halts. In order for such arguments to yield a useful conclusion, it is often necessary that the algorithm halts in a finite amount of time, or (even better) in a bounded amount of time.[10]

A basic strategy to ensure termination of an algorithm is to exploit a *monotonicity property*, or more precisely to show that some key quantity keeps increasing (or keeps decreasing) with each loop of the algorithm, while simultaneously staying bounded. (Or, as the economist Herbert Stein was fond of saying, "If something cannot go on forever, it must stop.")

Here are four common flavours of this monotonicity strategy:

- The *mass increment argument*. This is perhaps the most familiar way to ensure termination: make each improved object A' "heavier" than the previous one A by some nontrivial amount (e.g., by ensuring that the cardinality of A' is strictly greater than that of A, thus $|A'| \geq |A| + 1$). Dually, one can try to force the amount of "mass" remaining "outside" of A in some sense to decrease at every stage of the iteration. If there is a good upper bound on the "mass" of A that stays essentially fixed throughout the iteration process, and a lower bound on the mass increment at each stage, then the argument terminates. Many "greedy algorithm" arguments are of this type. The proof of the Hahn decomposition theorem (Theorem 1.2.2 of *Volume I*) also falls into this category. The general strategy here is to keep looking for useful pieces of mass outside of A, and add them to A to form A', thus exploiting the additivity properties of mass. Eventually no further usable mass remains to be added (i.e., A is *maximal* in some L^1 sense), and this should force some desirable property on A.
- The *density increment argument*. This is a variant of the mass increment argument, in which one increments the "density" of A rather than the "mass". For instance, A might be contained in some ambient space P, and one seeks to improve A to A' (and P to P') in such a way that the density of the new object in the new ambient space is better than that of the previous object (e.g.,

[10]In general, one cannot use infinitary iteration tools, such as *transfinite induction* or *Zorn's lemma* (Section 2.4 of *Volume I*), in combinatorial settings, because the iteration processes used to improve some target object A often degrade some other finitary quantity B in the process, and an infinite iteration would then have the undesirable effect of making B infinite.

$|A'|/|P'| \geq |A|/|P| + c$ for some $c > 0$). On the other hand, the density of A is clearly bounded above by 1. As long as one has a sufficiently good lower bound on the density increment at each stage, one can conclude an upper bound on the number of iterations in the algorithm. The prototypical example of this is Roth's proof of his theorem [**Ro1953**] that every set of integers of positive upper density contains an arithmetic progression of length three. The general strategy here is to keep looking for useful density fluctuations inside A and then "zoom in" to a region of increased density by reducing A and P appropriately. Eventually no further usable density fluctuation remains (i.e., A is *uniformly distributed*), and this should force some desirable property on A.

- The *energy increment argument.* This is an "L^2" analogue of the "L^1"-based mass increment argument (or the "L^∞"-based density increment argument), in which one seeks to increment the amount of "energy" that A captures from some reference object X, or (equivalently) to decrement the amount of energy of X which is still "orthogonal" to A. Here A and X are related somehow to a Hilbert space, and the energy involves the norm on that space. A classic example of this type of argument is the existence of orthogonal projections onto closed subspaces of a Hilbert space; this leads among other things to the construction of *conditional expectation* in measure theory, which then underlies a number of arguments in ergodic theory, as discussed for instance in Section 2.8 of *Poincaré's Legacies, Vol. I.* Another basic example is the standard proof of the *Szemerédi regularity lemma* (where the "energy" is often referred to as the "index"). These examples are related; see Section 2.2 for further discussion. The general strategy here is to keep looking for useful pieces of energy orthogonal to A, and add them to A to form A', thus exploiting square-additivity properties of energy, such as Pythagoras' theorem. Eventually, no further usable energy outside of A remains to be added (i.e., A is *maximal* in some L^2 sense), and this should force some desirable property on A.

- The *rank reduction argument.* Here, one seeks to make each new object A' to have a lower "rank", "dimension", or "order" than the previous one. A classic example here is the proof of the linear algebra fact that given any finite set of vectors, there exists a linearly independent subset which spans the same subspace; the proof of the more general *Steinitz exchange lemma* is in the same spirit. The general strategy here is to keep looking for "collisions" or "dependencies" within A, and use them to collapse A to an object

> A' of lower rank. Eventually, no further usable collisions within A remain, and this should force some desirable property on A.

Much of my work in additive combinatorics relies heavily on at least one of these types of arguments (and, in some cases, on a nested combination of two or more of them). Many arguments in nonlinear partial differential equations also have a similar flavour, relying on various *monotonicity formulae* for solutions to such equations, though the objective in PDE is usually slightly different, in that one wants to keep control of a solution as one approaches a singularity (or as some time or space coordinate goes off to infinity), rather than to ensure termination of an algorithm. (On the other hand, many arguments in the theory of *concentration compactness*, which is used heavily in PDE, do have the same algorithm-terminating flavour as the combinatorial arguments; see Section 2.1 of *Structure and Randomness* for more discussion.)

Recently, a new species of monotonicity argument was introduced by Moser [**Mo2009**], as the primary tool in his elegant new proof of the *Lovász local lemma.* This argument could be dubbed an *entropy compression argument* and only applies to *probabilistic algorithms* which require a certain collection R of random "bits" or other random choices as part of the input; thus each loop of the algorithm takes an object A (which may also have been generated randomly) and some portion of the random string R to (deterministically) create a better object A' (and a shorter random string R', formed by throwing away those bits of R that were used in the loop). The key point is to design the algorithm to be partially *reversible*, in the sense that given A' and R' and some additional data H' that logs the cumulative *history* of the algorithm up to this point, one can reconstruct A together with the remaining portion R not already contained in R'. Thus, each stage of the argument *compresses* the information-theoretic content of the string $A + R$ into the string $A' + R' + H'$ in a lossless fashion. However, a random variable such as $A + R$ cannot be compressed losslessly into a string of expected size smaller than the *Shannon entropy* of that variable. Thus, if one has a good lower bound on the entropy of $A + R$, and if the length of $A' + R' + H'$ is significantly less than that of $A + R$ (i.e., we need the marginal growth in the length of the history file H' per iteration to be less than the marginal amount of randomness used per iteration), then there is a limit as to how many times the algorithm can be run, much as there is a limit as to how many times a random data file can be compressed before no further length reduction occurs.

It is interesting to compare this method with the ones discussed earlier. In the previous methods, the failure of the algorithm to halt led to a new iteration of the object A which was "heavier", "denser", captured more "energy", or "lower rank" than the previous instance of A. Here, the failure of the algorithm to halt leads to new information that can be used to "compress" A (or more precisely, the full state $A+R$) into a smaller amount of space. I do not know yet of any application of this new type of termination strategy to the fields I work in, but one could imagine that it could eventually be of use (perhaps to show that solutions to PDE with sufficiently "random" initial data can avoid singularity formation?), so I thought I would discuss (a special case of) it here.

Rather than deal with the Lovász local lemma in full generality, I will work with a special case of this lemma involving the *k-satisfiability problem* (in *conjunctive normal form*). Here, one is given a set of *boolean variables* $x_1, \ldots, x_n$ together with their negations $\neg x_1, \ldots, \neg x_n$; we refer to the $2n$ variables and their negations collectively as *literals*. We fix an integer $k \geq 2$, and define a (length k) *clause* to be a *disjunction* of k literals, for instance

$$x_3 \vee \neg x_5 \vee x_9$$

is a clause of length three, which is true unless x_3 is false, x_5 is true, and x_9 is false. We define the *support* of a clause to be the set of variables that are involved in the clause, thus for instance $x_3 \vee \neg x_5 \vee x_9$ has support $\{x_3, x_5, x_9\}$. To avoid degeneracy, we assume that no clause uses a variable more than once (or equivalently, all supports have cardinality exactly k), thus for instance we do not consider $x_3 \vee x_3 \vee x_9$ or $x_3 \vee \neg x_3 \vee x_9$ to be clauses.

Note that the failure of a clause reveals complete information about all k of the boolean variables in the support; this will be an important fact later on.

The *k-satisfiability problem* is the following: given a set S of clauses of length k involving n boolean variables $x_1, \ldots, x_n$, is there a way to assign truth values to each of the $x_1, \ldots, x_n$, so that all of the clauses are simultaneously satisfied?

For general S, this problem is easy for $k = 2$ (essentially equivalent to the problem of 2-colouring a graph), but NP-complete for $k \geq 3$ (this is the famous *Cook-Levin theorem*). But the problem becomes simpler if one makes some more assumptions on the set S of clauses. For instance, if the clauses in S have disjoint supports, then they can be satisfied independently of each other, and so one easily has a positive answer to the satisfiability problem in this case. (Indeed, one only needs each clause in S to have *one* variable in its support that is disjoint from all the other supports in order to make this argument work.)

Now suppose that the clauses S are not completely disjoint but have a limited amount of overlap; thus *most* clauses in S have disjoint supports, but not all. With too much overlap, of course, one expects satisfability to fail (e.g., if S is the set of *all* length k clauses). But with a sufficiently small amount of overlap, one still has satisfiability:

Theorem 1.10.1 (Lovász local lemma, special case). *Suppose that S is a set of length k clauses, such that the support of each clause s in S intersects at most 2^{k-C} supports of clauses in S (including s itself), where C is a sufficiently large absolute constant. Then the clauses in S are simultaneously satisfiable.*

One of the reasons that this result is powerful is that the bounds here are uniform in the number n of variables. Apart from the loss of C, this result is sharp; consider for instance the set S of all 2^k clauses with support $\{x_1, \ldots, x_k\}$, which is clearly unsatisfiable.

The standard proof of this theorem proceeds by assigning each of the n boolean variables $x_1, \ldots, x_n$ a truth value $a_1, \ldots, a_n \in \{\text{true}, \text{false}\}$ independently at random (with each truth value occurring with an equal probability of $1/2$). Then each of the clauses in S has a positive zero probability of holding (in fact, the probability is $1-2^{-k}$). Furthermore, if E_s denotes the event that a clause $s \in S$ is satisfied, then the E_s are mostly independent of each other; indeed, each event E_s is independent of all but most 2^{k-C} other events $E_{s'}$. Applying the Lovász local lemma, one concludes that the E_s simultaneously hold with positive probability (if C is a little bit larger than $\log_2 e$), and the claim follows.

The textbook proof of the Lovász local lemma is short but nonconstructive; in particular, it does not easily offer any quick way to compute an actual satisfying assignment for $x_1, \ldots, x_n$, only saying that such an assignment exists. Moser's argument, by contrast, gives a simple and natural algorithm to locate such an assignment (and thus prove Theorem 1.10.1). (The constant C becomes 3 rather than $\log_2 e$, although the $\log_2 e$ bound has since been recovered in a paper of Moser and Tardos.)

As with the usual proof, one begins by randomly assigning truth values $a_1, \ldots, a_n \in \{\text{true}, \text{false}\}$ to $x_1, \ldots, x_n$; call this random assignment $A = (a_1, \ldots, a_n)$. If A satisfied all the clauses in S, we would be done. However, it is likely that there will be some nonempty subset T of clauses in S which are not satisfied by A.

We would now like to modify A in such a manner to reduce the number $|T|$ of violated clauses. If, for instance, we could always find a modification A' of A whose set T' of violated clauses was strictly smaller than T (assuming of course that T is nonempty), then we could iterate and be done (this is basically a mass decrement argument). One obvious way to try to achieve this is to pick a clause s in T that is violated by A, and modify the values of A on the support of s to create a modified set A' that satisfies s, which is easily accomplished; in fact, any nontrivial modification of A on the support will work here. In order to maximize the amount of entropy in the system (which is what one wants to do for an entropy compression argument), we will choose this modification of A' *randomly*; in particular, we will use k fresh random bits to replace the k bits of A in the support of s. (By doing so, there is a small probability (2^{-k}) that we in fact do not change A at all, but the argument is (very) slightly simpler if we do not bother to try to eliminate this case.)

If all the clauses had disjoint supports, then this strategy would work without difficulty. But when the supports are not disjoint, one has a problem: every time one modifies A to "fix" a clause s by modifying the variables on the support of s, one may cause other clauses s' whose supports overlap those of s to fail, thus potentially increasing the size of T by as much as $2^{k-C}-1$. One could then try fixing all the clauses which were broken by the first fix, but it appears that the number of clauses needed to repair could grow indefinitely with this procedure, and one might never terminate in a state in which all clauses are simultaneously satisfied.

The key observation of Moser, as alluded earlier, is that each failure of a clause s for an assignment A reveals k bits of information about A, namely the exact values that A assigns to the support of s. The plan is then to use each failure of a clause as a part of a *compression protocol* that compresses A (plus some other data) losslessly into a smaller amount of space. A crucial point is that at each stage of the process, the clause one is trying to fix is almost always going to be one that overlapped the clause that one had just previously fixed. Thus the total number of possibilities for each clause, given the previous clauses, is basically 2^{k-C}, which requires only $k-C$ bits of storage, compared with the k bits of entropy that have been eliminated. This is what is going to force the algorithm to terminate in finite time (with positive probability).

Let us make the details more precise. We will need the following objects:

- A truth assignment A of n truth values $a_1, \dots, a_n$, which is initially assigned randomly, but which will be modified as the algorithm progresses;
- A long random string R of bits, from which we will make future random choices, with each random bit being removed from R as it is read.

We also need a recursive algorithm Fix(s), which modifies the string A to satisfy a clause s in S (and, as a bonus, may also make A obey some other clauses in S that it did not previously satisfy). It is defined recursively:

- Step 1. If A already satisfies s, do nothing (i.e., leave A unchanged).
- Step 2. Otherwise, read off k random bits from R (thus shortening R by k bits), and use these to replace the k bits of A on the support of s in the obvious manner (ordering the support of s by some fixed ordering, and assigning the jth bit from R to the jth variable in the support for $1 \leq j \leq k$).
- Step 3. Next, find all the clauses s' in S whose supports intersect s, and which A now violates; this is a collection of at most 2^{k-C} clauses, possibly including s itself. Order these clauses s' in some arbitrary fashion, and then apply Fix(s') to each such clause in turn. (Thus the original algorithm Fix(s) is put "on hold" on some CPU stack while all the child processes Fix(s') are executed; once all of the child processes are complete, Fix(s) then terminates also.)

An easy induction shows that if Fix(s) terminates, then the resulting modification of A will satisfy s; and furthermore, any other clause s' in S which was already satisfied by A before Fix(s) was called, will continue to be satisfied by A after Fix(s) is called. Thus, Fix(s) can only serve to decrease the number of unsatisfied clauses T in S, and so one can fix all the clauses by calling Fix(s) once for each clause in T—provided that these algorithms all terminate.

Each time Step 2 of the Fix algorithm is called, the assignment A changes to a new assignment A', and the random string R changes to a shorter string R'. Is this process reversible? Yes, provided that one knows what clause s was being fixed by this instance of the algorithm. Indeed, if s, A', R' are known, then A can be recovered by changing the assignment of A' on the support of s to the only set of choices that violates s, while R can be recovered from R' by appending to R' the bits of A on the support of s.

This type of reversibility does not seem very useful for an entropy compression argument, because while R' is shorter than R by k bits, it requires about $\log |S|$ bits to store the clause s. So the map $A + R \mapsto A' + R' + s$ is only a compression if $\log |S| < k$, which is not what is being assumed here (and in any case the satisfiability of S in the case $\log |S| < k$ is trivial from the union bound).

The key trick is that while it does indeed take $\log |S|$ bits to store any given clause s, there is an economy of scale: after many recursive applications of the fix algorithm, the *marginal* amount of bits needed to store s drops to merely $k - C + O(1)$, which is less than k if C is large enough, and which will therefore make the entropy compression argument work.

Let us see why this is the case. Observe that the clauses s for which the above algorithm Fix(s) is called come in two categories. Firstly, there are those s which came from the original list T of failed clauses. Each of these will require $O(\log|S|)$ bits to store—but there are only $|T|$ of them. Since $|T| \leq |S|$, the net amount of storage space required for these clauses is $O(|S|\log|S|)$ at most. Actually, one can just store the subset T of S using $|S|$ bits (one for each element of S, to record whether it lies in T or not).

Of more interest is the other category of clauses s, in which $\mathrm{Fix}(s)$ is called recursively from some previously invoked call $\mathrm{Fix}(s')$ to the fix algorithm. But then s is one of the at most 2^{k-C} clauses in S whose support intersects that of s'. Thus one can encode s using s' and a number between 1 and 2^{k-C}, representing the position of s (with respect to some arbitrarily chosen fixed ordering of S) in the list of all clauses in S whose supports intersect that of s'. Let us call this number the *index* of the call $\mathrm{Fix}(s)$.

Now imagine that while the Fix routine is called, a running *log file* (or history) H of the routine is kept, which records s each time one of the original $|T|$ calls $\mathrm{Fix}(s)$ with $s \in T$ invoked, and it also records the index of any other call $\mathrm{Fix}(s)$ made during the recursive procedure. Finally, we assume that this log file records a termination symbol whenever a Fix routine terminates. By performing a *stack trace*, one sees that whenever a Fix routine is called, the clause s that is being repaired by that routine can be deduced from an inspection of the log file H up to that point.

As a consequence, at any intermediate stage in the process of all these Fix calls, the original state $A+R$ of the assignment and the random string of bits can be deduced from the current state $A'+R'$ of these objects, plus the history H' up to that point.

Now suppose for contradiction that S is not satisfiable; thus the stack of Fix calls can never completely terminate. We trace through this stack for M steps, where M is some large number to be chosen later. After these steps, the random string R has shortened by an amount of Mk; if we set R to initially have length Mk, then the string is now completely empty, $R' = \emptyset$. On the other hand, the history H' has size at most $O(|S|)+M(k-C+O(1))$, since it takes $|S|$ bits to store the initial clauses in T, $O(|S|)+O(M)$ bits to record all the instances when Step 1 occurs, and every subsequent call to Fix generates a $k-C$-bit number, plus possibly a termination symbol of size $O(1)$. Thus we have a lossless compression algorithm $A+R \mapsto A'+H'$ from $n+Mk$ completely random bits to $n+O(|S|)+M(k-C+O(1))$ bits (recall that A and R were chosen randomly and independently of each other). But since $n+Mk$ random bits cannot be compressed losslessly into any smaller space, we have the entropy bound

$$n + O(|S|) + M(k - C + O(1)) \geq n + Mk, \tag{1.49}$$

which leads to a contradiction if M is large enough (and if C is larger than an absolute constant). This proves Theorem 1.10.1.

Remark 1.10.2. Observe that the above argument in fact gives an explicit bound on M, and with a small bit of additional effort, it can be converted into a probabilistic algorithm that (with high probability) computes a satisfying assignment for S in time polynomial in $|S|$ and n.

Remark 1.10.3. One can replace the usage of randomness and Shannon entropy in the above argument with *Kolmogorov complexity* instead; thus, one sets $A + R$ to be a string of $n + Mk$ bits which cannot be computed by any algorithm of length $n + O(|S| \log |S|) + M(k - C + O(1))$, the existence of which is guaranteed as soon as (1.49) is violated. The proof now becomes deterministic, except of course for the problem of building the high-complexity string, which by their definition can only be constructed quickly by probabilistic methods.

Notes. This article first appeared at

`terrytao.wordpress.com/2009/08/05`,

but is based on an earlier blog post by Lance Fortnow at

`blog.computationalcomplexity.org/2009/06`.

Thanks to harrison, Heinrich, nh, and anonymous commenters for corrections.

There was some discussion online about the tightness of bounds in the argument.

1.11. The AKS primality test

The *Agrawal-Kayal-Saxena (AKS) primality test*, discovered in 2002, is the first provably deterministic algorithm to determine the primality of a given number with a run time which is guaranteed to be polynomial in the number of digits; thus, given a large number n, the algorithm will correctly determine whether that number is prime or not in time $O(\log^{O(1)} n)$. (Many previous primality testing algorithms existed, but they were either probabilistic in nature, had a running time slower than polynomial, or the correctness could not be guaranteed without additional hypotheses such as GRH.)

In this article I sketch the details of the test (and the proof that it works). (Of course, full details can be found in the original paper [**AgKaSa2004**], which is nine pages in length and almost entirely elementary in nature.) It relies on polynomial identities that are true modulo n when n is prime, but cannot hold for n nonprime as they would generate a large number of additional polynomial identities, eventually violating the *factor theorem* (which asserts that a polynomial identity of degree at most d can be obeyed by at most d values of the unknown). To remove some clutter in the notation, I have relied (somewhat loosely) on *asymptotic notation* in this article.

Our starting point is *Fermat's little theorem*, which asserts that

$$a^p = a \bmod p \tag{1.50}$$

for every prime p and every a. This theorem suggests an obvious primality test: to test whether a number n is prime, pick a few values of a and see whether $a^n = a \bmod n$. (Note that a^n can be computed in time $O(\log^{O(1)} n)$ for any fixed a by expressing n in binary, and repeatedly squaring a.) If the statement $a^n = a \bmod n$ fails for some a, then n would be composite. Unfortunately, the converse is not true: there exist nonprime numbers n, known as *Carmichael numbers*, for which $a^n = a \bmod n$ for all a coprime to n (561 is the first example). So Fermat's little theorem cannot be used, by itself, to establish primality for general n, because it is too weak to eliminate all nonprime numbers. (The situation improves though for more special types of n, such as Mersenne numbers; see Section 1.7 of *Poincaré's Legacies, Vol. I* for more discussion.)

However, there is a stronger version of Fermat's little theorem which does eliminate all nonprime numbers. Specifically, if p is prime and a is arbitrary, then one has the polynomial identity

$$(X + a)^p = X^p + a \bmod p, \tag{1.51}$$

where X is an indeterminate variable. (More formally, we have the identity $(X+a)^p = X^p+a$ in the ring $F_p[X]$ of polynomials of one variable X over the finite field F_p of p elements.) This identity (a manifestation of the *Frobenius endomorphism*) clearly implies (1.50) by setting $X = 0$; conversely, one can easily deduce (1.51) from (1.50) by expanding out $(X+a)^p$ using the *binomial theorem* and the observation that the binomial coefficients $\binom{p}{i} = \frac{p\cdot\dots\cdot(p-i+1)}{i!}$ are divisible by p for all $1 \leq i < p$. Conversely, if

$$(X + a)^n = X^n + a \bmod n \tag{1.52}$$

(i.e., $(X + a)^n = X^n + a$ in $(\mathbf{Z}/n\mathbf{Z})[X]$) for some a coprime to n, then by comparing coefficients using the binomial theorem, we see that $\binom{n}{i}$ is divisible by n for all $1 \leq i < n$. But if n is divisible by some smaller prime p, then by setting i equal to the largest power of p that divides n, one sees that $\binom{n}{i}$ is not divisible by enough powers of p to be divisible by n, a contradiction. Thus one can use (1.52) (for a single value of a coprime to n) to decide whether n is prime or not.

Unfortunately, this algorithm, while deterministic, is not polynomial-time, because the polynomial $(X+a)^n$ has $n+1$ coefficients and will therefore take at least $O(n)$ time to compute. However, one can speed up the process by descending to a quotient ring of $(\mathbf{Z}/n\mathbf{Z})[X]$, such as $F_p[X]/(X^r - 1)$ for some r. Clearly, if the identity $(X+a)^n = X^n + a$ holds in $(\mathbf{Z}/n\mathbf{Z})[X]$, then it will also hold in $(\mathbf{Z}/n\mathbf{Z})[X]/(X^r-1)$, thus

$$(X+a)^n = X^n + a \bmod n, X^r - 1. \tag{1.53}$$

The point of doing this is that (if r is not too large) the left-hand side of (1.53) can now be computed quickly (again by expanding n in binary and performing repeated squaring), because all polynomials can be reduced to be of degree less than r, rather than being as large as n. Indeed, if $r = O(\log^{O(1)} n)$, then one can test (1.53) in time $O(\log^{O(1)} n)$.

We are not done yet, because it could happen that (1.53) holds but (1.52) fails. But we have the following key theorem:

Theorem 1.11.1 (AKS theorem). *Suppose that* (1.53) *holds for all* $1 \leq a, r \leq O(\log^{O(1)} n)$, *and a is coprime to n. Then n is either a prime or a power of a prime.*

Of course, coprimality of a and n can be quickly tested using the *Euclidean algorithm*, and if coprimality fails, then n is of course composite. Also, it is easy to quickly test for the property that n is a power of an integer (just compute the roots $n^{1/k}$ for $1 \leq k \leq \log_2 n$), and such powers are clearly composite. From all this (and (1.51)), one soon sees that theorem gives rise to a deterministic polynomial-time test for primality. One can optimise the powers of $\log n$ in the bounds for a, r (as is done in [**AgKaSa2004**]), but we will not do so here to keep the exposition uncluttered.

Actually, we do not need (1.53) satisfied for all that many exponents r to make the theorem work; just one well-chosen r will do. More precisely, we have

Theorem 1.11.2 (AKS theorem, key step). *Let r be coprime to n, and such that n has order greater than $\log_2^2 n$ in the multiplicative group $(\mathbf{Z}/r\mathbf{Z})^\times$ (i.e., the residues $n^i \bmod r$ for $1 \leq i \leq \log^2 n$ are distinct). Suppose that for all $1 \leq a \leq O(r \log^{O(1)} n)$,* (1.53) *holds, and a is coprime to n. Then n is either a prime or a power of a prime.*

To find an r with the above properties we have

Lemma 1.11.3 (Existence of good r). *There exists $r = O(\log^{O(1)} n)$ coprime to n, such that n has order greater than $\log_2^2 n$ in $(\mathbf{Z}/r\mathbf{Z})^\times$.*

Proof. For each $1 \leq i \leq \log_2^2 n$, the number $n^i - 1$ has at most $O(\log^{O(1)} n)$ prime divisors (by the fundamental theorem of arithmetic). If one picks r to be the first prime not equal to any of these prime divisors, one obtains the claim. (One can use a crude version of the prime number theorem to get the upper bound on r.) □

It is clear that Theorem 1.11.1 follows from Theorem 1.11.2 and Lemma 1.11.3, so it suffices now to prove Theorem 1.11.2.

Suppose for contradiction that Theorem 1.11.2 fails. Then n is divisible by some smaller prime p, but is not a power of p. Since n is coprime to all numbers of size $O(\log^{O(1)} n)$, we know that p is not of polylogarithmic size, thus we may assume $p \geq \log^C n$ for any fixed C. As r is coprime to n, we see that r is not a multiple of p (indeed, one should view p as being much larger than r).

Let F be a field extension of F_p by a primitive rth root of unity X, thus $F = F_p[X]/h(X)$ for some factor $h(X)$ (in $F_p[X]$) of the rth *cyclotomic polynomial* $\Phi_r(X)$. From the hypothesis (1.53), we see that

$$(X + a)^n = X^n + a$$

in F for all $1 \leq a \leq A$, where $A = O(r \log^{O(1)} n)$. Note that n is coprime to every integer less than A, and thus $A < p$.

Meanwhile, from (1.51) one has

$$(X + a)^p = X^p + a$$

in F for all such a. The two equations give

$$(X^p + a)^{n/p} = (X^p)^{n/p} + a.$$

Note that the pth power X^p of a primitive rth root of unity X is again a primitive rth root of unity (and conversely, every primitive rth root arises in this fashion) and hence we also have

$$(X + a)^{n/p} = X^{n/p} + a$$

in F for all $1 \leq a \leq A$.

Inspired by this, we define a key concept: a positive integer m is said to be *introspective* if one has

$$(X + a)^m = X^m + a$$

in F for all $1 \leq a \leq A$, or equivalently if $(X + a)^m = \phi_m(X + a)$, where $\phi_m : F \to F$ is the ring homomorphism that sends X to X^m.

We have just shown that $p, n, n/p$ are all introspective; 1 is also trivially introspective. Furthermore, if m and m' are introspective, it is not hard to see that mm' is also introspective. Thus we in fact have a lot of introspective integers: any number of the form $p^i (n/p)^j$ for $i, j \geq 0$ is introspective.

It turns out in fact that it is not possible to create so many different introspective numbers, basically the presence of so many polynomial identities in the field would eventually violate the *factor theorem*. To see this, let $\mathcal{G} \subset F^\times$ be the multiplicative group generated by the quantities $X + a$ for $1 \leq a \leq A$. Observe that $z^m = \phi_m(z)$ for all $z \in \mathcal{G}$. We now show that this places incompatible lower and upper bounds on $\mathcal{G}$. We begin with the lower bound:

Proposition 1.11.4 (Lower bound on $\mathcal{G}$). $|\mathcal{G}| \geq 2^t$.

Proof. Let $P(X)$ be a product of less than t of the quantities $X+1, \ldots$, $X+A$ (allowing repetitions), then $P(X)$ lies in $\mathcal{G}$. Since $A \geq 2r \geq 2t$, there are certainly at least 2^t ways to pick such a product. So to establish the proposition, it suffices to show that all these products are distinct.

Suppose for contradiction that $P(X) = Q(X)$, where P, Q are different products of less than t of the $X+1, \ldots, X+A$. Then, for every introspective m, $P(X^m) = Q(X^m)$ as well (note that $P(X^m) = \phi_m(P(X))$). In particular, this shows that $X^{m_1}, \ldots, X^{m_t}$ are all roots of the polynomial $P - Q$. But this polynomial has degree less than t, and the $X^{m_1}, \ldots, X^{m_t}$ are distinct by hypothesis, and we obtain the desired contradiction by the factor theorem. □

Proposition 1.11.5 (Upper bound on $\mathcal{G}$). *Suppose that there are exactly t residue classes modulo r of the form $p^i(n/p)^j \bmod r$ for $i, j \geq 0$. Then $|\mathcal{G}| \leq n^{\sqrt{t}}$.*

Proof. By the pigeonhole principle, we must have a collision

$$p^i(n/p)^j = p^{i'}(n/p)^{j'} \bmod r$$

for some $0 \leq i, j, i', j' \leq \sqrt{t}$ with $(i,j) \neq (i',j')$. Setting $m := p^i(n/p)^j$ and $m' := p^{i'}(n/p)^{j'}$, we thus see that there are two distinct introspective numbers m, m' of size most $n^{\sqrt{t}}$ which are equal modulo r. (To ensure that m, m' are distinct, we use the hypothesis that n is not a power of p.) This implies that $\phi_m = \phi_{m'}$, and thus $z^m = z^{m'}$ for all $z \in \mathcal{G}$. But the polynomial $z^m - z^{m'}$ has degree at most $n^{\sqrt{t}}$, and the claim now follows from the factor theorem. □

Since n has order greater than $\log^2 n$ in $(\mathbf{Z}/r\mathbf{Z})^\times$, we see that the number t of residue classes r of the form $p^i(n/p)^j$ is at least $\log^2 n$. But then $2^t > n^{\sqrt{t}}$, and so Propositions 1.11.4 and 1.11.5 are incompatible.

Notes. This article first appeared at

`terrytao.wordpress.com/2009/08/11.`

Thanks to Leandro, theoreticalminimum and windfarmmusic for corrections.

A thorough discussion of the AKS algorithm can be found at [**Gr2005**].

1.12. The prime number theorem in arithmetic progressions, and dueling conspiracies

A fundamental problem in analytic number theory is to understand the distribution of the prime numbers $\{2, 3, 5, \ldots\}$. For technical reasons, it is convenient not to study the primes directly, but a proxy for the primes known as the *von Mangoldt function* $\Lambda : \mathbf{N} \to \mathbf{R}$, defined by setting $\Lambda(n)$ to equal $\log p$ when n is a prime p (or a power of that prime) and zero otherwise. The basic reason why the von Mangoldt function is useful is that it encodes the *fundamental theorem of arithmetic* (which in turn can be viewed as the defining property of the primes) very neatly via the identity

$$\log n = \sum_{d|n} \Lambda(d) \tag{1.54}$$

for every natural number n.

The most important result in this subject is the *prime number theorem*, which asserts that the number of prime numbers less than a large number x is equal to $(1 + o(1))\frac{x}{\log x}$:

$$\sum_{p \leq x} 1 = (1 + o(1))\frac{x}{\log x}.$$

Here, of course, $o(1)$ denotes a quantity that goes to zero as $x \to \infty$.

It is not hard to see (e.g., by *summation by parts*) that this is equivalent to the asymptotic

$$\sum_{n \leq x} \Lambda(n) = (1 + o(1))x \tag{1.55}$$

for the von Mangoldt function (the key point being that the squares, cubes, etc., of primes give a negligible contribution, so $\sum_{n \leq x} \Lambda(n)$ is essentially the same quantity as $\sum_{p \leq x} \log p$). Understanding the nature of the $o(1)$ term is a very important problem, with the conjectured optimal decay rate of $O(\sqrt{x} \log x)$ being equivalent to the *Riemann hypothesis*, but this will not be our concern here.

The prime number theorem has several important generalisations (for instance, there are analogues for other number fields such as the *Chebotarev density theorem*). One of the more elementary such generalisations is the *prime number theorem in arithmetic progressions*, which asserts that for fixed a and q with a coprime to q (thus $(a, q) = 1$), the number of primes less than x equal to a mod q is equal to $(1 + o_q(1))\frac{1}{\phi(q)}\frac{x}{\log x}$, where $\phi(q) :=$

$\#\{1 \leq a \leq q : (a,q) = 1\}$ is the *Euler totient function*:

$$\sum_{p \leq x : p = a \bmod q} 1 = (1 + o_q(1)) \frac{1}{\phi(q)} \frac{x}{\log x}.$$

(Of course, if a is not coprime to q, the number of primes less than x equal to a mod q is $O(1)$. The subscript q in the $o()$ and $O()$ notation denotes that the implied constants in that notation are allowed to depend on q.) This is a more quantitative version of *Dirichlet's theorem*, which asserts the weaker statement that the number of primes equal to $a \bmod q$ is infinite. This theorem is important in many applications in analytic number theory, for instance in *Vinogradov's theorem* that every sufficiently large odd number is the sum of three odd primes. (Imagine for instance if almost all of the primes were clustered in the residue class 2 mod 3, rather than 1 mod 3. Then almost all sums of three odd primes would be divisible by 3, leaving dangerously few sums left to cover the remaining two residue classes. This works similarly for other moduli than 3. It does not fully rule out the possibility that Vinogradov's theorem could still be true, but it does indicate why the prime number theorem in arithmetic progressions is a relevant tool in the proof of that theorem.)

As before, one can rewrite the prime number theorem in arithmetic progressions in terms of the von Mangoldt function as the equivalent form

$$\sum_{n \leq x : n = a \bmod q} \Lambda(n) = (1 + o_q(1)) \frac{1}{\phi(q)} x.$$

Philosophically, one of the main reasons why it is so hard to control the distribution of the primes is that we do not currently have very many tools with which one can rule out "conspiracies" between the primes, in which the primes (or the von Mangoldt function) decide to correlate with some structured object (and in particular, with a totally multiplicative function) which then visibly distorts the distribution of the primes. For instance, one could imagine a scenario in which the probability that a randomly chosen large integer n is prime is not asymptotic to $\frac{1}{\log n}$ (as is given by the prime number theorem), but instead it fluctuates depending on the phase of the complex number n^{it} for some fixed real number t. Thus, for instance, the probability might be significantly less than $1/\log n$ when $t \log n$ is close to an integer, and significantly more than $1/\log n$ when $t \log n$ is close to a half-integer. This would contradict the prime number theorem, and so this scenario would have to be eradicated somehow in the course of proving that theorem. In the language of *Dirichlet series*, this conspiracy is more commonly known as a zero of the Riemann zeta function at $1 + it$.

In the above scenario, the primality of a large integer n was somehow sensitive to asymptotic or "Archimedean" information about n, namely the approximate value of its logarithm. In modern terminology, this information reflects the local behaviour of n at the infinite *place* ∞. There are also potential consipracies in which the primality of n is sensitive to the local behaviour of n at finite places, and in particular to the residue class of $n \bmod q$ for some fixed modulus q. For instance, given a *Dirichlet character* $\chi : \mathbf{Z} \to \mathbf{C}$ of modulus q, i.e., a *completely multiplicative* function on the integers which is periodic of period q (and vanishes on those integers not coprime to q), one could imagine a scenario in which the probability that a randomly chosen large integer n is prime is large when $\chi(n)$ is close to $+1$ and small when $\chi(n)$ is close to -1, which would contradict the prime number theorem in arithmetic progressions. (Note the similarity between this scenario at q and the previous scenario at ∞; in particular, observe that the functions $n \to \chi(n)$ and $n \to n^{it}$ are both totally multiplicative.) In the language of Dirichlet series, this conspiracy is more commonly known as a zero of the *L-function* of χ at 1.

An especially difficult scenario to eliminate is that of *real characters*, such as the *Kronecker symbol* $\chi(n) = \left(\frac{n}{q}\right)$, in which numbers n which are quadratic nonresidues mod q are very likely to be prime, and quadratic residues mod q are unlikely to be prime. Indeed, there is a scenario of this form, the *Siegel zero* scenario, which we are still not able to eradicate (without assuming powerful conjectures such as the *Generalised Riemann Hypothesis (GRH)*), though fortunately Siegel zeroes are not quite strong enough to destroy the prime number theorem in arithmetic progressions.

It is difficult to prove that no conspiracy between the primes exist. However, it is not entirely impossible because we have been able to exploit two important phenomena. The first is that there is often an all-or-nothing dichotomy (somewhat resembling the *zero–one laws* in probability) regarding conspiracies: in the asymptotic limit, the primes can either conspire totally (or more precisely, anticonspire totally) with a multiplicative function or fail to conspire at all, but there is no middle ground. (In the language of Dirichlet series, this is reflected in the fact that zeroes of a meromorphic function can have order 1, or order 0 (i.e., are not zeroes after all), but cannot have an intermediate order between 0 and 1.) As a corollary of this fact, the prime numbers cannot conspire with two distinct multiplicative functions at once (by having a partial correlation with one and another partial correlation with another); thus one can use the existence of one conspiracy to exclude all the others. In other words, there is at most one conspiracy that can significantly distort the distribution of the primes. Unfortunately, this argument is *ineffective*, because it does not give any control at all on what that conspiracy is, or even if it exists in the first place!

But now one can use the second important phenomenon, which is that because of symmetries, one type of conspiracy can lead to another. For instance, because the von Mangoldt function is real-valued rather than complex-valued, we have conjugation symmetry; if the primes correlate with, say, n^{it}, then they must also correlate with n^{-it}. (In the language of Dirichlet series, this reflects the fact that the zeta function and L-functions enjoy symmetries with respect to reflection across the real axis (i.e., complex conjugation).) Combining this observation with the all-or-nothing dichotomy, we conclude that the primes cannot correlate with n^{it} for any nonzero t, which in fact leads directly to the prime number theorem (1.55), as we shall discuss below. Similarly, if the primes correlated with a Dirichlet character $\chi(n)$, then they would also correlate with the conjugate $\overline{\chi}(n)$, which is also inconsistent with the all-or-nothing dichotomy, except in the exceptional case when χ is real—which essentially means that χ is a quadratic character. In this one case (which is the only scenario which comes close to threatening the truth of the prime number theorem in arithmetic progressions), the above tricks fail, and one has to instead exploit the algebraic number theory properties of these characters instead, which has so far led to weaker results than in the nonreal case.

As mentioned previously in passing, these phenomena are usually presented using the language of Dirichlet series and complex analysis. This is a very slick and powerful way to do things, but here I would like to present the elementary approach to the same topics, which is slightly weaker but which I find to also be very instructive. (However, I will not be *too* dogmatic about keeping things elementary, if this comes at the expense of obscuring the key ideas; in particular, I will rely on multiplicative Fourier analysis (both at ∞ and at finite places) as a substitute for complex analysis in order to expedite various parts of the argument. Also, the emphasis here will be more on heuristics and intuition than on rigour.)

The material here is closely related to the theory of *pretentious characters* developed in [**GrSo2007**], as well as the earlier paper [**Gr1992**].

1.12.1. A heuristic elementary proof of the prime number theorem.

To motivate some of the later discussion, let us first give a highly nonrigorous *heuristic* elementary "proof" of the prime number theorem (1.55). Since we clearly have

$$\sum_{n \leq x} 1 = x + O(1),$$

one can view the prime number theorem as an assertion that the von Mangoldt function Λ "behaves like 1 on the average",

$$\Lambda(n) \approx 1, \tag{1.56}$$

where we will be deliberately vague as to what the "$\approx$" symbol means. (One can think of this symbol as denoting some sort of proximity in the *weak topology* or *vague topology*, after suitable normalisation.)

To see why one would expect (1.56) to be true, we take divisor sums of (1.56) to heuristically obtain

$$\sum_{d|n} \Lambda(d) \approx \sum_{d|n} 1. \tag{1.57}$$

By (1.54), the left-hand side is $\log n$; meanwhile, the right-hand side is the *divisor function* $\tau(n)$ of n, by definition. So we have a heuristic relationship between (1.56) and the informal approximation

$$\tau(n) \approx \log n. \tag{1.58}$$

In particular, we expect

$$\sum_{n \leq x} \tau(n) \approx \sum_{n \leq x} \log n. \tag{1.59}$$

The right-hand side of (1.59) can be approximated using the *integral test* as

$$\sum_{n \leq x} \log n = \int_1^x \log t \; dt + O(\log x) = x \log x - x + O(\log x) \tag{1.60}$$

(one can also use *Stirling's formula* to obtain a similar asymptotic). As for the left-hand side, we write $\tau(n) = \sum_{d|n} 1$ and then make the substitution $n = dm$ to obtain

$$\sum_{n \leq x} \tau(n) = \sum_{d,m: dm \leq x} 1.$$

The right-hand side is the number of lattice points underneath the hyperbola $dm = x$, and can be counted using the *Dirichlet hyperbola method*:

$$\sum_{d,m: dm \leq x} 1 = \sum_{d \leq \sqrt{x}} \sum_{m \leq x/d} 1 + \sum_{m \leq \sqrt{x}} \sum_{d \leq x/m} 1 - \sum_{d \leq \sqrt{x}} \sum_{m \leq \sqrt{x}} 1.$$

The third sum is equal to $(\sqrt{x} + O(1))^2 = x + O(\sqrt{x})$. The second sum is equal to the first. The first sum can be computed as

$$\sum_{d \leq \sqrt{x}} \sum_{m \leq x/d} 1 = \sum_{d \leq \sqrt{x}} (\frac{x}{d} + O(1)) = x \sum_{d \leq \sqrt{x}} \frac{1}{d} + O(1);$$

meanwhile, from the *integral test* and the definition of *Euler's constant* $\gamma = 0.577\cdots$ one has

$$\sum_{d\leq y}\frac{1}{d} = \log y + \gamma + O(1/y) \tag{1.61}$$

for any $y \geq 1$; combining all these estimates one obtains

$$\sum_{n\leq x}\tau(n) = x\log x + (2\gamma - 1)x + O(\sqrt{x}). \tag{1.62}$$

Comparing this with (1.60), we do see that $\tau(n)$ and $\log n$ are roughly equal "to top order" on average, thus giving some form of (1.58) and hence (1.57); if one could somehow invert the divisor sum operation, one could hope to get (1.56) and thus the prime number theorem.

(Looking at the next highest order terms in (1.60) and (1.62), we see that we expect $\tau(n)$ to in fact be slightly larger than $\log n$ on the average, and so $\Lambda(n)$ should be slightly less than 1 on the average. There is indeed a slight effect of this form; for instance, it is possible (using the prime number theorem) to prove

$$\sum_{d\leq y}\frac{\Lambda(d)}{d} = \log y - \gamma + o(1),$$

which should be compared with (1.61).)

One can partially translate the above discussion into the language of Dirichlet series, by transforming various arithmetical functions $f(n)$ to their associated Dirichlet series

$$F(s) := \sum_{n=1}^{\infty}\frac{f(n)}{n^s},$$

ignoring for now the issue of convergence of this series. By definition, the constant function 1 transforms to the Riemann zeta function $\zeta(s)$. Taking derivatives in s, we see (formally, at least) that if $f(n)$ has Dirichlet series $F(s)$, then $f(n)\log n$ has Dirichlet series $-F'(s)$; thus, for instance, $\log n$ has Dirichlet series $-\zeta'(s)$.

Most importantly, though, if $f(n), g(n)$ have Dirichlet series $F(s), G(s)$, respectively, then their *Dirichlet convolution* $f * g(n) := \sum_{d|n} f(d)g(\frac{n}{d})$ has Dirichlet series $F(s)G(s)$; this is closely related to the well-known ability of the Fourier transform to convert convolutions to pointwise multiplication. Thus, for instance, $\tau(n)$ has Dirichlet series $\zeta(s)^2$. Also, from (1.54) and the preceding discussion, we see that $\Lambda(n)$ has Dirichlet series $-\zeta'(s)/\zeta(s)$ (formally, at least). This already suggests that the von Mangoldt function will be sensitive to the zeroes of the zeta function.

An integral test computation closely related to (1.61) gives the asymptotic

$$\zeta(s) = \frac{1}{s-1} + \gamma + O(s-1)$$

for s close to one (and $\mathrm{Re}(s) > 1$, to avoid issues of convergence). This implies that the Dirichlet series $-\zeta'(s)/\zeta(s)$ for $\Lambda(n)$ has asymptotic

$$\frac{-\zeta'(s)}{\zeta(s)} = \frac{1}{s-1} - \gamma + O(s-1)$$

thus giving support to (1.56). Similarly, the Dirichlet series for $\log n$ and $\tau(n)$ have asymptotic

$$-\zeta'(s) = \frac{1}{(s-1)^2} + O(1)$$

and

$$\zeta(s)^2 = \frac{1}{(s-1)^2} + \frac{2\gamma}{s-1} + O(1)$$

which gives support to (1.58) (and is also consistent with (1.60), (1.62)).

Remark 1.12.1. One can connect the properties of Dirichlet series $F(s)$ more rigorously to asymptotics of partial sums $\sum_{n\leq x} f(n)$ by means of various transforms in Fourier analysis and complex analysis, in particular contour integration or the Hilbert transform, but this becomes somewhat technical and we will not do so here. I will remark, though, that asymptotics of $F(s)$ for s close to 1 are not enough, by themselves, to get really precise asymptotics for the sharply truncated partial sums $\sum_{n\leq x} f(n)$, for reasons related to the uncertainty principle; in order to control such sums one also needs to understand the behaviour of F far away from $s = 1$, and in particular for $s = 1 + it$ for large real t. On the other hand, the asymptotics for $F(s)$ for s near 1 are just about all one needs to control *smoothly* truncated partial sums such as $\sum_n f(n)\eta(n/x)$ for suitable cutoff functions η. Also, while Dirichlet series are very powerful tools, particularly with regards to understanding Dirichlet convolution identities, and controlling everything in terms of the zeroes and poles of such series, they do have the drawback that they do not easily encode such fundamental "physical space" facts as the pointwise inequalities $|\mu(n)| \leq 1$ and $\Lambda(n) \geq 0$, which are also an important aspect of the theory.

1.12.2. Almost primes.

One can hope to make the above heuristics precise by applying the *Möbius inversion formula*

$$1_{n=1} = \sum_{d|n} \mu(d),$$

where $\mu(d)$ is the *Möbius function*, defined as $(-1)^k$ when d is the product of k distinct primes for some $k \geq 0$, and zero otherwise. In terms of Dirichlet series, we thus see that μ has the Dirichlet series of $1/\zeta(s)$, and so can invert the divisor sum operation $f(n) \mapsto \sum_{d|n} f(d)$ (which corresponds to multiplication by $\zeta(s)$):

$$f(n) = \sum_{m|n} \mu(m)(\sum_{d|n/m} f(d)).$$

From (1.54) we then conclude

$$\Lambda(n) = \sum_{d|n} \mu(d) \log \frac{n}{d}, \tag{1.63}$$

while from $\tau(n) = \sum_{d|n} 1$ we have

$$1 = \sum_{d|n} \mu(d)\tau(\frac{n}{d}). \tag{1.64}$$

One can now hope to derive the prime number theorem (1.55) from the formulae (1.60) and (1.62). Unfortunately, this does not quite work: the prime number theorem is equivalent to the assertion

$$\sum_{n \leq x} (\Lambda(n) - 1) = o(x), \tag{1.65}$$

but if one inserts (1.63) and (1.64) into the left-hand side of (1.65), one obtains

$$\sum_{d \leq x} \mu(d) \sum_{m \leq x/d} (\log m - \tau(m)),$$

which, if one then inserts (1.60) and (1.62) and the trivial bound $\mu(d) = O(1)$, leads to

$$2Cx \sum_{d \leq x} \frac{\mu(d)}{d} + O(x).$$

Using the elementary inequality

$$|\sum_{d \leq x} \frac{\mu(d)}{d}| \leq 1 \tag{1.66}$$

(see [**Ta2010b**]), we only obtain a bound of $O(x)$ for (1.65) instead of $o(x)$. (A refinement of this argument, though, shows that the prime number theorem would follow if one had the asymptotic $\sum_{n \leq x} \mu(n) = o(x)$, which is in fact equivalent to the prime number theorem.)

We remark that if one computed $\sum_{n \leq x} \tau(n)$ or $\sum_{n \leq x} \Lambda(n)$ by the above methods, one would eventually be led to a variant of (1.66), namely

$$\sum_{d \leq x} \frac{\mu(d)}{d} \log \frac{x}{d} = O(1), \tag{1.67}$$

which is an estimate that will be useful later.

So we see that when trying to sum the von Mangoldt function Λ by elementary means, the error term $O(x)$ overwhelms the main term x. But there is a slight tweaking of the von Mangoldt function, the *second von Mangoldt function* Λ_2, that increases the size of the main term to $2x \log x$ while keeping the error term at $O(x)$, thus leading to a useful estimate; the price one pays for this is that this function is now a proxy for the *almost primes* rather than the primes. This function is defined by a variant of (1.63), namely

$$\Lambda_2(n) = \sum_{d|n} \mu(d) \log^2 \frac{n}{d}. \tag{1.68}$$

It is not hard to see that $\Lambda_2(n)$ vanishes once n has at least three distinct prime factors (basically because the quadratic function $x \mapsto x^2$ vanishes after being differentiated three or more times). Indeed, one can easily verify the identity

$$\Lambda_2(n) = \Lambda(n) \log n + \Lambda * \Lambda(n), \tag{1.69}$$

which corresponds to the Dirichlet series identity

$$\zeta''(s)/\zeta(s) = -(-\zeta'(s)/\zeta(s))' + (-\zeta'(s)/\zeta(s))^2;$$

the first term $\Lambda(n) \log n$ is mostly concentrated on primes, while the second term $\Lambda*\Lambda(n)$ is mostly concentrated on *semiprimes* (products of two distinct primes).

Now let us sum $\Lambda_2(n)$. In analogy with the previous discussion, we will do so by comparing the function $\log^2 n$ with something involving the divisor function. In view of (1.58), it is reasonable to try the approximation

$$\log^2 n \approx \tau(n) \log n;$$

from the identity

$$2 \log n = \sum_{d|n} \mu(d)\tau(\frac{n}{d}) \log \frac{n}{d} \tag{1.70}$$

(which corresponds to the Dirichlet series identity $-2\zeta'(s) = \frac{1}{\zeta(s)} - (\zeta^2(s))'$) we thus expect

$$\Lambda_2(n) \approx 2 \log n. \tag{1.71}$$

Now we make these heuristics more precise. From the integral test we have

$$\sum_{n \leq x} \log^2 n = x \log^2 x + C_1 x \log x + C_2 x + O(\log^2 x),$$

while from (1.62) and summation by parts one has

$$\sum_{n \leq x} \tau(n) \log n = x \log^2 x + C_3 x \log x + C_4 x + O(\sqrt{x} \log x),$$

where C_1, C_2, C_3, C_4 are explicit absolute constants whose exact value is not important here. Thus

$$\sum_{n \leq x} (\log^2 n - \tau(n) \log n) = C_5 x \log x + C_6 x + O(\sqrt{x} \log x) \tag{1.72}$$

for some other constants C_5, C_6.

Meanwhile, from (1.68), (1.70) one has

$$\sum_{n \leq x} (\Lambda_2(n) - 2\log(n)) = \sum_{d \leq x} \mu(d) \sum_{m \leq x/d} \log^2 n - \tau(n) \log n;$$

applying (1.72), (1.66), and (1.67) we see that the right-hand side is $O(x)$. Computing $\sum_{n \leq x} \log n$ by the integral test, we deduce the *Selberg symmetry formula*

$$\sum_{n \leq x} \Lambda_2(n) = 2x \log x + O(x). \tag{1.73}$$

One can view (1.73) as the "almost prime number theorem", the analogue of the prime number theorem for almost primes.

The fact that the almost primes have a relatively easy asymptotic, while the genuine primes do not, is a reflection of the *parity problem* in sieve theory; see Section 3.10 of *Structure and Randomness* for further discussion. The symmetry formula is however enough to get "within a factor of two" of the prime number theorem: if we discard the semiprimes $\Lambda * \Lambda$ from (1.69), we see that $\Lambda(n) \log n \leq \Lambda_2(n)$, and thus

$$\sum_{n \leq x} \Lambda(n) \log n \leq 2x \log x + O(x),$$

which by a summation by parts argument leads to

$$0 \leq \sum_{n \leq x} \Lambda(n) \leq 2x + O(\frac{x}{\log x}),$$

which is within a factor of two of (1.55) in some sense.

One can "twist" all of the above arguments by a Dirichlet character χ. For instance, (1.68) twists to

$$\Lambda_2(n)\chi(n) = \sum_{d|n} \mu(d)\chi(d) \log^2 \frac{n}{d} \chi(\frac{n}{d}).$$

On the other hand, if χ is a nonprincipal character of modulus q, then it has mean zero on any interval with length q, and it is then not hard to establish the asymptotic

$$\sum_{n \leq y} \log^2 n \chi(n) = O_q(\log^2 y).$$

This soon leads to the twisted version of (1.73):

$$\sum_{n \le x} \Lambda_2(n)\chi(n) = O_q(x). \tag{1.74}$$

Thus almost primes are asymptotically unbiased with respect to nonprincipal characters.

From the multiplicative Fourier analysis of Dirichlet characters modulo q (and the observation that Λ_2 is quite small on residue classes not coprime to q), one then has an "almost prime number theorem in arithmetic progressions":

$$\sum_{n \le x: n = a \bmod q} \Lambda_2(n) = \frac{2}{\phi(q)} x \log x + O_q(x).$$

As before, this lets us come within a factor of two of the actual prime number theorem in arithmetic progressions:

$$\sum_{n \le x: n = a \bmod q} \Lambda(n) \le \frac{2}{\phi(q)} x + O_q(\frac{x}{\log x}).$$

One can also twist things by the completely multiplicative function $n \mapsto n^{it}$, but with the caveat that the approximation $2 \log n$ to $\Lambda_2(n)$ can locally correlate with n^{it}. Thus for instance one has

$$\sum_{n \le x} (\Lambda_2(n) - 2 \log n)\chi(n) n^{it} = O_q(x)$$

for any fixed t and χ; in particular, if χ is nonprincipal, one has

$$\sum_{n \le x} \Lambda_2(n)\chi(n) n^{it} = O_q(x).$$

1.12.3. The all-or-nothing dichotomy. To summarise so far, the almost primes (as represented by Λ_2) are quite uniformly distributed. These almost primes can be split up into the primes (as represented by $\Lambda(n) \log n$) and the semiprimes (as represented by $\Lambda * \Lambda(n)$), thanks to (1.69).

One can rewrite (1.69) as a recursive formula for Λ:

$$\Lambda(n) = \frac{1}{\log n} \Lambda_2(n) - \frac{1}{\log n} \Lambda * \Lambda(n). \tag{1.75}$$

One can also twist this formula by a character χ and/or a completely multiplicative function $n \mapsto n^{it}$; thus for instance

$$\Lambda\chi(n) = \frac{1}{\log n} \Lambda_2\chi(n) - \frac{1}{\log n} \Lambda\chi * \Lambda\chi(n). \tag{1.76}$$

This recursion, combined with the uniform distribution properties on Λ_2, lead to various *all-or-nothing* dichotomies for Λ. Suppose, for instance, that

$\Lambda\chi$ behaves like a constant c on the average for some nonprincipal character χ:

$$\Lambda\chi(n) \approx c.$$

Then (from (1.58)) we expect $\Lambda\chi * \Lambda\chi$ to behave like $c^2 \log n$, thus

$$\frac{1}{\log n}\Lambda\chi * \Lambda\chi(n) \approx c^2.$$

On the other hand, from (1.74), $\frac{1}{\log n}\Lambda_2(n)$ is asymptotically uncorrelated with χ:

$$\frac{1}{\log n}\Lambda_2\chi \approx 0.$$

Putting all this together, one obtains

$$c \approx -c^2,$$

which suggests that c must be either close to 0, or close to -1.

Basically, the point is that there are only two equilibria for the recursion (1.76). One equilibrium occurs when Λ is asymptotically uncorrelated with χ; the other is when it is completely anticorrelated with χ, so that $\Lambda(n)$ is supported primarily on those n for which $\chi(n)$ is close to -1. Note in the latter case that $\chi(n) \approx -1$ for most primes n, and thus $\chi(n) \approx +1$ for most semiprimes n, thus leading to an equidistribution of $\chi(n)$ for almost primes (weighted by Λ_2). Any intermediate distribution of $\Lambda\chi$ would be inconsistent with the distribution of $\Lambda_2\chi$. (In terms of Dirichlet series, this assertion corresponds to the fact that the L-function of χ either has a zero of order 1, or a zero of order 0 (i.e., not a zero at all) at $s = 1$.)

A similar phenomenon occurs when twisting Λ by n^{it}; basically, the average value of $(\Lambda(n) - 1)n^{it}$ must asymptotically either be close to 0, or close to -1; no other asymptotic ends up being compatible with the distribution of $(\Lambda_2(n) - 2\log n)n^{it}$. (Again, this corresponds to the fact that the Riemann zeta function has a zero of order 1 or 0 at $1+it$.) More generally, the average value of $(\Lambda(n) - 1)\chi(n)n^{it}$ must asymptotically approach either 0 or -1.

Remark 1.12.2. One can make the above heuristics precise either by using Dirichlet series (and analytic continuation and the theory of zeroes of meromorphic functions) or by smoothing out arithmetic functions such as $\Lambda\chi$ by a suitable multiplicative convolution with a mollifier (as is basically done in elementary proofs of the prime number theorem); see also [**GrSo2007**] for a closely related theory. We will not pursue these details here, however.

1.12.4. Dueling conspiracies. In the previous section we have seen (heuristically, at least), that the von Mangoldt function $\Lambda(n)$ (or more precisely, $\Lambda(n) - 1$) will either have no correlation or a maximal amount of anticorrelation with a completely multiplicative function such as $\chi(n)$, n^{it}, or $\chi(n)n^{it}$. On the other hand, it is not possible for this function to maximally anticorrelate (or to *conspire*) with two such functions; thus the presence of one conspiracy excludes the presence of all others.

Suppose for instance that we had two distinct nonprincipal characters χ, χ' for which one had maximal anticorrelation:

$$\Lambda(n)\chi(n), \Lambda(n)\chi'(n) \approx -1.$$

One could then combine the two statements to obtain

$$\Lambda(n)(\chi(n) + \chi'(n)) \approx -2.$$

Meanwhile, $\frac{1}{\log n}\Lambda_2(n)$ does not correlate with either χ or χ'. It will be convenient to exploit this to normalise Λ, obtaining

$$(\Lambda(n) - \frac{1}{2\log n}\Lambda_2(n))(\chi(n) + \chi'(n)) \approx -2.$$

(Note from (1.56) and (1.71) that we expect $\Lambda(n) - \frac{1}{2\log n}\Lambda_2(n)$ to have mean zero.)

On the other hand, since $0 \leq \Lambda(n)\log n \leq \Lambda_2(n)$, one has

$$|\Lambda(n) - \frac{1}{2\log n}\Lambda_2(n)| \leq \frac{1}{2\log n}\Lambda_2(n),$$

and hence by the triangle inequality

$$\Lambda_2(n)|\chi(n) + \chi'(n)| \gtrapprox 4\log n,$$

in the sense that averages of the left-hand side should be at least as large as averages of the right-hand side. From this, (1.71), and Cauchy-Schwarz, one thus expects

$$\Lambda_2(n)|\chi(n) + \chi'(n)|^2 \gtrapprox 8\log n.$$

But if one expands out the left-hand side using (1.71) and (1.74), one only ends up with $4\log n + O_q(1)$ on the average, a contradiction for n sufficiently large.

Remark 1.12.3. The above argument belongs to a family of L^2-based arguments which go by various names (*almost orthogonality*, TT^*, *large sieve*, etc.) The L^2 argument can more generally be used to establish square-summability estimates on averages such as $\frac{1}{x}\sum_{n\leq x}\Lambda(n)\chi(n)$ as χ varies, but we will not make this precise here.

As one consequence of the above arguments, one can show that $\Lambda(n)$ cannot maximally anticorrelate with any nonreal character χ, since (by the reality of Λ) it would then also maximally anticorrelate with the complex conjugate $\overline{\chi}$, which is distinct from χ. A similar argument shows that $\Lambda(n)$ cannot maximally anticorrelate with n^{it} for any nonzero t, a fact which can soon lead to the prime number theorem either by Dirichlet series methods, by Fourier-analytic means, or by elementary means. (Sketch of Fourier-analytic proof: L^2 methods provide L^2-type bounds on the averages of $\Lambda(n)n^{it}$ in t, while the above arguments show that these averages are also small in L^∞. Applying (1.75) a few times to take advantage of the smoothing effects of convolution, one eventually concludes that these averages can be made arbitrarily small in L^1 asymptotically, at which point the prime number theorem follows from Fourier inversion.)

Remark 1.12.4. There is a slightly different argument of an L^1 nature rather than an L^2 nature (i.e., using tools such as the triangle inequality, union bound, etc.) that can also achieve similar results. For instance, suppose that $\Lambda(n)$ maximally anticorrelates with χ and χ'. Then $\chi(n), \chi'(n) \approx -1$ for most primes n, which implies that $\chi\chi'(n) \approx +1$ for most primes n, which is incompatible with the all-or-nothing dichotomy unless $\chi\chi'$ is principal. This is an alternate way to exclude correlation with nonreal characters. Similarly, if $\Lambda(n)n^{it} \approx -1$, then $\Lambda(n)n^{2it} \approx +1$, which is also incompatible with the zero–one law; this is essentially the method underlying the standard proof of the prime number theorem (which relates $\zeta(1+it)$ with $\zeta(1+2it)$).

1.12.5. Quadratic characters.

The one difficult scenario to eliminate is that of maximal anticorrelation with a real nonprincipal (i.e., quadratic) character χ, thus

$$\Lambda(n)\chi(n) \approx -1.$$

This scenario implies that the quantity

$$L(1,\chi) := \sum_{n=1}^\infty \frac{\chi(n)}{n}$$

vanishes. Indeed, if one starts with the identity

$$\log n\chi(n) = \sum_{d|n} \Lambda\chi(d)\chi(\frac{n}{d})$$

and sums in n, one sees that

$$\sum_{n\leq x} \log n\chi(n) = \sum_{d,m:dm\leq x} \Lambda\chi(d)\chi(m).$$

The left-hand side is $O_q(\log x)$ by the mean zero and periodicity properties of χ. To estimate the right-hand side, we use the hyperbola method and rewrite it as

$$\sum_{m\leq M}\chi(m)\sum_{d\leq x/m}\Lambda\chi(d)+\sum_{d\leq x/M}\Lambda\chi(d)\sum_{M<m\leq x/d}\chi(m)$$

for some parameter M (sufficiently slowly growing in x) to be optimised later. Writing $\sum_{d\leq x/m}\Lambda\chi(d)=(-1+o_q(1))x/m$ and $\sum_{M<m\leq x/d}\chi(m)=O_q(1)$, we can express this as

$$x(\sum_{m\leq M}\frac{\chi(m)}{m}+o_q(1))+O_q(x/M);$$

sending $x\to\infty$ (and $M\to\infty$ at a slower rate), we conclude $L(1,\chi)=0$ as required.

It is remarkably difficult to show that $L(1,\chi)$ does not, in fact, vanish. One way to do this is to use the *class number formula*, that relates this quantity to the class number of the quadratic number field $\mathbf{Q}(\sqrt{-d})$ associated to the conductor d of χ, together with some related number-theoretic quantities. A more elementary (but significantly weaker) method proceeds by using the easily verified fact that the convolution $1*\chi$ is nonnegative and is at least 1 on the squares; this should be interpreted as a fact from algebraic number theory, and basically corresponds to the fact that the number of representations of an integer n as the norm x^2+dy^2 of an integer in $\mathbf{Z}(\sqrt{d})$ (or more generally, as the norm of an ideal in that ring) is nonnegative, and is at least 1 on the squares. In particular we have

$$\sum_{n\leq x}\frac{1*\chi(n)}{\sqrt{n}}\geq\frac{1}{2}\log x+O(1).$$

On the other hand from the hyperbola method we can express the left-hand side as

$$\sum_{d\leq\sqrt{x}}\frac{\chi(d)}{\sqrt{d}}\sum_{m\leq x/d}\frac{1}{\sqrt{m}}+\sum_{m<\sqrt{x}}\frac{1}{\sqrt{m}}\sum_{\sqrt{x}<d\leq x/m}\frac{\chi(d)}{\sqrt{d}}. \tag{1.77}$$

From the mean zero and periodicity properties of χ we have $\sum_{\sqrt{x}<d\leq x/m}\frac{\chi(d)}{\sqrt{d}}=O_q(x^{-1/4})$, so the second term in (1.77) is $O_q(1)$. Meanwhile, from the *midpoint rule*, $\sum_{m\leq y}\frac{1}{\sqrt{m}}=2\sqrt{y}-\frac{3}{2}+O(1/\sqrt{y})$, and so the first term in (1.77) is

$$2\sqrt{x}\sum_{d\leq\sqrt{x}}\frac{\chi(d)}{d}+O(|\sum_{d\leq\sqrt{x}}\frac{\chi(d)}{\sqrt{d}}|)+O(1)=2\sqrt{x}L(1,\chi)+O(1).$$

Putting all this together, we have

$$\frac{1}{2}\log x+O(1)\leq 2\sqrt{x}L(1,\chi)+O_q(1),$$

which leads to a contradiction as $x \to \infty$ if $L(1,\chi)$ vanishes.

Note in fact that the above argument shows that $L(1,\chi)$ is positive. If one carefully computes the dependence of the above argument on the modulus q, one obtains a lower bound of the form $L(1,\chi) \geq \exp(-q^{1/2+o(1)})$, which is quite poor. Using a nontrivial improvement on the error term in counting lattice points under the hyperbola (or better still, by smoothing the sum $\sum_{n\leq x}$), one can improve this a bit, to $L(1,\chi) \geq q^{-O(1)}$. In contrast, the class number method gives a bound $L(1,\chi) \geq q^{-1/2+o(1)}$.

We can improve this even further for all but at most one real primitive character χ:

Theorem 1.12.5 (Siegel's theorem). *For every $\varepsilon > 0$, one has $L(1,\chi) \gg_\varepsilon q^{-\varepsilon}$ for all but at most one real primitive character χ, where the implied constant is effective and q is the modulus of χ.*

Throwing in this (hypothetical) one exceptional character, we conclude that $L(1,\chi) \gg_\varepsilon q^{-\varepsilon}$ for *all* real primitive characters χ, but now the implied constant is ineffective, which is the usual way in which Siegel's theorem is formulated (but the above nearly effective refinement can be obtained by the same methods). It is a major open problem in the subject to eliminate this exceptional character and recover an effective estimate for some $\varepsilon < 1/2$.

Proof. Let $\varepsilon > 0$ (which we can assume to be small), and let $c > 0$ be a small number depending (effectively) on ε to be chosen later. Our task is to show that $L(1,\chi) \geq cq^{-\varepsilon}$ for all but at most one primitive real character χ. Note we may assume q is large (effectively) depending on ε, as the claim follows from the previous bounds on $L(1,\chi)$ otherwise.

Suppose then for contradiction that $L(1,\chi) < cq^{-\varepsilon}$ and $L(1,\chi') < c(q')^{-\varepsilon}$ for two distinct primitive real characters χ, χ' of (large) modulus q, q', respectively.

We begin by modifying the proof that $L(1,\chi)$ was positive, which relied (among other things) on the observation that $1 * \chi$, and equals 1 at 1. In particular, one has

$$\sum_{n\leq x} \frac{1 * \chi(n)}{n^s} \geq 1 \tag{1.78}$$

for any $x \geq 1$ and any real s. (One can get slightly better bounds by exploiting that $1 * \chi$ is also at least 1 on square numbers, as before, but this is really only useful for $s \leq 1/2$, and we are now going to take s much closer to 1.)

On the other hand, one has the asymptotics

$$\sum_{n\le x}\frac{1}{n^s}=\zeta(s)+\frac{x^{1-s}}{1-s}+O(x^{-s})$$

for any real s close (but not equal) to 1, and similarly

$$\sum_{n\le x}\frac{\chi(n)}{n^s}=L(s,\chi)+O(q^{O(1)}x^{-s})$$

for any real s close to 1; similarly for $\chi',\chi\chi'$. From the hyperbola method, we can then conclude

$$\sum_{n\le x}\frac{1*\chi(n)}{n^s}=\zeta(s)L(s,\chi)+\frac{x^{1-s}}{1-s}L(1,\chi)+O(q^{O(1)}x^{0.5-s}) \tag{1.79}$$

for all real s sufficiently close to 1. Indeed, one can expand the left-hand side of (1.79) as

$$\sum_{d\le\sqrt{x}}\frac{\chi(d)}{d^s}\sum_{m\le x/d}\frac{1}{m^s}+\sum_{m<\sqrt{x}}\frac{1}{m^s}\sum_{\sqrt{x}<d\le x/m}\frac{\chi(d)}{d^s},$$

and the claim then follows from the previous asymptotics. (One can improve the error term by smoothing the summation, but we will not need to do so here.)

Now set $x=Cq^C$ for a large absolute constant C. If $0.99\le s<1$, then the error term in $O(q^{O(1)}x^{0.5-s})$ is then at most $1/2$ (say) if C is large enough. We conclude from (1.78) that

$$\zeta(s)L(s,\chi)\ge\frac{1}{2}-O(\frac{q^{O(1-s)}}{1-s}L(1,\chi))$$

for $0.99\le s<1$. Since $L(1,\chi)\le cq^{-\varepsilon}$ and c is assumed small (depending on ε), this implies that $\zeta(s)L(s,\chi)$ is positive in the range

$$L(1,\chi)\ll 1-s\ll\varepsilon$$

(this can be seen by checking the cases $1-s\le 1/\log q$ and $1-s>1/\log q$ separately). On the other hand, $\zeta(s)L(s,\chi)$ has a simple pole at $s=1$ with positive residue and is thus negative for $s<1$ extremely close to 1. By the intermediate value theorem, we conclude that $L(s,\chi)$ has a zero for some $s=1-O(L(1,\chi))$. Conversely, it is not difficult (using summation by parts) to show that $L'(s,\chi)=O(\log^2 q)$ for $s=1-O(1/\log q)$, and so by the mean value theorem we see that the zero of $L(s,\chi)$ must also obey $1-s\gg L(1,\chi)/\log^2 q$. Thus $L(s,\chi)$ has a zero for some $s<1$ with

$$L(1,\chi)/\log^2 q\ll 1-s\ll L(1,\chi). \tag{1.80}$$

Similarly, $L(s',\chi')$ has a zero for some $s'<1$ with

$$L(1,\chi')/\log^2 q'\ll 1-s'\ll L(1,\chi'). \tag{1.81}$$

Now, we consider the function

$$f := 1 * \chi * \chi' * \chi\chi'.$$

One can also show that f is nonnegative and equals 1 at 1, thus

$$\sum_{n \leq x} \frac{f(n)}{n^s} \geq 1.$$

(The algebraic number theory interpretation of this positivity is that $f(n)$ is the number of representations of n as the norm of an ideal in the *biquadratic field* generated by $\sqrt{q}$ and $\sqrt{q'}$.)

Also, by (a more complicated version of) the derivation of (1.79), one has

$$\sum_{n \leq x} \frac{f(n)}{n^s} = \zeta(s)L(s,\chi)L(s,\chi')L(s,\chi\chi')$$
$$+ \frac{x^{1-s}}{1-s} L(1,\chi)L(1,\chi')L(1,\chi\chi') + O((qq')^{O(1)} x^{0.9-s})$$

(say). Arguing as before, we conclude that

$$\zeta(s)L(s,\chi)L(s,\chi')L(s,\chi\chi') \geq \frac{1}{2} - O(\frac{(qq')^{O(1-s)}}{1-s} L(1,\chi)L(1,\chi')L(1,\chi\chi'))$$

for $0.99 \leq s < 1$. Using the bound $L(1,\chi\chi') \ll \log(qq')$ (which can be established by summation by parts), we conclude that $\zeta(s)L(s,\chi)L(s,\chi')L(s,\chi\chi')$ is positive in the range

$$L(1,\chi)L(1,\chi') \log(qq') \ll 1 - s \ll \varepsilon.$$

Since we already know $L(s,\chi)$ and $L(s',\chi')$ have zeroes for some s, s' obeying (1.80) and (1.81)

$$\frac{L(1,\chi)}{\log^2 q}, \frac{L(1,\chi')}{\log^2 q'} \ll L(1,\chi)L(1,\chi') \log(qq');$$

taking geometric means and rearranging, we obtain

$$L(1,\chi)L(1,\chi') \gg \log(qq')^{-O(1)}.$$

But this contradicts the hypotheses $L(1,\chi) \leq cq^{-\varepsilon}$, $L(1,\chi') \leq c(q')^{-\varepsilon}$ if c is small enough. □

Remark 1.12.6. Siegel's theorem leads to a version of the prime number theorem in arithmetic progressions known as the *Siegel-Walfisz theorem*. As with Siegel's theorem, the bounds are ineffective unless one is allowed to exclude a single exceptional modulus q (and its multiples), in which case one has a modified prime number theorem which favours the quadratic non-residues mod q; see [**Gr1992**].

Remark 1.12.7. One can improve the effective bounds in Siegel's theorem if one is allowed to exclude a larger set of bad moduli. For instance, the arguments in Section 1.12.4 allow one to establish a bound of the form $L(1,\chi) \gg \log^{-O(1)} q$ after excluding at most one q in each hyper-dyadic range $2^{100^k} \leq q \leq 2^{100^{k+1}}$ for each k; one can of course replace 100 by other exponents here, but at the cost of worsening the $O(1)$ term. (This is essentially an observation of Landau.)

Notes. This article first appeared at

`terrytao.wordpress.com/2009/09/24.`

Thanks to anonymous commenters for corrections.

David Speyer noted the connection between Siegel's theorem and the classification of imaginary quadratic fields with unique factorisation.

1.13. Mazur's swindle

Let d be a natural number. A basic operation in the topology of oriented, connected, compact, d-dimensional manifolds (hereby referred to simply as *manifolds* for short) is that of *connected sum*: given two manifolds M, N, the connected sum $M\#N$ is formed by removing a small ball from each manifold and then gluing the boundary together (in the orientation-preserving manner). This gives another oriented, connected, compact manifold, and the exact nature of the balls removed and their gluing is not relevant for topological purposes (any two such procedures give homeomorphic manifolds). It is easy to see that this operation is associative and commutative up to homeomorphism, thus $M\#N \cong N\#M$ and $(M\#N)\#O \cong M\#(N\#O)$, where we use $M \cong N$ to denote the assertion that M is homeomorphic to N.

(It is important that the orientation is preserved; if, for instance, $d = 3$, and M is a 3-manifold which is *chiral* (thus $M \not\cong -M$, where $-M$ is the orientation reversal of M), then the connect sum $M\#M$ of M with itself is also chiral (by the *prime decomposition*; in fact one does not even need the irreducibility hypothesis for this claim), but $M\# -M$ is not. A typical example of an irreducible chiral manifold is the complement of a *trefoil knot*. Thanks to Danny Calegari for this example.)

The d-dimensional sphere S^d is an identity (up to homeomorphism) of connect sum: $M\#S^d \cong M$ for any M. A basic result in the subject is that the sphere is itself irreducible:

Theorem 1.13.1 (Irreducibility of the sphere). *If $S^d \cong M\#N$, then $M, N \cong S^d$.*

For $d = 1$ (curves), this theorem is trivial because the only connected 1-manifolds are homeomorphic to circles. For $d = 2$ (surfaces), the theorem is also easy by considering the *genus* of $M, N, M\#N$. For $d = 3$ the result follows from the *prime decomposition*. But for higher d, these ad hoc methods no longer work. Nevertheless, there is an elegant proof of Theorem 1.13.1, due to Mazur [**Ma1959**], which is known as *Mazur's swindle*. The reason for this name should become clear when one sees the proof, which I reproduce below.

Suppose $M\#N \cong S^d$. Now consider the infinite connected sum

$$(M\#N)\#(M\#N)\#(M\#N)\#\cdots.$$

This is an infinite connected sum of spheres and can thus be viewed as a half-open cylinder, which is topologically equivalent to a sphere with a small ball removed. Alternatively, one can contract the boundary at infinity to a point to recover the sphere S^d. On the other hand, by using the associativity of connected sum (which will still work for the infinite connected sum, if one thinks about it carefully), the above manifold is also homeomorphic to

$$M\#(N\#M)\#(N\#M)\#\cdots.$$

which is the connected sum of M with an infinite sequence of spheres or, equivalently, M with a small ball removed. Contracting the small balls to a point, we conclude that $M \cong S^d$, and a similar argument gives $N \cong S^d$.

A typical corollary of Theorem 1.13.1 is a generalisation of the *Jordan curve theorem*: any *locally flat* embedded copy of S^{d-1} in S^d divides the sphere S^d into two regions homeomorphic to balls B^d. (Some sort of regularity hypothesis, such as local flatness, is essential, thanks to the counterexample of the *Alexander horned sphere*. If one assumes smoothness instead of local flatness, the problem is known as the *Schönflies problem*, and is apparently quite subtle, especially in the four-dimensional case $d = 4$.)

One can ask whether there is a way to prove Theorem 1.13.1 for general d without recourse to the infinite sum swindle. I do not know the complete answer to this, but some evidence against this hope can be seen by noting that if one works in the smooth category instead of the topological category (i.e., working with smooth manifolds, and only equating manifolds that are diffeomorphic and not merely homeomorphic), then the *exotic spheres* in five and higher dimensions provide a counterexample to the smooth version of Theorem 1.13.1: it is possible to find two exotic spheres whose connected sum is diffeomorphic to the standard sphere. (Indeed, in five and higher dimensions, the exotic sphere structures on S^d form a finite abelian group under connected sum, with the standard sphere being the identity element. The situation in four dimensions is much less well understood.) The problem with the swindle here is that the homeomorphism generated by the infinite number of applications of the associativity law is not smooth when one identifies the boundary with a point.

The basic idea of the swindle—grouping an alternating infinite sum in two different ways—also appears in a few other contexts. Most classically, it is used to show that the sum $1-1+1-1+\cdots$ does not converge in any sense which is consistent with the infinite associative law, since this would then imply that $1=0$; indeed, one can view the swindle as a dichotomy between the infinite associative law and the presence of nontrivial cancellation. (In the topological manifold category, one has the former but not the latter, whereas in the case of $1-1+1-1+\cdots$, one has the latter but not the former.) The *alternating series test* can also be viewed as a variant of the swindle.

Another variant of the swindle arises in the proof of the *Cantor-Bernstein-Schröder theorem*. Suppose one has two sets A, B, together with injections from A to B and from B to A. The first injection leads to an identification $B \cong C \uplus A$ for some set C, while the second injection leads to an identification $A \cong D \uplus B$. Iterating this leads to identifications

$$A \cong (D \uplus C \uplus D \uplus \cdots) \uplus X$$

and

$$B \cong (C \uplus D \uplus C \uplus \cdots) \uplus X$$

for some additional set X. Using the identification $D \uplus C \cong C \uplus D$ then yields an explicit bijection between A and B.

Notes. This article first appeared at

`terrytao.wordpress.com/2009/10/05.`

Thanks to Jan, Peter, and an anonymous commenter for corrections.

Thanks again to Danny Calegari for telling me about the swindle while we were both waiting to catch an airplane.

Several commenters provided further examples of swindle-type arguments. Scott Morrison noted that Mazur's argument also shows that nontrivial knots do not have inverses: one cannot untie a knot by tying another one. Qiaochu Yuan provided a swindle argument that showed that $GL(H)$ is contractible for any infinite-dimensional Hilbert space H. In a similar spirit, Pace Nielsen recalled the Eilenberg swindle that shows that for every projective module P, there exists a free module F with $P \oplus F \equiv F$. Tim Gowers also mentioned Pelczynski's decomposition method in the theory of Banach spaces as a similar argument.

1.14. Grothendieck's definition of a group

In his wonderful article [**Th1994**], Bill Thurston describes (among many other topics) how one's understanding of given concept in mathematics (such as that of the derivative) can be vastly enriched by viewing it simultaneously from many subtly different perspectives. In the case of the derivative, he gives seven standard such perspectives (infinitesimal, symbolic, logical, geometric, rate, approximation, microscopic) and then mentions a much later perspective in the sequence (as describing a flat connection for a graph).

One can of course do something similar for many other fundamental notions in mathematics. For instance, the notion of a *group* G can be thought of in a number of (closely related) ways, such as the following:

(0) **Motivating examples**: A group is an abstraction of the operations of addition/subtraction or multiplication/division in arithmetic or linear algebra, or of composition/inversion of transformations.

(1) **Universal algebraic**: A group is a set G with an identity element e, a unary inverse operation $\cdot^{-1} : G \to G$, and a binary multiplication operation $\cdot : G \times G \to G$ obeying the relations (or axioms) $e \cdot x = x \cdot e = x$, $x \cdot x^{-1} = x^{-1} \cdot x = e$, $(x \cdot y) \cdot z = x \cdot (y \cdot z)$ for all $x, y, z \in G$.

(2) **Symmetric**: A group is all the ways in which one can transform a space V to itself while preserving some object or structure O on this space.

(3) **Representation theoretic**: A group is identifiable with a collection of transformations on a space V which is closed under composition and inverse, and contains the identity transformation.

(4) **Presentation theoretic**: A group can be generated by a collection of generators subject to some number of relations.

(5) **Topological**: A group is the fundamental group $\pi_1(X)$ of a connected topological space X.

(6) **Dynamic**: A group represents the passage of time (or of some other variable(s) of motion or action) on a (reversible) dynamical system.

(7) **Category theoretic**: A group is a category with one object in which all morphisms have inverses.

(8) **Quantum**: A group is the classical limit $q \to 0$ of a quantum group.

- etc.

One can view a large part of group theory (and related subjects, such as representation theory) as exploring the interconnections between a variety of these perspectives. As one's understanding of the subject matures, many of these formerly distinct perspectives slowly merge into a single unified perspective.

From a recent talk by Ezra Getzler, I learned a more sophisticated perspective on a group that is somewhat analogous to Thurston's example of a sophisticated perspective on a derivative (and coincidentally, flat connections play a central role in both):

(37) **Sheaf theoretic**: A group is identifiable with a (set-valued) sheaf on the category of simplicial complexes such that the morphisms associated to collapses of d-simplices are bijective for $d > 1$ (and merely surjective for $d \leq 1$).

This interpretation of the group concept is apparently due to Grothendieck, though it is motivated also by homotopy theory. One of the key advantages of this interpretation is that it generalises easily to the notion of an *n-group* (simply by replacing 1 with n in (37)), whereas the other interpretations listed earlier require a certain amount of subtlety in order to generalise correctly (in particular, they usually themselves require higher-order notions, such as *n-categories*).

The connection of (37) with any of the other perspectives of a group is elementary, but not immediately obvious; I enjoyed working out exactly what the connection was and thought it might be of interest to some readers here.

1.14.1. Flat connections.

To see the relationship between (37) and more traditional concepts of a group, such as (1), we will begin by recalling the machinery of flat connections.

Let G be a group, and let X be a topological space. A *principal G-connection* ω on X can be thought of as an assignment of a group element $\omega(\gamma) \in G$ to every path γ in X which obeys the following four properties:

- *Invariance under reparameterisation.* If γ' is a reparameterisation of γ, then $\omega(\gamma) = \omega(\gamma')$.
- *Identity.* If γ is a constant path, then $\omega(\gamma)$ is the identity element.
- *Inverse.* If $-\gamma$ is the reversal of a path γ, then $\omega(-\gamma)$ is the inverse of $\omega(\gamma)$.
- *Groupoid homomorphism.* If γ_2 starts where γ_1 ends (so that one can define the concatenation $\gamma_1+\gamma_2$), then $\omega(\gamma_1+\gamma_2) = \omega(\gamma_2)\omega(\gamma_1)$. (Depending on one's conventions, one may wish to reverse the order of the group multiplication on the right-hand side.)

Intuitively, $\omega(\gamma)$ represents a way to use the group G to connect (or "parallel transport") the fibre at the initial point of γ to the fibre at the final point; see Section 1.4 of *Poincaré's Legacies, Vol. II* for more discussion. Note that the identity property is redundant, being implied by the other three properties.

We say that a connection ω is *flat* if $\omega(\gamma)$ is the identity element for every "short" closed loop γ, thus strengthening the identity property. One could define "short" rigorously (e.g., one could use "*contractible*" as a substitute), but we will prefer here to leave the concept intentionally vague.

Typically, one studies connections when the structure group G and the base space X are continuous rather than discrete. However, there is a combinatorial model for connections which is suitable for discrete groups, in which the base space X is now an *(abstract) simplicial complex* Δ—a vertex set V, together with a number of *simplices* in V, by which we mean ordered $d+1$-tuples $(x_0, \ldots, x_d)$ of distinct vertices in V for various integers d (with d being the *dimension* of the simplex $(x_0, \ldots, x_d)$). In our definition of a simplicial complex, we add the requirement that if a simplex lies in the complex, then all faces of that simplex (formed by removing one of the vertices, but leaving the order of the remaining vertices unchanged) also lie in the complex. We also assume a well-defined *orientation*, in the sense that every $d+1$-tuple $\{x_0, \ldots, x_d\}$ is represented by at most one simplex (thus, for instance, a complex cannot contain both an edge $(0, 1)$ and its reversal $(1, 0)$). Though it will not matter too much here, one can think of the vertex set V here as being restricted to be finite.

A *path* γ in a simplicial complex Δ is then a sequence of 1-simplices (x_i, x_{i+1}) or their formal reverses $-(x_i, x_{i+1})$, with the final point of each 1-simplex being the initial point of the next. If G is a (discrete) group, a *principal G-connection* ω on Δ is then an assignment of a group element $\omega(\gamma) \in G$ to each such path γ, obeying the groupoid homomorphism property and the inverse property (and hence the identity property). Note that the reparameterisation property is no longer needed in this abstract combinatorial model. Note that a connection can be determined by the group elements $\omega(b \leftarrow a)$ it assigns to each 1-simplex (a, b). (I have written the simplex $b \leftarrow a$ from right to left, as this makes the composition law cleaner.)

So far, only the 1-skeleton (i.e., the simplices of dimension at most 1) of the complex have been used. But one can use the 2-skeleton to define the notion of a *flat* connection: we say that a principal G-connection ω on Δ is flat if the boundary of every 2-simplex (a, b, c), oriented appropriately, is assigned the identity element, or more precisely that

$$\omega(c \leftarrow a)^{-1}\omega(c \leftarrow b)\omega(b \leftarrow a) = e,$$

or in other words that

$$\omega(c \leftarrow a) = \omega(c \leftarrow b)\omega(b \leftarrow a);$$

thus, in this context, a "short loop" means a loop that is the boundary of a 2-simplex. Note that this corresponds closely to the topological concept of a flat connection when applied to, say, a triangulated manifold.

Fix a group G. Given any simplicial complex Δ, let $\mathcal{O}(\Delta)$ be the set of flat connections on Δ. One can get some feeling for this set by considering some basic examples:

- If Δ is a single 0-dimensional simplex (i.e., a point), then there is only the trivial path, which must be assigned the identity element e of the group. Thus, in this case, $\mathcal{O}(\Delta)$ can be identified with $\{e\}$.
- If Δ is a 1-dimensional simplex, say $(0, 1)$, then the path from 0 to 1 can be assigned an arbitrary group element $\omega(1 \leftarrow 0) \in G$, and this is the only degree of freedom in the connection. So in this case, $\mathcal{O}(\Delta)$ can be identified with G.
- Now suppose Δ is a 2-dimensional simplex, say $(0, 1, 2)$. Then the group elements $\omega(1 \leftarrow 0)$ and $\omega(2 \leftarrow 1)$ are arbitrary elements of G, but $\omega(2 \leftarrow 0)$ is constrained to equal $\omega(2 \leftarrow 1)\omega(1 \leftarrow 0)$. This determines the entire flat connection, so $\mathcal{O}(\Delta)$ can be identified with G^2.
- Generalising this example, if Δ is a k-dimensional simplex, then $\mathcal{O}(\Delta)$ can be identified with G^k.

An important operation one can do on flat connections is that of *pullback*. Let $\phi : \Delta \to \Delta'$ be a *morphism* from one simplicial complex Δ to another Δ'; by this, we mean a map from the vertex set of Δ to the vertex set of Δ' such that every simplex in Δ maps to a simplex in Δ' in an order-preserving manner (though note that ϕ is allowed to be noninjective, so that the dimension of the simplex can decrease by mapping adjacent vertices to the same vertex). Given such a morphism, and given a flat connection ω' on Δ', one can then pull back that connection to yield a flat connection $\phi^*\omega'$ on Δ, defined by the formula

$$\phi^*\omega'(w \leftarrow v) := \omega'(\phi(w) \leftarrow \phi(v))$$

for any 1-simplex (v, w) in Δ, with the convention that $\omega'(u \leftarrow u)$ is the identity for any u. It is easy to see that this is still a flat connection. Also, if $\phi : \Delta \to \Delta'$ and $\psi : \Delta' \to \Delta''$ are morphisms, then the operations of pullback by ψ and then by ϕ compose to equal the operation of pullback by $\psi \circ \phi$: $\phi^*\psi^* = (\psi \circ \phi)^*$. In the language of category theory, pullback is a contravariant functor from the category of simplicial complexes to the category of sets (with each simplicial complex being mapped to its set of flat connections).

A special case of a morphism is an *inclusion morphism* $\iota : \Delta \to \Delta'$ to a simplicial complex Δ' from a subcomplex Δ. The associated pullback operation is the *restriction* operation, that maps a flat connection ω' on Δ' to its restriction $\omega' \downharpoonright_\Delta$ to Δ.

1.14.2. Sheaves. We currently have a set $\mathcal{O}(\Delta)$ of flat connections assigned to each simplicial complex Δ, together with pullback maps (and in particular, restriction maps) connecting these sets to each other. One can easily observe that this system of structures obeys the following axioms:

- *Identity.* There is only one flat connection on a point.
- *Locality.* If $\Delta = \Delta_1 \cup \Delta_2$ is the union of two simplicial complexes, then a flat connection on Δ is determined by its restrictions to Δ_1 and Δ_2. In other words, the map $\omega \mapsto (\omega \downharpoonright_{\Delta_1}, \omega \downharpoonright_{\Delta_2})$ is an injection from $\mathcal{O}(\Delta)$ to $\mathcal{O}(\Delta_1) \times \mathcal{O}(\Delta_2)$.
- *Gluing.* If $\Delta = \Delta_1 \cup \Delta_2$ and ω_1, ω_2 are flat connections on Δ_1, Δ_2 which agree when restricted to $\Delta_1 \cap \Delta_2$ (and if the orientations of Δ_1, Δ_2 on the intersection $\Delta_1 \cap \Delta_2$ agree), then there exists a flat connection ω on Δ which agrees with ω_1, ω_2 on Δ_1, Δ_2. (Note that this gluing of ω_1 and ω_2 is unique, by the previous axiom. It is important that the orientations match; we cannot glue $(0, 1)$ to $(1, 0)$, for instance.)

One can consider more abstract assignments of sets to simplicial complexes, together with pullback maps, which obey these three axioms. A system which obeys the first two axioms is known as a *presheaf*, while a system that obeys all three is known as a *sheaf*. (One can also consider presheaves and sheaves on more general topological spaces than simplicial complexes, for instance the spaces of smooth (or continuous, or holomorphic, etc.) functions (or forms, sections, etc.) on open subsets of a manifold form a sheaf.)

Thus, flat connections associated to a group G form a sheaf. But flat connections form a special type of sheaf that obeys an additional property (listed above as (37)). To explain this property, we first consider a key example when $\Delta = (0, 1, 2)$ is the standard 2-simplex (together with subsimplices) and Δ' is the subcomplex formed by removing the 2-face $(0, 1, 2)$ and the 1-face $(0, 2)$, leaving only the 1-faces $(0, 1), (1, 2)$ and the 0-faces $0, 1, 2$. Then of course every flat connection on Δ restricts to a flat connection on Δ'. But the flatness property ensures that this restriction is invertible: given a flat connection on Δ', there exists a unique extension of this connection back to

Δ. This is nothing more than the property, remarked earlier, that to specify a flat connection on the 2-simplex $(0, 1, 2)$, it suffices to know what the connection is doing on $(0, 1)$ and $(1, 2)$, as the behaviour on the remaining edge can then be deduced from the group law; conversely, any specification of the connection on those two 1-simplices determines a connection on the remainder of the 2-simplex.

This observation can be generalised. Given any simplicial complex Δ, define a k-dimensional *collapse* Δ' of Δ to be a simplicial complex obtained from Δ by removing the interior of a k-simplex, together with one of its faces; thus for instance the complex consisting of $(0, 1), (1, 2)$ (and subsimplices) is a 2-dimensional collapse of the 2-simplex $(0, 1, 2)$ (and subsimplices). We then see that the sheaf of flat connections obeys an additional axiom:

- *Grothendieck's axiom.* If Δ' is a k-dimensional collapse of Δ, then the restriction map from $\mathcal{O}(\Delta)$ to $\mathcal{O}(\Delta')$ is surjective for all k and bijective for $k \geq 2$.

This axiom is trivial for $k = 0$. For $k = 1$, it is true because if an edge (and one of its vertices) can be removed from a complex, then it is not the boundary of any 2-simplex, and the value of a flat connection on that edge is thus completely unconstrained. (In any event, the $k = 1$ case of this axiom can be deduced from the sheaf axioms.) For $k = 2$, it follows because if one can remove a 2-simplex and one of its edges from a complex, then the edge is not the boundary of any other 2-simplex and thus the connection on that edge only constrained to precisely be the product of the connection on the other two edges of the 2-simplex. For $k = 3$, it follows because if one removes a 3-simplex and one of its 2-simplex faces, the constraint associated to that 2-simplex is implied by the constraints coming from the other three faces of the 3-simplex (I recommend drawing a tetrahedron and chasing some loops around to see this), and so one retains bijectivity. For $k \geq 4$, the axiom becomes trivial again because the k-simplices and $k - 1$-simplices have no impact on the definition of a flat connection.

Grothendieck's beautiful observation is that the converse holds: if a (concrete) sheaf $\Delta \mapsto \mathcal{O}(\Delta)$ obeys Grothendieck's axiom, then it is equivalent to the sheaf of flat connections of some group G defined canonically from the sheaf. Let us see how this works. Suppose we have a sheaf $\Delta \mapsto \mathcal{O}(\Delta)$, which is concrete in the sense that $\mathcal{O}(\Delta)$ is a set, and the morphisms between these sets are given by functions. In analogy with the preceding discussion, we will refer to elements of $\mathcal{O}(\Delta)$ as (abstract) flat connections, though a priori we do not assume there is a group structure behind these connections.

By the sheaf axioms there is only one flat connection on a point, which we will call the trivial connection. Now consider the space $\mathcal{O}(0, 1)$ of flat connections on the standard 1-simplex $(0, 1)$. If the sheaf was indeed the sheaf of flat connections on a group G, then $\mathcal{O}(0,1)$ is canonically identifiable with G. Inspired by this, we will *define* G to equal the space $\mathcal{O}(0,1)$ of flat connections on $(0,1)$. The flat connections on any other 1-simplex (u,v) can then be placed in one-to-one correspondence with elements of G by the morphism $u \mapsto 0, v \mapsto 1$, so flat connections on (u,v) can be viewed as being *equivalent* to an element of G.

At present, G is merely a set, not a group. To make it into a group, we need to introduce an identity element, an inverse operation, and a multiplication operation, and verify the group axioms.

To obtain an identity element, we look at the morphism from $(0,1)$ to a point, and pull back the trivial connection on that point to obtain a flat connection e on $(0,1)$, which we will declare to be the identity element. (Note from the functorial nature of pullback that it does not matter which point we choose for this.)

Now we define the multiplication operation. Let $g, h \in G$, then g and h are flat connections on $(0,1)$. By using the morphism $i \mapsto i-1$ from $(1,2)$ to $(0,1)$, we can pull back h to $(1,2)$ to create a flat connection $\tilde{h}$ on $(1,2)$ that is equivalent to h. The restriction of g and $\tilde{h}$ to the point 1 is trivial, so by the gluing axiom we can glue g and $\tilde{h}$ to a flat connection on the complex $(0,1),(1,2)$. By Grothendieck's axiom, one can then uniquely extend this connection to the 2-simplex $(0,1,2)$, which can then be restricted to the edge $(0,2)$. Mapping this edge back to $(0,1)$, we obtain an element of G, which we will define to be hg.

This operation is closed. To verify the identity property, observe that if $g \in G$, then by starting with the simplex $(0,1,2)$ and pulling back g under the morphism that sends 2 to 1 but is the identity on $0,1$, we obtain a flat connection on $(0,1,2)$ which is equal to g on $(0,1)$, equivalent to the identity on $(1,2)$, and is equivalent to g on $(0,2)$ (after identifying $(0,2)$ with $(0,1)$). From the definition of group multiplication, this shows that $eg = g$; a similar argument (using a slightly different morphism from $(0,1,2)$ to $(0,1)$) gives $ge = g$.

Now we establish associativity. Let $f, g, h \in G$. Using the definition of multiplication, we can create a flat connection on the 2-simplex $(0,1,2)$ which equals h on $(0,1)$, is equivalent to g on $(1,2)$, and is equivalent to gh on $(0,2)$. We then glue on the edge $(2,3)$ and extend the flat connection to be equivalent to f on $(2,3)$. Using Grothendieck's axiom and the definition of multiplication, we can then extend the flat connection to the 2-simplex $(0,2,3)$ to be equivalent to $f(gh)$ on $(0,3)$. By another use of that axiom, we can also extend the flat connection to the 2-simplex $(1,2,3)$, to be equivalent to fg on $(1,3)$. Now that we have three of the four faces of the 3-simplex $(0,1,2,3)$, we can now apply the $k=3$ case of Grothendieck's axiom and extend the flat connection to the entire 3-simplex, and in particular to the 2-simplex $(0, 1, 3)$. Using the definition of multiplication again, we thus see that $f(gh) = (fg)h$, giving associativity.

Next, we establish the inverse property. It will suffice to establish the existence of a left inverse and a right inverse for each group element, since the associativity property will then guarantee that these two inverses equal each other. We shall just establish the left-inverse property, as the right inverse is analogous. Let $g \in G$ be arbitrary. By the gluing axiom, one can form a flat connection on the complex $(0, 1), (0, 2)$ which equals g on $(0, 1)$ and is equivalent to the identity on $(0, 2)$. By Grothendieck's axiom, this can be extended to a flat connection on $(0, 1, 2)$; the restriction of this connection to $(1, 2)$ is equivalent to some element of G, which we define to be g^{-1}. By construction, $g^{-1}g = e$ as required.

We have just shown that G is a group. The last thing to do is to show that this abstract sheaf $\mathcal{O}$ can be indeed identified with the sheaf of G-flat connections. This is fairly straightforward: given an abstract flat connection on a complex, the restriction of that connection to any edge is equivalent to an element of G. To verify that this genuinely determines a G-connection on that complex, we need to verify that if (u, v) and (v, u) are both in the complex, that the group elements g, h assigned to these edges invert each other. But we can pull back $(u, v), (v, u)$ to the 2-simplex $(0, 1, 2)$ by mapping $0, 2$ to u and 1 to v, creating a flat connection that is equal to g on $(0, 1)$, equivalent to h on $(1, 2)$, and equivalent to the identity on $(0, 2)$; by definition of multiplication or inverse we conclude that g, h invert each other as desired.

Thus the abstract connection defines a G-connection. From the definition of multiplication it is also clear that every 2-simplex in the complex imposes the right relation on the three elements of G associated to the edges of that simplex that makes the G-connection flat. Thus we have a canonical way to associate a G-flat connection to each abstract flat connection; the only remaining things to do are to verify that this association is bijective.

We induct on the size of the complex. If the complex is not a single simplex, the claim follows from the induction hypothesis by viewing the complex as the union of two (possibly overlapping) smaller complexes and using the gluing and locality axioms. So we may assume that the complex consists of a single simplex. If the simplex is 0 or 1-dimensional the claim is easy; for $k \geq 2$ the claim follows from Grothendieck's axiom (which applies both for the abstract flat connections (by hypothesis) and for G-flat connections (as verified earlier)) and the induction hypothesis.

Notes. This article first appeared at

`terrytao.wordpress.com/2009/10/19.`

Thanks to Lior, Raj, and anonymous commenters for corrections.

Raj and Ben Wieland noted the close connection to the Kan extension property.

1.15. The "no self-defeating object" argument

A fundamental tool in any mathematician's toolkit is that of *reductio ad absurdum*: showing that a statement X is false by assuming first that X is true, and showing that this leads to a logical contradiction. A particularly pure example of *reductio ad absurdum* occurs when establishing the nonexistence of a hypothetically overpowered object or structure X, by showing that X's powers are "self-defeating": the very existence of X and its powers can be used (by some clever trick) to construct a counterexample to that power. Perhaps the most well-known example of a self-defeating object comes from the *omnipotence paradox* in philosophy: "Can an omnipotent being create a rock so heavy that He cannot lift it?" More generally, a large number of other paradoxes in logic or philosophy can be reinterpreted as a proof that a certain overpowered object or structure does not exist.

In mathematics, perhaps the first example of a self-defeating object one encounters is that of a largest natural number:

Proposition 1.15.1 (No largest natural number)**.** *There does not exist a natural number N which is larger than all other natural numbers.*

Proof. Suppose for contradiction that there was such a largest natural number N. Then $N + 1$ is also a natural number which is strictly larger than N, contradicting the hypothesis that N is the largest natural number. □

Note the argument does not apply to the *extended natural number system* in which one adjoins an additional object ∞ beyond the natural numbers, because $\infty + 1$ is defined equal to ∞. However, the above argument does show that the existence of a largest number is not compatible with the *Peano axioms.*

This argument, by the way, is perhaps the only mathematical argument I know of which is routinely taught to primary school children *by other primary school children*, thanks to the schoolyard game of naming the largest number. It is arguably one's first exposure to a mathematical *nonexistence result*, which seems innocuous at first but can be surprisingly deep, as such results preclude in advance all future attempts to establish existence of that object, no matter how much effort or ingenuity is poured into this task.

One sees this in a typical instance of the above schoolyard game; one player tries furiously to cleverly construct some impressively huge number N, but no matter how much effort is expended in doing so, the player is defeated by the simple response "... plus one!" (unless, of course, N is infinite, ill defined, or otherwise not a natural number).

It is not only individual objects (such as natural numbers) which can be self-defeating; structures (such as orderings or enumerations) can also be self-defeating. (In modern set theory, one considers structures to themselves be a kind of object, and so the distinction between the two concepts is often blurred.) Here is one example (related to, but subtly different from, the previous one):

Proposition 1.15.2 (The natural numbers cannot be finitely enumerated). *The natural numbers* $\mathbf{N} = \{0, 1, 2, 3, \ldots\}$ *cannot be written as* $\{a_1, \ldots, a_n\}$ *for any finite collection* $a_1, \ldots, a_n$ *of natural numbers.*

Proof. Suppose for contradiction that such an enumeration $\mathbf{N}=\{a_1, \ldots, a_n\}$ existed. Then consider the number $a_1 + \cdots + a_n + 1$; this is a natural number, but it is larger than (and hence not equal to) any of the natural numbers $a_1, \ldots, a_n$, contradicting the hypothesis that $\mathbf{N}$ is enumerated by $a_1, \ldots, a_n$. □

Here, it is the *enumeration* which is self-defeating, rather than any individual natural number such as a_1 or a_n. (For this article, we allow enumerations to contain repetitions.)

The above argument may seem trivial, but a slight modification of it already gives an important result, namely *Euclid's theorem*:

Proposition 1.15.3 (The primes cannot be finitely enumerated). *The prime numbers* $\mathcal{P} = \{2, 3, 5, 7, \ldots\}$ *cannot be written as* $\{p_1, \ldots, p_n\}$ *for any finite collection of prime numbers.*

Proof. Suppose for contradiction that such an enumeration $\mathcal{P}=\{p_1, \ldots, p_n\}$ existed. Then consider the natural number $p_1 \times \cdots \times p_n + 1$; this is a natural number larger than 1 which is not divisible by any of the primes $p_1, \ldots, p_n$. But, by the *fundamental theorem of arithmetic* (or by the method of *infinite descent*, which is another classic application of *reductio ad absurdum*), every natural number larger than 1 must be divisible by some prime, contradicting the hypothesis that $\mathcal{P}$ is enumerated by $p_1, \ldots, p_n$. □

Remark 1.15.4. Continuing the number-theoretic theme, the "dueling conspiracies" arguments in Section 1.12.4 can also be viewed as an instance of this type of "no self-defeating object"; for instance, a zero of the Riemann zeta function at $1 + it$ implies that the primes anticorrelate almost completely with the multiplicative function n^{it}, which is self-defeating because it also implies complete anticorrelation with n^{-it}, and the two are incompatible. Thus we see that the *prime number theorem* (a much stronger version of Proposition 1.15.3) also emerges from this type of argument.

In this article I would like to collect several other well-known examples of this type of "no self-defeating object" argument. Each of these is well studied and probably quite familiar to many of you, but I feel that by collecting them all in one place, the commonality of theme between these arguments becomes more apparent. (For instance, as we shall see, many well-known "paradoxes" in logic and philosophy can be interpreted mathematically as rigorous "no self-defeating object" arguments.)

1.15.1. Set theory. Many famous foundational results in set theory come from a "no self-defeating object" argument. (Here, we shall be implicitly using a standard axiomatic framework for set theory, such as *Zermelo-Frankel set theory*; the situation becomes different for other set theories, much as results such as Proposition 1.15.1 change if one uses other number systems such as the extended natural numbers.) The basic idea here is that any sufficiently overpowered set-theoretic object is capable of encoding a version of the *liar paradox* ("this sentence is false", or more generally a statement which can be shown to be logically equivalent to its negation) and thus lead to a contradiction. Consider for instance this variant of *Russell's paradox*:

Proposition 1.15.5 (No universal set). *There does not exist a set which contains all sets (including itself).*

Proof. Suppose for contradiction that there existed a universal set X which contained all sets. Using the *axiom schema of specification*, one can then construct the set

$$Y := \{A \in X : A \notin A\}$$

of all sets in the universe which did not contain themselves. As X is universal, Y is contained in X. But then, by definition of Y, one sees that $Y \in Y$ if and only if $Y \notin Y$, a contradiction. □

Remark 1.15.6. As a corollary, there also does not exist any set Z which contains all *other* sets (not including itself), because the set $X := Z \cup \{Z\}$ would then be universal.

One can "localise" the above argument to a smaller domain than the entire universe, leading to the important

Proposition 1.15.7 (Cantor's theorem). *Let X be a set. Then the power set $2^X := \{A : A \subset X\}$ of X cannot be enumerated by X, i.e., one cannot write $2^X := \{A_x : x \in X\}$ for some collection $(A_x)_{x \in X}$ of subsets of X.*

Proof. Suppose for contradiction that there existed a set X whose power set 2^X could be enumerated as $\{A_x : x \in X\}$ for some $(A_x)_{x \in X}$. Using the axiom schema of specification, one can then construct the set

$$Y := \{x \in X : x \notin A_x\}.$$

The set Y is an element of the power set 2^X. As 2^X is enumerated by $\{A_x : x \in X\}$, we have $Y = A_y$ for some $y \in X$. But then by the definition of Y, one sees that $y \in A_y$ if and only if $y \notin A_y$, a contradiction. □

As is well known, one can adapt Cantor's argument to the real line, showing that the reals are uncountable:

Proposition 1.15.8 (The real numbers cannot be countably enumerated). *The real numbers* $\mathbf{R}$ *cannot be written as* $\{x_n : n \in \mathbf{N}\}$ *for any countable collection* $x_1, x_2, \ldots$ *of real numbers.*

Proof. Suppose for contradiction that there existed a countable enumeration of $\mathbf{R}$ by a sequence $x_1, x_2, \ldots$ of real numbers. Consider the decimal expansion of each of these numbers. Note that, due to the well-known "$0.999\cdots = 1.000\cdots$" issue, the decimal expansion may be nonunique, but each real number x_n has at most two decimal expansions. For each n, let $a_n \in \{0, 1, \ldots, 9\}$ be a digit which is not equal to the nth digit of any of the decimal expansions of x_n; this is always possible because there are ten digits to choose from and at most two decimal expansions of x_n. (One can avoid any implicit invocation of the *axiom of choice* here by setting a_n to be (say) the *least* digit which is not equal to the nth digit of any of the decimal expansions of x_n.) Then the real number given by the decimal expansion $0.a_1a_2a_3\cdots$ differs in the nth digit from any of the decimal expansions of x_n for each n, and so is distinct from every x_n, a contradiction. □

Remark 1.15.9. One can of course deduce Proposition 1.15.8 directly from Proposition 1.15.7 by using the decimal representation to embed $2^{\mathbf{N}}$ into $\mathbf{R}$. But notice how the two arguments have a slightly different (though closely related) basis; the former argument proceeds by encoding the liar paradox, while the second proceeds by exploiting Cantor's diagonal argument. The two perspectives are indeed a little different: for instance, Cantor's diagonal argument can also be modified to establish the *Arzelá-Ascoli theorem*, whereas I do not see any obvious way to prove that theorem by encoding the liar paradox.

Remark 1.15.10. It is an interesting psychological phenomenon that Proposition 1.15.8 is often considered far more unintuitive than any of the other propositions here, despite being in the same family of arguments; most people have no objection to the fact that every effort to finitely enumerate the natural numbers, for instance, is doomed to failure, but for some reason it is much harder to let go of the belief that, at some point, some sufficiently ingenious person will work out a way to countably enumerate the real numbers. I am not exactly sure why this disparity exists, but I suspect it is at least partly due to the fact that the rigorous construction of the real numbers is quite sophisticated and often not presented properly until the advanced undergraduate level. (Or perhaps it is because we do not play the game "enumerate the real numbers" often enough in schoolyards.)

Remark 1.15.11. One can also use the diagonal argument to show that any reasonable notion of a "constructible real number" must itself be nonconstructive (this is related to the *interesting number paradox*). Part of the problem is that the question of determining whether a proposed construction of a real number is actually well defined is a variant of the *halting problem*, which we will discuss below.

While a genuinely universal set is not possible in standard set theory, one at least has the notion of an *ordinal*, which contains all the ordinals less than it. (In the discussion below, we assume familiarity with the theory of ordinals.) One can modify the above arguments concerning sets to give analogous results about the ordinals. For instance:

Proposition 1.15.12 (Ordinals do not form a set). *There does not exist a set that contains all the ordinals.*

Proof. Suppose for contradiction that such a set existed. By the axiom schema of specification, one can then find a set Ω which consists precisely of all the ordinals and nothing else. But then $\Omega \cup \{\Omega\}$ is an ordinal which is not contained in Ω (by the *axiom of foundation*, also known as the *axiom of regularity*), a contradiction. □

Remark 1.15.13. This proposition (together with the theory of ordinals) can be used to give a quick proof of *Zorn's lemma*: see Section 2.4 of *Volume I* for further discussion. Notice the similarity between this argument and the proof of Proposition 1.15.1.

Remark 1.15.14. Once one has Zorn's lemma, one can show that various other classes of mathematical objects do not form sets. Consider for instance the class of all vector spaces. Observe that any chain of (real) vector spaces (ordered by inclusion) has an upper bound (namely the union or limit of these spaces). Thus, if the class of all vector spaces was a set, then Zorn's lemma would imply the existence of a maximal vector space V. But one can simply adjoin an additional element not already in V (e.g., $\{V\}$) to V, and contradict this maximality. (An alternate proof: every object is an element of some vector space, and in particular every set is an element of some vector space. If the class of all vector spaces formed a set, then by the *axiom of* *union*, we see that union of all vector spaces is a set also, contradicting Proposition 1.15.5.)

One can localise the above argument to smaller cardinalities, for instance:

Proposition 1.15.15 (Countable ordinals are uncountable). *There does not exist a countable enumeration $\omega_1, \omega_2, \ldots$ of the countable ordinals. (Here we consider finite sets and countably infinite sets to both be countable.)*

Proof. Suppose for contradiction that there exists a countable enumeration $\omega_1, \omega_2, \ldots$ of the countable ordinals. Then the set $\Omega := \bigcup_n \omega_n$ is also a countable ordinal, as is the set $\Omega \cup \{\Omega\}$. But $\Omega \cup \{\Omega\}$ is not equal to any of the ω_n (by the axiom of foundation), a contradiction. $\square$

Remark 1.15.16. One can show the existence of uncountable ordinals (e.g., by considering all the well-orderings of subsets of the natural numbers, up to isomorphism), and then there exists a least uncountable ordinal Ω. By construction, this ordinal consists precisely of all the countable ordinals, but is itself uncountable, much as $\mathbf{N}$ consists precisely of all the finite natural numbers, but is itself infinite (Proposition 1.15.2). The least uncountable ordinal is notorious, among other things, for providing a host of counterexamples to various intuitively plausible assertions in point set topology, and in particular in showing that the topology of sufficiently uncountable spaces cannot always be adequately explored by countable objects such as sequences.

Remark 1.15.17. The existence of the least uncountable ordinal can explain why one cannot contradict Cantor's theorem on the uncountability of the reals simply by iterating the diagonal argument (or any other algorithm) in an attempt to "exhaust" the reals. From *transfinite induction* we see that the diagonal argument allows one to assign a different real number to each countable ordinal, but this does not establish countability of the reals, because the set of all countable ordinals is itself uncountable. (This is similar to how one cannot contradict Proposition 1.15.5 by iterating the $N \to N+1$ map, as the set of all finite natural numbers is itself infinite.) In any event, even once one reaches the first uncountable ordinal, one may not yet have completely exhausted the reals; for instance, using the diagonal argument given in the proof of Proposition 1.15.8, only the real numbers in the interval $[0, 1]$ will ever be enumerated by this procedure. (Also, the question of whether *all* real numbers in $[0, 1]$ can be enumerated by the iterated diagonal algorithm requires the *continuum hypothesis*, and even with this hypothesis I am not sure whether the statement is decidable.)

1.15.2. Logic. The "no self-defeating object" argument leads to a number of important nonexistence results in logic. Again, the basic idea is to show that a sufficiently overpowered logical structure will eventually lead to the existence of a self-contradictory statement, such as the liar paradox. To state examples of this properly, one unfortunately has to invest a fair amount of time in first carefully setting up the language and theory of logic. I will not do so here, and instead use informal English sentences as a proxy for precise logical statements to convey a taste (but not a completely rigorous description) of these logical results here.

The liar paradox itself—the inability to assign a consistent truth value to "this sentence is false"—can be viewed as an argument demonstrating that there is no consistent way to interpret (i.e., assign a truth value to) sentences, when the sentences are (a) allowed to be self-referential, and (b) allowed to invoke the very notion of truth given by this interpretation. One's first impulse is to say that the difficulty here lies more with (a) than with (b), but there is a clever trick, known as *Quining* (or *indirect self-reference*), which allows one to modify the liar paradox to produce a non-self-referential statement to which one still cannot assign a consistent truth value. The idea is to work not with fully formed sentences S, which have a single truth value, but instead with *predicates* S, whose truth value depends on a variable x in some range. For instance, S may be "x is two characters long", and the range of x may be the set of strings (i.e., finite sequences of characters). Then for every string T, the statement $S(T)$ (formed by replacing every appearance of x in S with T) is either true or false. For instance, $S($"*ab*"$)$ is true, but $S($"*abc*"$)$ is false. Crucially, predicates are themselves strings, and can thus be fed into themselves as input; for instance, $S(S)$ is false. If however U is the predicate "x is sixty-five characters long", observe that $U(U)$ is true.

Now consider the *Quine predicate* Q given by

"x is a predicate whose range is the set of strings, and $x(x)$ is false"

whose range is the set of strings. Thus, for any string T, $Q(T)$ is the sentence

"T is a predicate whose range is the set of strings, and $T(T)$ is false."

This predicate is defined nonrecursively, but the sentence $Q(Q)$ captures the essence of the liar paradox: it is true if and only if it is false. This shows that there is no consistent way to interpret sentences in which the sentences are allowed to come from predicates, are allowed to use the concept of a string, and also are allowed to use the concept of truth as given by that interpretation.

Note that the proof of Proposition 1.15.5 is basically the set-theoretic analogue of the above argument, with the connection being that one can identify a predicate $T(x)$ with the set $\{x : T(x) \text{ true}\}$.

By making one small modification to the above argument—replacing the notion of truth with the related notion of provability—one obtains the celebrated *Gödel's (second) incompleteness theorem*:

Theorem 1.15.18 (Gödel's incompleteness theorem, informal statement). *No consistent logical system which has the notion of a string, can provide a proof of its own logical consistency. (Note that a proof can be viewed as a certain type of string.)*

Remark 1.15.19. Because one can encode strings in numerical form (e.g., using the *ASCII code*), it is also true (informally speaking) that no consistent logical system which has the notion of a natural number can provide a proof of its own logical consistency.

Proof (Informal sketch only). Suppose for contradiction that one had a consistent logical system inside of which its consistency could be proven. Now let Q be the predicate given by

x is a predicate whose range is the set of strings, and $x(x)$ is not provable

and whose range is the set of strings. Define the *Gödel sentence* G to be the string $G := Q(Q)$. Then G is logically equivalent to the assertion "G is not provable". Thus, if G were false, then G would be provable, which (by the consistency of the system) implies that G is true, a contradiction; thus, G must be true. Because the system is provably consistent, the above argument can be placed inside the system itself, to *prove* inside that system that G must be true; thus G is provable and G is then false, a contradiction. (It becomes quite necessary to carefully distinguish the notions of truth and provability (both inside a system and externally to that system) in order to get this argument straight!) □

Remark 1.15.20. It is not hard to show that a consistent logical system which can model the standard natural numbers cannot *disprove* its own consistency either (i.e., it cannot establish the statement that one can deduce a contradiction from the axioms in the systems in n steps for some natural number n); thus the consistency of such a system is undecidable within that system. Thus this theorem strengthens the (more well-known) first Gödel incompleteness theory, which asserts the existence of undecidable statements inside a consistent logical system which contains the concept of a string (or a natural number). On the other hand, the incompleteness theorem does not preclude the possibility that the consistency of a weak theory could be proven in a strictly stronger theory (e.g., the consistency of Peano arithmetic is provable in Zermelo-Frankel set theory).

Remark 1.15.21. One can use the incompleteness theorem to establish the undecidability of other overpowered problems. For instance, *Matiyasevich's* *theorem* demonstrates that the problem of determining the solvability of a system of Diophantine equations is, in general, undecidable, because one can encode statements such as the consistency of set theory inside such a system.

1.15.3. Computability. One can adapt these arguments in logic to analogous arguments in the theory of computation; the basic idea here is to show that a sufficiently overpowered computer program cannot exist, by feeding the source code for that program into the program itself (or some slight modification thereof) to create a contradiction. As with logic, a properly rigorous formalisation of the theory of computation would require a fair amount of preliminary machinery, for instance to define the concept of Turing machine (or some other universal computer), and so I will once again use informal English sentences as an informal substitute for a precise programming language.

A fundamental "no self-defeating object" argument in the subject, analogous to the other liar paradox type arguments encountered previously, is the *Turing halting theorem*:

Theorem 1.15.22 (Turing halting theorem). *There does not exist a program P which takes a string S as input, and determines in finite time whether S is a program (with no input) that halts in finite time.*

Proof. Suppose for contradiction that such a program P existed. Then one could easily modify P to create a variant program Q, which takes a string S as input, and halts if and only if S is a program (with S itself as input) that does not halt in finite time. Indeed, all Q has to do is call P with the string $S(S)$, defined as the program (with no input) formed by declaring S to be the input string for the program S. If P determines that $S(S)$ does not halt, then Q halts; otherwise, if P determines that $S(S)$ does halt, then Q performs an infinite loop and does not halt. Then observe that $Q(Q)$ will halt if and only if it does not halt, a contradiction. □

Remark 1.15.23. As one can imagine from the proofs, this result is closely related to, but not quite identical with, the Gödel incompleteness theorem. That latter theorem implies that the question of whether a given program halts or not in general is undecidable (consider a program designed to search for proofs of the inconsistency of set theory). By contrast, the halting theorem (roughly speaking) shows that this question is *uncomputable* (i.e., there is no algorithm to decide halting in general) rather than *undecidable* (i.e., there are programs whose halting can neither be proven nor disproven).

On the other hand, the halting theorem can be used to establish the incompleteness theorem. Indeed, suppose that all statements in a suitably strong and consistent logical system were either provable or disprovable.

Then one could build a program that determined whether an input string S, when run as a program, halts in finite time, simply by searching for all proofs or disproofs of the statement "S halts in finite time"; this program is guaranteed to terminate with a correct answer by hypothesis.

Remark 1.15.24. While it is not possible for the halting problem for a given computing language to be computable in that language, it is certainly possible that it is computable in a strictly stronger language. When that is the case, one can then invoke *Newcomb's paradox* to argue that the weaker language does not have unlimited "free will" in some sense.

Remark 1.15.25. In the language of *recursion theory*, the halting theorem asserts that the set of programs that do not halt is not a *decidable set* (or a *recursive set*). In fact, one can make the slightly stronger assertion that the set of programs that do not halt is not even a *semidecidable set* (or a *recursively enumerable set*), i.e., there is no algorithm which takes a program as input and halts in finite time if and only if the input program does not halt. This is because the complementary set of programs that do halt is clearly semidecidable (one simply runs the program until it halts, running forever if it does not), and so if the set of programs that do not halt is also semidecidable, then it is decidable (by running both algorithms in parallel; this observation is a special case of *Post's theorem*).

Remark 1.15.26. One can use the halting theorem to exclude overly general theories for certain types of mathematical objects. For instance, one cannot hope to find an algorithm to determine the existence of smooth solutions to arbitrary nonlinear partial differential equations, because it is possible to simulate a Turing machine using the laws of classical physics, which in turn can be modeled using (a moderately complicated system of) nonlinear PDE. Instead, progress in nonlinear PDE has instead proceeded by focusing on much more specific classes of such PDE (e.g., elliptic PDE, parabolic PDE, hyperbolic PDE, gauge theories, etc.)

One can place the halting theorem in a more "quantitative" form. Call a function $f : \mathbf{N} \to \mathbf{N}$ *computable* if there exists a computer program which, when given a natural number n as input, returns $f(n)$ as output in finite time. Define the *Busy Beaver function* $BB : \mathbf{N} \to \mathbf{N}$ by setting $BB(n)$ to equal the largest output of any program of at most n characters in length (and taking no input), which halts in finite time. Note that there are only finitely many such programs for any given n, so $BB(n)$ is well defined. On the other hand, it is uncomputable, even to upper bound:

Proposition 1.15.27. *There does not exist a computable function f such that one has $BB(n) \leq f(n)$ for all n.*

Proof. Suppose for contradiction that there existed a computable function $f(n)$ such that $BB(n) \leq f(n)$ for all n. We can use this to contradict the halting theorem, as follows. First observe that once the Busy Beaver function can be upper bounded by a computable function, then for any n, the run time of any halting program of length at most n can also be upper bounded by a computable function. This is because if a program of length n halts in finite time, then a trivial modification of that program (of length larger than n, but by a computable factor) is capable of outputting the run time of that program (by keeping track of a suitable "clock" variable, for instance). Applying the upper bound for Busy Beaver to that increased length, one obtains the bound on run time.

Now, to determine whether a given program S halts in finite time or not, one simply simulates (runs) that program for time up to the computable upper bound of the possible running time of S, given by the length of S. If the program has not halted by then, then it never will. This provides a program P obeying the hypotheses of the halting theorem, a contradiction. □

Remark 1.15.28. A variant of the argument shows that $BB(n)$ grows faster than any computable function: thus if f is computable, then $BB(n) > f(n)$ for all sufficiently large n. We leave the proof of this result as an exercise to the reader.

Remark 1.15.29. Sadly, the most important unsolved problem in complexity theory, namely the $P \neq NP$ *problem*, does not seem to be susceptible to the "no self-defeating object" argument, basically because such arguments tend to be *relativisable*, whereas the $P \neq NP$ problem is not; see Section 1.9 for more discussion. On the other hand, one has the curious feature that many proposed *proofs* that $P \neq NP$ appear to be self-defeating; this is most strikingly captured in the celebrated work of Razborov and Rudich [**RaRu1997**], who showed (very roughly speaking) that any sufficiently "natural" proof that $P \neq NP$ could be used to disprove the existence of an object closely related to the belief that $P \neq NP$, namely the existence of pseudo-random number generators. (I am told, though, that diagonalisation arguments can be used to prove some other inclusions or noninclusions in complexity theory that are not subject to the relativisation barrier, though I do not know the details.)

1.15.4. Game theory. Another basic example of the "no self-defeating object" argument arises from game theory, namely the *strategy stealing argument.* Consider for instance a generalised version of naughts and crosses (tic-tac-toe), in which two players take turns placing naughts and crosses on some game board (not necessarily square, and not necessarily two-dimensional), with the naughts player going first, until a certain pattern of all naughts or all crosses is obtained as a subpattern, with the naughts player winning if the pattern is all naughts, and the crosses player winning if the pattern is all crosses. (If all positions are filled without either pattern occurring, the game is a draw.) We assume that the winning patterns for the cross player are exactly the same as the winning patterns for the naughts player (but with naughts replaced by crosses, of course).

Proposition 1.15.30. *In any generalised version of naughts and crosses, there is no strategy for the second player (i.e., the crosses player) which is guaranteed to ensure victory.*

Proof. Suppose for contradiction that the second player had such a winning strategy W. The first player can then *steal* that strategy by placing a naught arbitrarily on the board, and then pretending to be the second player and using W accordingly. Note that occasionally, the W strategy will compel the naughts player to place a naught on the square that he or she has already occupied, but in such cases the naughts player may simply place the naught somewhere else instead. (It is not possible that the naughts player would run out of places, thus forcing a draw, because this would imply that W could lead to a draw as well, a contradiction.) If we denote this stolen strategy by W', then W' guarantees a win for the naughts player; playing the W' strategy for the naughts player against the W strategy for the crosses player, we obtain a contradiction. □

Remark 1.15.31. The key point here is that in naughts and crosses games, it is possible to play a *harmless move*—a move which gives up the turn of play, but does not actually decrease one's chance of winning. In games such as chess, there does not appear to be any analogue of the harmless move, and so it is not known whether black actually has a strategy guaranteed to win or not in chess, though it is suspected that this is not the case.

Remark 1.15.32. The *Hales-Jewett theorem* shows that for any fixed board length, an n-dimensional game of naughts and crosses is unable to end in a draw if n is sufficiently large. An induction argument shows that for any two-player game that terminates in bounded time in which draws are impossible, one player must have a guaranteed winning strategy; by the above proposition, this strategy must be a win for the naughts player. Note, however, that Proposition 1.15.30 provides no information as to *how* to locate this winning strategy, other than that this strategy belongs to the naughts player. Nevertheless, this gives a second example in which the "no self-defeating object" argument can be used to ensure the *existence* of some object, rather than the *nonexistence* of an object. (The first example was the prime number theorem, discussed earlier.)

The strategy-stealing argument can be applied to real-world economics and finance, though as with any other application of mathematics to the real world, one has to be careful as to the implicit assumptions one is making about reality and how it conforms to one's mathematical model when doing so. For instance, one can argue that in any market or other economic system in which the net amount of money is approximately constant, it is not possible to locate a universal trading strategy which is guaranteed to make money for the user of that strategy, since if everyone applied that strategy, then the net amount of money in the system would increase, a contradiction. Note however that there are many loopholes here; it may be that the strategy is difficult to copy or relies on exploiting some other group of participants who are unaware or unable to use the strategy and would then lose money (though in such a case, the strategy is not truly universal as it would stop working once enough people used it). Unfortunately, there can be strong psychological factors that can cause people to override the conclusions of such strategy-stealing arguments with their own rationalisations, as can be seen, for instance, in the perennial popularity of pyramid schemes, or to a lesser extent, market bubbles (though one has to be careful about applying the strategy-stealing argument in the latter case, since it is possible to have net wealth creation through external factors such as advances in technology).

Note also that the strategy-stealing argument also limits the universal predictive power of *technical analysis* to provide predictions other than that the prices obey a *martingale*, though again there are loopholes in the case of markets that are illiquid or highly volatile.

1.15.5. Physics. In a similar vein, one can try to adapt the "no self-defeating object" argument from mathematics to physics, but again one has to be much more careful with various physical and metaphysical assumptions that may be implicit in one's argument. For instance, one can argue that under the laws of special relativity, it is not possible to construct a totally immovable object. The argument would be that if one could construct an immovable object O in one inertial reference frame, then by the *principle of relativity* it should be possible to construct an object O' which is immovable in another inertial reference frame which is moving with respect to the first; setting the two on a collision course, we obtain the classic contradiction between an irresistible force and an immovable object. Note however that there are several loopholes here which allow one to avoid contradiction; for instance, the two objects O, O' could simply pass through each other without interacting.

In a somewhat similar vein, using the laws of special relativity, one can argue that it is not possible to systematically generate and detect *tachyon particles* (particles traveling faster than the speed of light) because these could be used to transmit localised information faster than the speed of light and then (by the principle of relativity) to send localised information back into the past from one location to a distant one. Setting up a second tachyon beam to reflect this information back to the original location, one could then send localised information back to one's own past (rather than to the past of an observer at a distant location), allowing one to set up a classic *grandfather paradox*. However, as before, there are a large number of loopholes in this argument which could let one avoid contradiction; for instance, if the apparatus needed to set up the tachyon beam is larger than the distance the beam travels (as is for instance the case in *Mexican wave*-type tachyon beams), then there is no causality paradox. Another loophole occurs if the tachyon beam is not fully localised but propagates in space-time in a manner to interfere with the second tachyon beam. A third loophole occurs if the universe exhibits quantum behaviour (in particular, the ability to exist in entangled states) instead of nonquantum behaviour, which allows for such superluminal mechanisms as wave function collapse to occur without any threat to causality or the principle of relativity. A fourth loophole occurs if the effects of relativistic gravity (i.e., general relativity) become significant. Nevertheless, the paradoxical effect of time travel is so strong that this physical argument is a fairly convincing way to rule out many commonly imagined types of faster-than-light travel or communication (and we have a number of other arguments also that exclude more modes of faster-than-light behaviour, though this is an entire article topic in its own right).

Notes. This article first appeared at

`terrytao.wordpress.com/2009/10/27.`

Thanks to Seva Lev and an anonymous commenter for corrections.

1.16. From Bose-Einstein condensates to the nonlinear Schrödinger equation

The *Schrödinger equation*

$$i\hbar\partial_t|\psi\rangle = H|\psi\rangle$$

is the fundamental equation of motion for (nonrelativistic) quantum mechanics, modeling both one-particle systems and N-particle systems for $N > 1$. Remarkably, despite being a *linear* equation, solutions $|\psi\rangle$ to this equation can be governed by a *nonlinear* equation in the large particle limit $N \to \infty$. In particular, when modeling a *Bose-Einstein condensate* with a suitably scaled interaction potential V in the large particle limit, the solution can be governed by the *cubic nonlinear Schrödinger equation*

$$i\partial_t \phi = \Delta\phi + \lambda|\phi|^2\phi. \tag{1.82}$$

I recently attended a talk by Natasa Pavlovic on the rigorous derivation of this type of limiting behaviour, which was initiated by the pioneering work of Hepp and Spohn and which has now attracted a vast recent literature. The rigorous details here are rather sophisticated, but the heuristic explanation of the phenomenon is fairly simple, and actually rather pretty in my opinion, involving the foundational quantum mechanics of N-particle systems. I am recording this heuristic derivation here, partly for my own benefit, but perhaps it will be of interest to some readers.

This discussion will be purely formal, in the sense that (important) analytic issues such as differentiability, existence and uniqueness, etc., will be largely ignored.

1.16.1. A quick review of classical mechanics. The phenomena discussed here are purely quantum mechanical in nature, but to motivate the quantum mechanical discussion, it is helpful to first quickly review the more familiar (and more conceptually intuitive) classical situation.

Classical mechanics can be formulated in a number of essentially equivalent ways: *Newtonian*, *Hamiltonian*, and *Lagrangian*. The formalism of Hamiltonian mechanics for a given physical system can be summarised briefly as follows:

- The physical system has a *phase space* Ω of states $\vec{x}$ (which is often parameterised by position variables q and momentum variables p). Mathematically, it has the structure of a *symplectic manifold*, with some *symplectic form* ω (which would be $\omega = dp \wedge dq$ if one had position and momentum coordinates available).
- The complete state of the system at any given time t is given (in the case of *pure states*) by a point $\vec{x}(t)$ in the phase space Ω.
- Every physical observable (e.g., energy, momentum, position, etc.) A is associated to a function (also called A) mapping the phase space Ω to the range of the observable (e.g., for real observables, A would be a function from Ω to $\mathbf{R}$). If one measures the observable A at time t, one will obtain the measurement $A(x(t))$.
- There is a special observable, the *Hamiltonian* $H : \Omega \to \mathbf{R}$, which governs the evolution of the state $\vec{x}(t)$ through time, via *Hamilton's equations of motion*. If one has position and momentum coordinates $\vec{x}(t) = (q_i(t), p_i(t))_{i=1}^n$, these equations are given by the

formulae

$$\partial_t p_i = -\frac{\partial H}{\partial q_i}; \quad \partial_t q_i = \frac{\partial H}{\partial p_i};$$

more abstractly, just from the symplectic form ω on the phase space, the equations of motion can be written as

$$\partial_t \vec{x}(t) = -\nabla_\omega H(\vec{x}(t)), \tag{1.83}$$

where $\nabla_\omega H$ is the symplectic gradient of H.

Hamilton's equation of motion can also be expressed in a dual form in terms of observables A, as *Poisson's equation of motion*

$$\partial_t A(\vec{x}(t)) = -\{H, A\}(\vec{x}(t))$$

for any observable A, where $\{H, A\} := \nabla_\omega H \cdot \nabla A$ is the *Poisson bracket.* One can express Poisson's equation more abstractly as

$$\partial_t A = -\{H, A\}. \tag{1.84}$$

In the above formalism, we are assuming that the system is in a *pure state* at each time t, which means that it only occupies a single point $\vec{x}(t)$ in phase space. One can also consider *mixed states* in which the state of the system at a time t is not fully known, but is instead given by a *probability distribution* $\rho(t, \vec{x})\, dx$ on phase space. The act of measuring an observable A at a time t will thus no longer be deterministic, but will itself be a random variable, whose expectation $\langle A \rangle$ is given by

$$\langle A \rangle(t) = \int_\Omega A(\vec{x})\rho(t, \vec{x})\, d\vec{x}. \tag{1.85}$$

The equation of motion of a mixed state ρ is given by the *advection equation*

$$\partial_t \rho = \operatorname{div}(\rho \nabla_\omega H)$$

using the same vector field $-\nabla_\omega H$ that appears in (1.83); this equation can also be derived from (1.84) and (1.85) and a duality argument.

Pure states can be viewed as the special case of mixed states in which the probability distribution $\rho(t, \vec{x})\, d\vec{x}$ is a *Dirac mass*[11] $\delta_{\vec{x}(t)}(\vec{x})$. One can thus think of mixed states as continuous averages of pure states, or equivalently the space of mixed states is the convex hull of the space of pure states.

Suppose one had a 2-particle system, in which the joint phase space $\Omega = \Omega_1 \times \Omega_2$ is the product of the two one-particle phase spaces. A pure joint state is then a point $x = (\vec{x}_1, \vec{x}_2)$ in Ω, where $\vec{x}_1$ represents the state

[11] We ignore for now the formal issues of how to perform operations such as derivatives on Dirac masses. This can be accomplished using the theory of distributions in Section 1.13 of *Volume I* or equivalently by working in the dual setting of observables, but this is not our concern here.

of the first particle, and $\vec{x}_2$ is the state of the second particle. If the joint Hamiltonian $H : \Omega \to \mathbf{R}$ split as

$$H(\vec{x}_1, \vec{x}_2) = H_1(\vec{x}_1) + H_2(\vec{x}_2),$$

then the equations of motion for the first and second particles would be completely *decoupled*, with no interactions between the two particles. However, in practice, the joint Hamiltonian contains coupling terms between $\vec{x}_1, \vec{x}_2$ that prevents one from totally decoupling the system. For instance, one may have

$$H(\vec{x}_1, \vec{x}_2) = \frac{|p_1|^2}{2m_1} + \frac{|p_2|^2}{2m_2} + V(q_1 - q_2),$$

where $\vec{x}_1 = (q_1, p_1)$, $\vec{x}_2 = (q_2, p_2)$ are written using position coordinates q_i and momentum coordinates p_i, $m_1, m_2 > 0$ are constants (representing mass), and $V(q_1 - q_2)$ is some *interaction potential* that depends on the spatial separation $q_1 - q_2$ between the two particles.

In a similar spirit, a mixed joint state is a joint probability distribution $\rho(\vec{x}_1, \vec{x}_2)\, d\vec{x}_1 d\vec{x}_2$ on the product state space. To recover the (mixed) state of an individual particle, one must consider a *marginal distribution*, such as

$$\rho_1(\vec{x}_1) := \int_{\Omega_2} \rho(\vec{x}_1, \vec{x}_2)\, d\vec{x}_2$$

(for the first particle) or

$$\rho_2(\vec{x}_2) := \int_{\Omega_1} \rho(\vec{x}_1, \vec{x}_2)\, d\vec{x}_1$$

(for the second particle). Similarly, for N-particle systems, if the joint distribution of N distinct particles is given by $\rho(\vec{x}_1, \ldots, \vec{x}_N)\, d\vec{x}_1 \cdots d\vec{x}_N$, then the distribution of the first particle (say) is given by

$$\rho_1(\vec{x}_1) = \int_{\Omega_2 \times \cdots \times \Omega_N} \rho(\vec{x}_1, \vec{x}_2, \ldots, \vec{x}_N)\, d\vec{x}_2 \cdots d\vec{x}_N,$$

the distribution of the first two particles is given by

$$\rho_{12}(\vec{x}_1, \vec{x}_2) = \int_{\Omega_3 \times \cdots \times \Omega_N} \rho(\vec{x}_1, \vec{x}_2, \ldots, \vec{x}_N)\, d\vec{x}_3 \cdots d\vec{x}_N,$$

and so forth.

A typical Hamiltonian in this case may take the form

$$H(\vec{x}_1, \ldots, \vec{x}_n) = \sum_{j=1}^{N} \frac{|p_j|^2}{2m_j} + \sum_{1 \leq j < k \leq N} V_{jk}(q_j - q_k),$$

which is a combination of single-particle Hamiltonians H_j and interaction perturbations. If the momenta p_j and masses m_j are normalised to be of size $O(1)$, and the potential V_{jk} has an average value (i.e., an L^1-norm) of $O(1)$

also, then the former sum has size $O(N)$ and the latter sum has size $O(N^2)$, so the latter will dominate. In order to balance the two components and get a more interesting limiting dynamics when $N \to \infty$, we shall therefore insert a normalising factor of $\frac{1}{N}$ on the right-hand side, giving a Hamiltonian

$$H(\vec{x}_1, \ldots, \vec{x}_n) = \sum_{j=1}^{N} \frac{|p_j|^2}{2m_j} + \frac{1}{N} \sum_{1 \le j < k \le N} V_{jk}(q_j - q_k).$$

Now imagine a system of N *indistinguishable* particles. By this, we mean that all the state spaces $\Omega_1 = \cdots = \Omega_N$ are identical, and all observables (including the Hamiltonian) are symmetric functions of the product space $\Omega = \Omega_1^N$ (i.e., invariant under the action of the symmetric group S_N). In such a case, one may as well average over this group (since this does not affect any physical observable) and assume that all mixed states ρ are also symmetric. (One cost of doing this, though, is that one has to largely give up pure states $(\vec{x}_1, \ldots, \vec{x}_N)$, since such states will not be symmetric except in the very exceptional case $\vec{x}_1 = \cdots = \vec{x}_N$.)

A typical example of a symmetric Hamiltonian is

$$H(\vec{x}_1, \ldots, \vec{x}_n) = \sum_{j=1}^{N} \frac{|p_j|^2}{2m} + \frac{1}{N} \sum_{1 \le j < k \le N} V(q_j - q_k),$$

where V is even (thus all particles have the same individual Hamiltonian, and interact with the other particles using the same interaction potential). In many physical systems, it is natural to consider only *short-range* interaction potentials, in which the interaction between q_j and q_k is localised to the region $q_j - q_k = O(r)$ for some small r. We model this by considering Hamiltonians of the form

$$H(\vec{x}_1, \ldots, \vec{x}_n) = \sum_{j=1}^{N} H(\vec{x}_j) + \frac{1}{N} \sum_{1 \le j < k \le N} \frac{1}{r^d} V\left(\frac{\vec{x}_j - \vec{x}_k}{r}\right),$$

where d is the ambient dimension of each particle (thus in physical models, d would usually be 3); the factor of $\frac{1}{r^d}$ is a normalisation factor designed to keep the L^1-norm of the interaction potential of size $O(1)$. It turns out that an interesting limit occurs when r goes to zero as N goes to infinity by some power law $r = N^{-\beta}$. Imagine for instance N particles of "radius" r bouncing around in a box, which is a basic model for classical gases.

An important example of a symmetric mixed state is a *factored* state

$$\rho(\vec{x}_1, \ldots, \vec{x}_N) = \rho_1(\vec{x}_1) \cdots \rho_1(\vec{x}_N),$$

where ρ_1 is a single-particle probability density function; thus ρ is the tensor product of N copies of ρ_1. If there are no interaction terms in the Hamiltonian, then Hamilton's equation of motion will preserve the property of being a factored state (with ρ_1 evolving according to the one-particle equation); but with interactions, the factored nature may be lost over time.

1.16.2. A quick review of quantum mechanics. Now we turn to quantum mechanics. This theory is fundamentally rather different in nature than classical mechanics (in the sense that the basic objects, such as states and observables, are a different type of mathematical object than in the classical case), but they share many features in common also, particularly those relating to the Hamiltonian and other observables. (This relationship is made more precise via the *correspondence principle*, and more precise still using *semiclassical analysis*.)

The formalism of quantum mechanics for a given physical system can be summarised briefly as follows:

- The physical system has a *phase space* $\mathbf{H}$ of states $|\psi\rangle$ (which is often parameterised as a complex-valued function of the position space). Mathematically, it has the structure of a complex *Hilbert space*, which is traditionally manipulated using *Dirac's bra-ket notation*.
- The complete *state* of the system at any given time t is given (in the case of *pure states*) by a unit vector $|\psi(t)\rangle$ in the phase space $\mathbf{H}$.
- Every physical observable A is associated to a linear operator on $\mathbf{H}$; real-valued observables are associated to self-adjoint linear operators. If one measures the observable A at time t, one will obtain the random variable whose expectation $\langle A\rangle$ is given by $\langle\psi(t)|A|\psi(t)\rangle$. (The full distribution of A is given by the *spectral measure* of A relative to $|\psi(t)\rangle$.)
- There is a special observable, the *Hamiltonian* $H : \mathbf{H} \to \mathbf{H}$, which governs the evolution of the state $|\psi(t)\rangle$ through time, via *Schrödinger's equations of motion*

$$i\hbar\partial_t|\psi(t)\rangle = H|\psi(t)\rangle. \tag{1.86}$$

Schrödinger's equation of motion can also be expressed in a dual form in terms of observables A as *Heisenberg's equation of motion*

$$\partial_t\langle\psi|A|\psi\rangle = \frac{i}{\hbar}\langle\psi|[H,A]|\psi\rangle,$$

or more abstractly as

$$\partial_t A = \frac{i}{\hbar}[H,A], \tag{1.87}$$

where $[,]$ is the *commutator* or *Lie bracket* (compare with (1.84)).

The states $|\psi\rangle$ are pure states, analogous to the pure states x in Hamiltonian mechanics. One also has *mixed states* ρ in quantum mechanics. Whereas in classical mechanics, a mixed state ρ is a probability distribution (a nonnegative function of total mass $\int_\Omega \rho = 1$), in quantum mechanics a mixed state is a nonnegative (i.e., *positive semi-definite*) *operator* ρ on $\mathbf{H}$ of total *trace* $\operatorname{tr} \rho = 1$. If one measures an observable A at a mixed state ρ, one obtains a random variable with expectation $\operatorname{tr} A\rho$. From (1.87) and duality, one can infer that the correct equation of motion for mixed states must be given by

$$\partial_t \rho = \frac{i}{\hbar}[H, \rho]. \tag{1.88}$$

One can view pure states as the special case of mixed states which are rank one projections,

$$\rho = |\psi\rangle\langle\psi|.$$

Morally speaking, the space of mixed states is the convex hull of the space of pure states (just as in the classical case), though things are a little trickier than this when the phase space $\mathbf{H}$ is infinite dimensional, due to the presence of continuous spectrum in the *spectral theorem.*

Pure states suffer from a *phase ambiguity*: a phase rotation $e^{i\theta}|\psi\rangle$ of a pure state $|\psi\rangle$ leads to the same mixed state, and the two states cannot be distinguished by any physical observable.

In a single particle system, modeling a (scalar) quantum particle in a d-dimensional position space $\mathbf{R}^d$, one can identify the Hilbert space $\mathbf{H}$ with $L^2(\mathbf{R}^d \to \mathbf{C})$ and can describe the pure state $|\psi\rangle$ as a *wave function* $\psi : \mathbf{R}^d \to \mathbf{C}$, which is normalised as

$$\int_{\mathbf{R}^d} |\psi(x)|^2 \, dx = 1$$

as $|\psi\rangle$ has to be a unit vector. (If the quantum particle has additional features such as *spin*, then one needs a fancier wave function, but let us ignore this for now.) A mixed state is then a function $\rho : \mathbf{R}^d \times \mathbf{R}^d \to \mathbf{C}$ which is Hermitian (i.e., $\rho(x, x') = \overline{\rho(x', x)}$) and positive definite, with unit trace $\int_{\mathbf{R}^d} \rho(x, x) \, dx = 1$; a pure state ψ corresponds to the mixed state $\rho(x, x') = \psi(x)\overline{\psi(x')}$.

A typical Hamiltonian in this setting is given by the operator

$$H\psi(x) := \frac{|p|^2}{2m}\psi(x) + V(x)\psi(x),$$

where $m > 0$ is a constant, p is the *momentum operator* $p := -i\hbar\nabla_x$, ∇_x is the gradient in the x variable (so $|p|^2 = -\hbar^2\Delta_x$, where Δ_x is the Laplacian; note that ∇_x is skew-adjoint and should thus be thought of as being imaginary rather than real), and $V : \mathbf{R}^d \to \mathbf{R}$ is some potential.

Physically, this depicts a particle of mass m in a potential well given by the potential V.

Now suppose one has an N-particle system of scalar particles. A pure state of such a system can then be given by an N-particle wave function $\psi : (\mathbf{R}^d)^N \to \mathbf{C}$, normalised so that

$$\int_{(\mathbf{R}^d)^N} |\psi(x_1, \ldots, x_N)|^2 \, dx_1 \cdots dx_N = 1$$

and a mixed state is a Hermitian positive semi-definite function $\rho : (\mathbf{R}^d)^N \times (\mathbf{R}^d)^N \to \mathbf{C}$ with trace

$$\int_{(\mathbf{R}^d)^N} \rho(x_1, \ldots, x_N; x_1, \ldots, x_N) \, dx_1 \cdots dx_N = 1,$$

with a pure state ψ being identified with the mixed state

$$\rho(x_1, \ldots, x_N; x_1', \ldots, x_N') := \psi(x_1, \ldots, x_N)\overline{\psi(x_1', \ldots, x_N')}.$$

In classical mechanics, the state of a single particle was the marginal distribution of the joint state. In quantum mechanics, the state of a single particle is instead obtained as the *partial trace* of the joint state. For instance, the state of the first particle is given as

$$\rho_1(x_1; x_1') := \int_{(\mathbf{R}^d)^{N-1}} \rho(x_1, x_2, \ldots, x_N; x_1', x_2, \ldots, x_N) \, dx_2 \cdots dx_N,$$

the state of the first two particles is given as

$$\rho_{12}(x_1, x_2; x_1', x_2') := \int_{(\mathbf{R}^d)^{N-2}} \rho(x_1, x_2, x_3, \ldots, x_N; \\ x_1', x_2', x_3, \ldots, x_N) \, dx_3 \cdots dx_N,$$

and so forth. (These formulae can be justified by considering observables of the joint state that only affect, say, the first two position coordinates x_1, x_2 and using duality.)

A typical Hamiltonian in this setting is given by the operator

$$H\psi(x_1, \ldots, x_N) = \sum_{j=1}^{N} \frac{|p_j|^2}{2m_j} \psi(x_1, \ldots, x_N) \\ + \frac{1}{N} \sum_{1 \leq j < k \leq N} V_{jk}(x_j - x_k), \psi(x_1, \ldots, x_N)$$

where we normalise just as in the classical case, and $p_j := -i\hbar\nabla_{x_j}$.

An interesting feature of quantum mechanics—not present in the classical world—is that even if the N-particle system is in a pure state, individual particles may be in a mixed state: the partial trace of a pure state need not remain pure. Because of this, when considering a subsystem of a larger system, one cannot always assume that the subsystem is in a pure state, but must work instead with mixed states throughout, unless there is some reason (e.g., a lack of coupling) to assume that pure states are somehow preserved.

Now consider a system of N indistinguishable quantum particles. As in the classical case, this means that all observables (including the Hamiltonian) for the joint system are invariant with respect to the action of the symmetric group S_N. Because of this, one may as well assume that the (mixed) state of the joint system is also symmetric with respect to this action. In the special case when the particles are *bosons*, one can also assume that pure states $|\psi\rangle$ are also symmetric with respect to this action (in contrast to *fermions*, where the action on pure states is antisymmetric). A typical Hamiltonian in this setting is given by the operator

$$\begin{aligned} H\psi(x_1,\dots,x_N) &= \sum_{j=1}^N \frac{|p_j|^2}{2m}\psi(x_1,\dots,x_N) \\ &\quad + \frac{1}{N}\sum_{1\le j<k\le N} V(x_j-x_k)\psi(x_1,\dots,x_N) \end{aligned}$$

for some even potential V; if one wants to model short-range interactions, one might instead pick the variant

$$\begin{aligned} H\psi(x_1,\dots,x_N) &= \sum_{j=1}^N \frac{|p_j|^2}{2m}\psi(x_1,\dots,x_N) \\ &\quad + \frac{1}{N}\sum_{1\le j<k\le N} r^d V\Big(\frac{x_j-x_k}{r}\Big)\psi(x_1,\dots,x_N) \end{aligned} \tag{1.89}$$

for some $r > 0$. This is a typical model for an N-particle *Bose-Einstein condensate.* (Longer-range models can lead to more nonlocal variants of the nonlinear Schrödinger equation (NLS) for the limiting equation, such as the *Hartree equation.*)

1.16.3. NLS. Suppose we have a Bose-Einstein condensate given by a (symmetric) mixed state

$$\rho(t,x_1,\dots,x_N;x_1',\dots,x_N')$$

evolving according to the equation of motion (1.88) using the Hamiltonian (1.89). One can take a partial trace of the equation of motion (1.88) to obtain an equation for the state $\rho_1(t,x_1;x_1')$ of the first particle (note from symmetry that all the other particles will have the same state function). If one does take this trace, one soon finds that the equation of motion becomes

$$\begin{aligned}\partial_t \rho_1(t, x_1; x_1') = \frac{i}{\hbar}\Big[&(\frac{|p_1|^2}{2m} - \frac{|p_1'|^2}{2m})\rho_1(t, x_1; x_1') \\ &+ \frac{1}{N}\sum_{j=2}^{N}\int_{\mathbf{R}^d}\frac{1}{r^d}[V(\frac{x_1 - x_j}{r}) \\ &- V(\frac{x_1' - x_j}{r})]\rho_{1j}(t, x_1, x_j; x_1', x_j)\, dx_j\Big],\end{aligned}$$

where ρ_{1j} is the partial trace to the $1, j$ particles. Using symmetry, we see that all the summands in the j summation are identical, so we can simplify this as

$$\begin{aligned}\partial_t \rho_1(t, x_1; x_1') = \frac{i}{\hbar}\Big[&(\frac{|p_1|^2}{2m} - \frac{|p_1'|^2}{2m})\rho_1(t, x_1; x_1') \\ &+ \frac{N-1}{N}\int_{\mathbf{R}^d}\frac{1}{r^d}[V(\frac{x_1 - x_2}{r}) \\ &- V(\frac{x_1' - x_2}{r})]\rho_{12}(t, x_1, x_2; x_1', x_2)\, dx_2\Big].\end{aligned}$$

This does not completely describe the dynamics of ρ_1, as one also needs an equation for ρ_{12}. But one can repeat the same argument to get an equation for ρ_{12} involving ρ_{123}, and so forth, leading to a system of equations known as the *BBGKY hierarchy*. For simplicity we shall just look at the first equation in this hierarchy.

Let us now formally take two limits in the above equation, sending the number of particles N to infinity and the interaction scale r to zero. The effect of sending N to infinity should simply be to eliminate the $\frac{N-1}{N}$ factor. The effect of sending r to zero should be to send $\frac{1}{r^d}V(\frac{x}{r})$ to the Dirac mass $\lambda\delta(x)$, where $\lambda := \int_{\mathbf{R}^d} V$ is the total mass of V. *Formally* performing these two limits, one is led to the equation

$$\begin{aligned}\partial_t \rho_1(t, x_1; x_1') = \frac{i}{\hbar}\Big[&(\frac{|p_1|^2}{2m} - \frac{|p_1'|^2}{2m})\rho_1(t, x_1; x_1') \\ &+ \lambda(\rho_{12}(t, x_1, x_1; x_1', x_1) - \rho_{12}(t, x_1, x_1'; x_1', x_1'))\Big].\end{aligned}$$

One can perform a similar formal limiting procedure for the other equations in the BBGKY hierarchy, obtaining a system of equations known as the *Gross-Pitaevskii hierarchy.*

We next make an important simplifying assumption, which is that in the limit $N \to \infty$ any two particles in this system become *decoupled,* meaning that the two-particle mixed state factors as the tensor product of two one-particle states:

$$\rho_{12}(t, x_1, x_2; x_1', x_2) \approx \rho_1(t, x_1; x_1')\rho_1(t, x_2; x_2').$$

One can view this as a *mean field approximation*, modeling the interaction of one particle x_1 with all the other particles by the mean field ρ_1.

Making this assumption, the previous equation simplifies to

$$\partial_t \rho_1(t, x_1; x_1') = \frac{i}{\hbar}\Big[\big(\frac{|p_1|^2}{2m} - \frac{|p_1'|^2}{2m}\big) + \lambda(\rho_1(t, x_1; x_1) - \rho_1(t, x_1'; x_1'))\Big]\rho_1(t, x_1; x_1').$$

If we assume, furthermore, that ρ_1 is a pure state, thus

$$\rho_1(t, x_1; x_1') = \psi(t, x_1)\overline{\psi(t, x_1')},$$

then (up to the phase ambiguity mentioned earlier) $\psi(t, x)$ obeys the *Gross-Pitaevskii equation*

$$\partial_t \psi(t, x) = \frac{i}{\hbar}\Big[\frac{|p|^2}{2m} + \lambda|\psi(t, x)|^2\Big]\psi(t, x),$$

which (up to some factors of $\hbar$ and m, which can be renormalised away) is essentially (1.82).

An alternate derivation of (1.82), using a slight variant of the above mean field approximation, comes from studying the Hamiltonian (1.89). Let us make the (very strong) assumption that at some fixed time t one is in a completely factored pure state

$$\psi(x_1, \dots, x_N) = \psi_1(x_1) \cdots \psi_1(x_N),$$

where ψ_1 is a one-particle wave function, in particular obeying the normalisation

$$\int_{\mathbf{R}^d} |\psi_1(x)|^2 \, dx = 1.$$

(This is an unrealistically strong version of the mean field approximation. In practice, one only needs the two-particle partial traces to be completely factored for the discussion below.) The expected value of the Hamiltonian,

$$\langle \psi | H | \psi \rangle = \int_{(\mathbf{R}^d)^N} \psi(x_1, \dots, x_N)\overline{H\psi(x_1, \dots, x_N)} \, dx_1 \cdots dx_N,$$

can then be simplified as

$$N \int_{\mathbf{R}^d} \psi_1(x)\overline{\frac{|p_1|^2}{2m}\psi_1(x)} \, dx + \frac{N-1}{2}\int_{\mathbf{R}^d \times \mathbf{R}^d} r^{-d} V\big(\frac{x_1 - x_2}{r}\big)|\psi_1(x_1)|^2|\psi_1(x_2)| \, dx_1 dx_2.$$

Again sending $r \to 0$, this formally becomes

$$N \int_{\mathbf{R}^d} \psi_1(x)\overline{\frac{|p_1|^2}{2m}\psi_1(x)} \, dx + \frac{N-1}{2}\lambda \int_{\mathbf{R}^d \times \mathbf{R}^d} |\psi_1(x_1)|^4 \, dx_1,$$

which in the limit $N \to \infty$ is asymptotically

$$N \int_{\mathbf{R}^d} \psi_1(x) \overline{\frac{|p_1|^2}{2m} \psi_1(x)} + \frac{\lambda}{2} |\psi_1(x_1)|^4 \, dx_1.$$

Up to some normalisations, this is the Hamiltonian for the NLS equation (1.82).

There has been much progress recently in making the above derivations precise; see e.g., [**Sc2006**], [**KlMa2008**], [**KiScSt2008**], [**ChPa2009**]. A key step is to show that the Gross-Pitaevskii hierarchy necessarily preserves the property of being a completely factored state. This requires a uniqueness theory for this hierarchy, which is surprisingly delicate, due to the fact that it is a system of infinitely many coupled equations over an unbounded number of variables.

Remark 1.16.1. Interestingly, the above heuristic derivation only works when the interaction scale r is much larger than N^{-1}. For $r \sim N^{-1}$, the coupling constant λ acquires a nonlinear correction, becoming essentially the *scattering length* of the potential rather than its mean. (Thanks to Bob Jerrard for pointing out this subtlety.)

Notes. This article first appeared at

`terrytao.wordpress.com/2009/11/26.`

Thanks to CJ, liuyao, Mio, and M.S. for corrections.

Bob Jerrard provided a heuristic argument as to why the coupling constant becomes nonlinear in the regime $r \sim N^{-1}$.

Chapter 2

Technical articles

2.1. Polymath1 and three new proofs of the density Hales-Jewett theorem

During the first few months of 2009, I was involved in the *Polymath1 project*, a massively collaborative mathematical project whose purpose was to investigate the viability of various approaches to proving the *density Hales-Jewett theorem*. For simplicity I will focus attention here on the model case $k = 3$ of a three-letter alphabet, in which case the theorem reads as follows:

Theorem 2.1.1 ($k = 3$ density Hales-Jewett theorem). *Let* $0 < \delta \leq 1$. *Then if* n *is a sufficiently large integer, any subset* A *of the cube* $[3]^n = \{1,2,3\}^n$ *of density* $|A|/3^n$ *at least* δ *contains at least one combinatorial line* $\{\ell(1), \ell(2), \ell(3)\}$, *where* $\ell \in \{1,2,3,x\}^n \backslash [3]^n$ *is a string of* 1*'s,* 2*'s,* 3*'s, and* x*'s containing at least one "wildcard"* x*, and* $\ell(i)$ *is the string formed from* ℓ *by replacing all* x*'s with* i*'s.*

The full density Hales-Jewett theorem is the same statement, but with [3] replaced by $[k]$ for some $k \geq 1$. (The case $k = 1$ is trivial, and the case $k = 2$ follows from *Sperner's theorem.*) As a result of the project, three new proofs of this theorem were established, at least one of which has extended [**Po2009**] to cover the case of general k.

This theorem was first proven in [**FuKa1989**] by Furstenberg and Katznelson, by first converting it to a statement in ergodic theory; the original paper with the Furstenberg-Katznelson argument was for the $k = 3$ case only, and it gave only part of the proof in detail. But in a subsequent paper [**FuKa1991**] a full proof in general k was provided. The remaining components of the original $k = 3$ argument were later completed in unpublished notes of McCutcheon.[1] One of the new proofs is essentially a finitary translation of this $k = 3$ argument; in principle one could also finitise the significantly more complicated argument of Furstenberg and Katznelson for general k, but this has not been properly carried out yet (the other two proofs are likely to generalise much more easily to higher k). The result is considered quite deep; for instance, the general k case of the density Hales-Jewett theorem already implies *Szemerédi's theorem*, which is a highly nontrivial theorem in its own right, as a special case.

Another of the proofs is based primarily on the density increment method that goes back to Roth, and also incorporates some ideas from a paper of Ajtai and Szemerédi [**AjSz1974**] establishing what we have called the *corners theorem* (and which is also implied by the $k = 3$ case of the density Hales-Jewett theorem). A key new idea involved studying the correlations of the original set A with special subsets of $[3]^n$, such as ij*-insensitive sets*, or intersections of ij-insensitive and ik-insensitive sets.

[1] `http://www.msci.memphis.edu/~randall/preprints/HJk3.pdf`

This correlations idea inspired a new ergodic proof of the density Hales-Jewett theorem for all values of k by Austin [**Au2009b**], which is in the spirit of the *triangle removal lemma* (or hypergraph removal lemma) proofs of *Roth's theorem* (or the *multidimensional Szemerédi theorem*). A finitary translation of this argument in the $k = 3$ case has been sketched out; I believe it also extends in a relatively straightforward manner to the higher k case (in analogy with some proofs of the hypergraph removal lemma).

2.1.1. Simpler cases of the density Hales-Jewett theorem.

In order to motivate the known proofs of the density Hales-Jewett theorem, it is instructive to consider some simpler theorems which are implied by this theorem. The first is the *corners theorem* of Ajtai and Szemerédi:

Theorem 2.1.2 (Corners theorem). *Let $0 < \delta \leq 1$. Then if n is a sufficiently large integer, any subset A of the square $[n]^2$ of density $|A|/n^2$ at least δ contains at least one right-angled triangle (or "corner") $\{(x,y), (x+r,y), (x,y+r)\}$ with $r \neq 0$.*

The $k = 3$ density Hales-Jewett theorem implies the corners theorem; this is proven by utilising the map $\phi : [3]^n \to [n]^2$ from the cube to the square, defined by mapping a string $x \in [3]^n$ to a pair (a,b), where a,b are the number of 1's and 2's, respectively in x. The key point is that ϕ maps combinatorial lines to corners. (Strictly speaking, this mapping only establishes the corners theorem for dense subsets of $[n/3 - \sqrt{n}, n/3 + \sqrt{n}]^2$, but it is not difficult to obtain the general case from this by replacing n by n^2 and using translation invariance.)

The corners theorem is also closely related to the problem of finding dense sets of points in a triangular grid without any equilateral triangles, a problem which we have called *Fujimura's problem.*

The corners theorem in turn implies

Theorem 2.1.3 (Roth's theorem). *Let $0 < \delta \leq 1$. Then if n is a sufficiently large integer, any subset A of the interval $[n]$ of density $|A|/n$ at least δ contains at least one arithmetic progression $a, a+r, a+2r$ of length three.*

Roth's theorem can be deduced from the corners theorem by considering the map $\psi : [n]^2 \to [3n]$ defined by $\psi(a,b) := a + 2b$; the key point is that ψ maps corners to arithmetic progressions of length three.

There are higher k analogues of these implications; the general k version of the density Hales-Jewett theorem implies a general k version of the corners theorem known as the multidimensional Szemerédi theorem, which in term implies a general version of Roth's theorem known as *Szemerédi's theorem.*

2.1.2. The density increment argument. The strategy of the density increment argument, which goes back to Roth's proof [**Ro1953**] of Theorem 2.1.3, is to perform a downward induction on the density δ. Indeed, the theorem is obvious for high enough values of δ; for instance, if $\delta > 2/3$, then partitioning the cube $[3]^n$ into lines and applying the pigeonhole principle will already give a combinatorial line. So the idea is to deduce the claim for a fixed density δ from that of a higher density δ.

A key concept here is that of an m-dimensional *combinatorial subspace* of $[3]^n$—a set of the form $\phi([3]^m)$, where $\phi \in \{1, 2, 3, *_1, \ldots, *_m\}^n$ is a string formed using the base alphabet and m wildcards $*_1, \ldots, *_m$ (with each wildcard appearing at least once), and $\phi(a_1 \cdots a_m)$ is the string formed by substituting a_i for $*_i$ for each i. (Thus, for instance, a combinatorial line is a combinatorial subspace of dimension 1.) The identification ϕ between $[3]^m$ and the combinatorial space $\phi([3]^m)$ maps combinatorial lines to combinatorial lines. Thus, to prove Theorem 2.1.1, it suffices to show

Proposition 2.1.4 (Lack of lines implies density increment)**.** *Let* $0 < \delta \leq 1$. *Then if* n *is a sufficiently large integer and* $A \subset [3]^n$ *has density at least* δ *and has no combinatorial lines, then there exists an* m*-dimensional subspace* $\phi([3]^m)$ *of* $[3]^n$ *on which* A *has density at least* $\delta + c(\delta)$, *where* $c(\delta) > 0$ *depends only on* δ *(and is bounded away from zero on any compact range of* δ*), and* $m \geq m_0(n, \delta)$ *for some function* $m_0(n, \delta)$ *that goes to infinity as* $n \to \infty$ *for fixed* δ.

It is easy to see that Proposition 2.1.4 implies Theorem 2.1.1 (for instance, one could consider the infimum of all δ for which the theorem holds and show that having this infimum nonzero would lead to a contradiction).

Now we have to figure out how to get that density increment. The original argument of Roth relied on Fourier analysis, which in turn relies on an underlying translation-invariant structure which is not present in the density Hales-Jewett setting. (Arithmetic progressions are translation invariant, but combinatorial lines are not.) It turns out that one can proceed instead by adapting a (modification of) an argument of Ajtai and Szemerédi, which gave the first proof of Theorem 2.1.2.

The (modified) Ajtai-Szemerédi argument uses the density increment method, assuming that A has no right-angled triangles and showing that A has an increased density on a *subgrid*—a product $P \times Q$ of fairly long arithmetic progressions with the same spacing. The argument proceeds in two stages, which we describe slightly informally (in particular, glossing over some technical details regarding quantitative parameters such as ε) as follows:

- Step 1. If $A \subset [n]^2$ is dense but has no right-angled triangles, then A has an increased density on a Cartesian product $U \times V$ of dense sets $U, V \subset [n]$ (which are not necessarily arithmetic progressions).
- Step 2. Any Cartesian product $U \times V$ in $[n]^2$ can be partitioned into reasonably large grids $P \times Q$, plus a remainder term of small density.

From Step 1, Step 2, and the pigeonhole principle we obtain the desired density increment of A on a grid $P \times Q$, and then the density increment argument gives us the corners theorem.

Step 1 is actually quite easy. If A is dense, then it must also be dense on some diagonal $D = \{(x, y) : x + y = const\}$, by the pigeonhole principle. Let U and V denote the rows and columns that $A \cap D$ occupies. Every pair of points in $A \cap D$ forms the hypotenuse of some corner, whose third vertex lies in $U \times V$. Thus, if A has no corners, then A must avoid all the points formed by $U \times V$ (except for those of the diagonal D). Thus A has a significant density *decrease* on the Cartesian product $U \times V$. Dividing the remainder $[n]^2 \backslash (U \times V)$ into three further Cartesian products $U \times ([n]\backslash V)$, $([n]\backslash U) \times V$, $([n]\backslash U) \times ([n]\backslash V)$ and using the pigeonhole principle, we obtain the claim (after redefining U, V appropriately).

Step 2 can be obtained by iterating a one-dimensional version:

- Step 2a. Any set $U \subset [n]$ can be partitioned into reasonably long arithmetic progressions P, plus a remainder term of small density.

Indeed, from Step 2a, one can partition $U \times [n]$ into products $P \times [n]$ (plus a small remainder), which can be easily repartitioned into grids $P \times Q$ (plus a small remainder). This partitions $U \times V$ into sets $P \times (V \cap Q)$ (plus a small remainder). Applying Step 2a again, each $V \cap Q$ can be partitioned further into progressions Q' (plus a small remainder), which allows us to partition each $P \times (V \cap Q)$ into grids $P' \times Q'$ (plus a small remainder).

So all one has left to do is establish Step 2a. But this can be done by the greedy algorithm: locate one long arithmetic progression P in U and remove it from U, then locate another to remove, and so forth until no further long progressions remain in the set. But *Szemerédi's theorem* then tells us the remaining set has low density, and one is done!

This argument has the apparent disadvantage of requiring a deep theorem (Szemerédi's theorem) in order to complete the proof. However, interestingly enough, when one adapts the argument to the density Hales-Jewett theorem, one gets to replace Szemerédi's theorem by a more elementary result—one which in fact follows from the (easy) $k = 2$ version of the density Hales-Jewett theorem, i.e., *Sperner's theorem.*

We first need to understand the analogue of the Cartesian products $U \times V$. Note that $U \times V$ is the intersection of a *vertically insensitive set* $U \times [n]$ and a *horizontally insensitive set* $[n] \times V$. By vertically insensitive we mean that membership of a point (x, y) in that set is unaffected if one moves that point in a vertical direction, and similarly for horizontally insensitive. In a similar fashion, define a 12-*insensitive set* to be a subset of $[3]^n$, membership in which is unaffected if one flips a coordinate from a 1 to a 2 or vice versa (e.g., if 1223 lies in the set, then so must 1213, 1113, 2113, etc.) Similarly, define the notion of a 13-*insensitive set.* We then define a *complexity* 1 *set* to be the intersection $E_{12} \cap E_{13}$ of a 12-insensitive set E_{12} and a 13-insensitive set E_{13}; these are analogous to the Cartesian products $U \times V$.

(For technical reasons, one actually has to deal with *local* versions of insensitive sets and complexity 1 sets, in which one is only allowed to flip a moderately small number of the n coordinates rather than all of them. But to simplify the discussion, let me ignore this (important) detail, which is also a major issue to be addressed in the other two proofs of this theorem.)

The analogues of Steps 1 and 2 for the density Hales-Jewett theorem are then

- Step 1. If $A \subset [3]^n$ is dense but has no combinatorial lines, then A has an increased density on a (local) complexity 1 set $E_{12} \cap E_{13}$.
- Step 2. Any (local) complexity 1 set $E_{12} \cap E_{13} \subset [3]^n$ can be partitioned into moderately large combinatorial subspaces (plus a small remainder).

We can sketch how Step 1 works as follows. Given any $x \in [3]^n$, let $\pi_{1 \to 2}(x)$ denote the string formed by replacing all 1's with 2's, e.g., $\pi_{1 \to 2}(1321) = 2322$. Similarly, define $\pi_{1 \to 3}(x)$. Observe that $x, \pi_{1 \to 2}(x)$, $\pi_{1 \to 3}(x)$ forms a combinatorial line (except in the rare case when x does not contain any 1's). Thus if we let $E_{12} := \{x : \pi_{1 \to 2}(x) \in A\}$, $E_{13} := \{x : \pi_{1 \to 3}(x) \in A\}$, we see that A must avoid essentially all of $E_{12} \cap E_{13}$. On the other hand, observe that E_{12} and E_{13} are 12-insensitive and 13-insensitive sets, respectively. Taking complements and using the same sort of pigeonhole argument as before, we obtain the claim. (Actually, this argument does not quite work because E_{12}, E_{13} could be very sparse; this problem can be fixed, but requires one to use local complexity 1 sets rather than global ones and also to introduce the concept of *equal-slices measure*; I will not discuss these issues here.)

Step 2 can be reduced, much as before, to the following analogue of Step 2a:

- Step 2a. Any 12-insensitive set $E_{12} \subset [3]^n$ can be partitioned into moderately large combinatorial subspaces (plus a small remainder).

Identifying the letters 1 and 2 together, one can quotient $[3]^n$ down to $[2]^n$; the preimages of this projection are precisely the 12-insensitive sets. Because of this, Step 2a is basically equivalent (modulo some technicalities about measure) to

- Step 2a′. Any $E \subset [2]^n$ can be partitioned into moderately large combinatorial subspaces (plus a small remainder).

By the greedy algorithm, we will be able to accomplish this step if we can show that every dense subset of $[2]^n$ contains moderately large subspaces. But this turns out to be possible by carefully iterating Sperner's theorem (which shows that every dense subset of $[2]^n$ contains combinatorial lines).

This proof of Theorem 2.1.1 extends without major difficulty to the case of higher k; see [**Po2009**].

2.1.3. The triangle removal argument. The *triangle removal lemma* of Ruzsa and Szemerédi [**RuSz1978**] is a graph-theoretic result which implies the corners theorem (and hence Roth's theorem). It asserts the following:

Lemma 2.1.5 (Triangle removal lemma). *For every $\varepsilon > 0$ there exists $\delta > 0$ such that if a graph G on n vertices has fewer than δn^3 triangles, then the triangles can be deleted entirely by removing at most εn^2 edges.*

Let us see how the triangle removal lemma implies the corners theorem. A corner is, of course, already a triangle in the geometric sense, but we need to convert it to a triangle in the graph-theoretic sense, as follows. Let A be a subset of $[n]^2$ with no corners; the aim is to show that A has small density. Let V_h be the set of all horizontal lines in $[n]^2$, V_v the set of vertical lines, and V_d the set of diagonal lines (thus all three sets have size about n). We create a tripartite graph G on the vertex sets $V_h \cup V_v \cup V_d$ by joining a horizontal line $h \in V_h$ to a vertical line $v \in V_v$ whenever h and v intersect at a point in A, and similarly connecting V_h or V_v to V_d. Observe that a triangle in G corresponds either to a corner in A, or to a "degenerate" corner in which the horizontal, vertical, and diagonal line are all concurrent. In particular, there are very few triangles in G, which can then be deleted by removing a small number of edges from G by the triangle removal lemma. But each edge removed can delete at most one degenerate corner, and the number of degenerate corners is $|A|$, and so $|A|$ is small as required.

All known proofs of the triangle removal lemma proceed by some version of the following three steps:

- *Regularity lemma step.* Applying tools such as the *Szemerédi regularity lemma*, one can partition the graph G into components G_{ij} between cells V_i, V_j of vertices, such that most of the G_{ij} are "pseudo-random". One way to define what pseudo-random means is to view each graph component G_{ij} as a subset of the Cartesian product $V_i \times V_j$, in which case G_{ij} is pseudo-random if it does not have a significant density increment on any smaller Cartesian product $V'_i \times V'_j$ of nontrivial size.

- *Counting lemma step.* By exploiting the pseudo-randomness property, one shows that if G has a triple G_{ij}, G_{jk}, G_{ki} of dense pseudo-random graphs between cells V_i, V_j, V_k of nontrivial size, then this triple must generate a large number of triangles; hence, if G has very few triangles, then one cannot find such a triple of dense pseudo-random graphs.

- *Cleaning step.* If one then removes all components of G which are too sparse or insufficiently pseudo-random, one can thus eliminate all triangles.

Pulling this argument back to the corners theorem, we see that cells such as V_i, V_j, V_k will correspond either to horizontally insensitive sets, vertically insensitive sets, or diagonally insensitive sets. Thus this proof of the corners theorem proceeds by partitioning $[n]^2$ in three different ways into insensitive sets in such a way that A is pseudo-random with respect to many of the cells created by any two of these partitions, counting the corners generated by any triple of large cells in which A is pseudo-random and dense, and cleaning out all the other cells.

It turns out that a variant of this argument can give Theorem 2.1.1; this was in fact the original approach studied by the Polymath1 project, though it was only after a detour through ergodic theory (as well as the development of the density-increment argument discussed above) that the triangle-removal approach could be properly executed. In particular, an ergodic argument based on the infinitary analogue of the triangle removal lemma (and its hypergraph generalisations) was developed by Austin [**Au2009b**], which then inspired the combinatorial version sketched here.

The analogue of the vertex cells V_i are given by certain 12-insensitive sets E^a_{12}, 13-insensitive sets E^b_{13}, and 23-insensitive sets E^c_{23}. Roughly speaking, a set $A \subset [3]^n$ would be said to be pseudo-random with respect to a cell $E^a_{12} \cap E^b_{13}$ if $A \cap E^a_{12} \cap E^b_{13}$ has no further density increment on any smaller cell $E'_{12} \cap E'_{13}$ with E'_{12} a 12-insensitive subset of E^a_{12}, and E'_{13} a 13-insensitive subset of E^b_{13}. (This is an oversimplification, glossing over an important refinement of the concept of pseudo-randomness involving the discrepancy between global densities in $[3]^n$ and local densities in subspaces of $[3]^n$.) There is a similar notion of A being pseudo-random with respect to a cell $E^b_{13} \cap E^c_{23}$ or $E^c_{23} \cap E^a_{12}$.

We briefly describe the regularity lemma step. By modifying the proof of the regularity lemma, one can obtain three partitions

$$[3]^n = E_{12}^1 \cup \cdots \cup E_{12}^{M_{12}} = E_{13}^1 \cup \cdots \cup E_{13}^{M_{13}} = E_{23}^1 \cup \cdots \cup E_{23}^{M_{23}}$$

of $[3]^n$ 12-insensitive, 13-insensitive, and 23-insensitive components, respectively, where M_{12}, M_{13}, M_{23} are not too large, and A is pseudo-random with respect to most cells $E_{12}^a \cap E_{13}^b$, $E_{13}^b \cap E_{23}^c$, and $E_{23}^c \cap E_{12}^a$.

In order for the counting step to work, one also needs an additional "stationarity" reduction, which is difficult to state precisely, but roughly speaking asserts that the "local" statistics of sets such as E_{12}^a on medium-dimensional subspaces are close to the corresponding "global" statistics of such sets; this can be achieved by an additional pigeon holing argument. We will gloss over this issue, pretending that there is no distinction between local statistics and global statistics. (Thus, for instance, if E_{12}^a has large global density in $[3]^n$, we shall assume that E_{12}^a also has large density on most medium-sized subspaces of $[3]^n$.)

Now for the counting lemma step. Suppose we can find a, b, c such that the cells $E_{12}^a, E_{13}^b, E_{23}^c$ are large, and that A intersects $E_{12}^a \cap E_{13}^b$, $E_{13}^b \cap E_{23}^c$, and $E_{23}^c \cap E_{12}^a$ in a dense pseudo-random manner. We claim that this will force A to have a large number of combinatorial lines ℓ, with $\ell(1)$ in $A \cap E_{12}^a \cap E_{13}^b$, $\ell(2)$ in $A \cap E_{23}^c \cap E_{12}^a$, and $\ell(3)$ in $A \cap E_{13}^b \cap E_{23}^c$. Because of the dense pseudo-random nature of A in these cells, it turns out that it will suffice to show that there are a lot of lines $\ell(1)$ with $\ell(1) \in E_{12}^a \cap E_{13}^b$, $\ell(2) \in E_{23}^c \cap E_{12}^a$, and $\ell(3) \in E_{13}^b \cap E_{23}^c$.

One way to generate a line ℓ is by taking the triple $\{x, \pi_{1\to2}(x), \pi_{1\to3}(x)\}$, where $x \in [3]^n$ is a generic point. (Actually, as we will see below, we would have to have a subspace of $[3]^n$ before using this recipe to generate lines.) Then we need to find many x obeying the constraints

$$x \in E_{12}^a \cap E_{13}^b, \quad \pi_{1\to2}(x) \in E_{23}^c \cap E_{12}^a, \quad \pi_{1\to3}(x) \in E_{13}^b \cap E_{23}^c.$$

Because of the various insensitivity properties, many of these conditions are redundant, and we can simplify to

$$x \in E_{12}^a \cap E_{13}^b, \quad \pi_{1\to2}(x) \in E_{23}^c.$$

Now note that the property "$\pi_{1\to2}(x) \in E_{23}^c$" is 123-insensitive; it is simultaneously 12-insensitive, 23-insensitive, and 13-insensitive. As E_{23}^c is assumed to be large, there will be large combinatorial subspaces on which (a suitably localised version of) this property "$\pi_{1\to2}(x) \in E_{23}^c$" will be always true. Localising to this space (taking advantage of the stationarity properties alluded to earlier), we are now looking for solutions to

$$x \in E_{12}^a \cap E_{13}^b.$$

We will pick x to be of the form $\pi_{2\to 1}(y)$ for some y. We can then rewrite the constraints on y as

$$y \in E^a_{12}, \quad \pi_{2\to 1}(y) \in E^b_{13}.$$

The property "$\pi_{2\to 1}(y) \in E^b_{13}$" is 123-invariant, and E^b_{13} is large, so by arguing as before we can pass to a large subspace where this property is always true. The largeness of E^a_{12} then gives us a large number of solutions.

Taking contrapositives, we conclude that if A in fact has no combinatorial lines, then there do not exist any triple $E^a_{12}, E^b_{13}, E^c_{23}$ of large cells with respect to which A is dense and pseudo-random. This forces A to be confined either to very small cells, or to very sparse subsets of cells, or to the rare cells which fail to be pseudo-random. None of these cases can contribute much to the density of A, and so A itself is very sparse, contradicting the hypothesis in Theorem 2.1.1 that A is dense (this is the cleaning step). This concludes the sketch of the triangle-removal proof of this theorem.

The ergodic version of this argument in [**Au2009b**] works for all values of k, so I expect the combinatorial version to do so as well.

2.1.4. The finitary Furstenberg-Katznelson argument. Furstenberg and Katznelson gave the first proof of Theorem 2.1.1, in [**FuKa1989**], by translating it into a recurrence statement about a certain type of stationary process indexed by an infinite cube $[3]^\omega := \bigcup_{n=1}^\infty [3]^n$. This argument was inspired by a long string of other successful proofs of density Ramsey theorems via ergodic means, starting with the initial paper of Furstenberg [**Fu1977**] giving an ergodic theory proof of Szemerédi's theorem. The latter proof was transcribed into a finitary language in [**Ta2006b**], so it was reasonable to expect that the Furstenberg-Katznelson argument could similarly be translated into a combinatorial framework.

Let us first briefly describe the original strategy of Furstenberg to establish Roth's theorem, but phrased in informal and vaguely combinatorial language. The basic task is to get a nontrivial lower bound on averages of the form

$$\mathbf{E}_{a,r} f(a) f(a+r) f(a+2r), \tag{2.1}$$

where we will be a bit vague about what a, r are ranging over, and where f is some nonnegative function of positive mean. It is then natural to study more general averages of the form

$$\mathbf{E}_{a,r} f(a) g(a+r) h(a+2r). \tag{2.2}$$

Now, it turns out that certain types of functions f, g, h give a negligible contribution to expressions such as (2.2). In particular, if f is *weakly mixing*, which roughly means that the pair correlations

$$\mathbf{E}_a f(a)f(a+r)$$

are small for most r, then the average (2.2) is small no matter what g, h are (so long as they are bounded). This can be established by some applications of the Cauchy-Schwarz inequality (or its close cousin, the *van der Corput lemma*). As a consequence of this, all weakly mixing components of f can essentially be discarded when considering an average such as (2.1).

After getting rid of the weakly mixing components, what is left? Being weakly mixing is like saying that almost all the shifts $f(\cdot + r)$ of f are close to orthogonal to each other. At the other extreme is that of *periodicity*—the shifts $f(\cdot + r)$ periodically recur to become equal to f again. There is a slightly more general notion of *almost periodicity*—roughly, this means that the shifts $f(\cdot + r)$ do not have to recur exactly to f again, but they are forced to range in a precompact set, which basically means that for every $\varepsilon > 0$, that $f(\cdot + r)$ lies within ε (in some suitable norm) of some finite-dimensional space. A good example of an almost periodic function is an *eigenfunction*, in which we have $f(a + r) = \lambda_r f(a)$ for each r and some quantity λ_r independent of a (e.g., one can take $f(a) = e^{2\pi i\alpha a}$ for some $\alpha \in \mathbf{R}$). In this case, the finite-dimensional space is simply the scalar multiples of $f(a)$ (and one can even take $\varepsilon = 0$ in this special case).

It is easy to see that nontrivial almost periodic functions are not weakly mixing; more generally, any function which correlates nontrivially with an almost periodic function can also be seen to not be weakly mixing. In the converse direction, it is also fairly easy to show that any function which is not weakly mixing must have nontrivial correlation with an almost periodic function. Because of this, it turns out that one can basically decompose *any* function into almost periodic and weakly mixing components. For the purposes of getting lower bounds on (2.1), this allows us to essentially reduce matters to the special case when f is almost periodic. But then the shifts $f(\cdot + r)$ are almost ranging in a finite-dimensional set, which allows one to essentially assign each shift r a colour from a finite range of colours. If one then applies the *van der Waerden theorem*, one can find many arithmetic progressions $a, a + r, a + 2r$ which have the same colour, and this can be used to give a nontrivial lower bound on (2.1). (Thus we see that the role of a compactness property such as almost periodicity is to reduce density Ramsey theorems to colouring Ramsey theorems.)

This type of argument can be extended to more advanced recurrence theorems, but certain things become more complicated. For instance, suppose one wanted to count progressions of length 4. This amounts to lower bounding expressions such as

$$\mathbf{E}_{a,r} f(a) f(a+r) f(a+2r) f(a+3r). \tag{2.3}$$

It turns out that f being weakly mixing is no longer enough to give a negligible contribution to expressions such as (2.3). For that, one needs the stronger property of being *weakly mixing relative to almost periodic functions*. Roughly speaking, this means that for most r, the expression $f(\cdot)f(\cdot+r)$ is not merely of small mean (which is what weak mixing would mean), but that this expression furthermore does not correlate strongly with any almost periodic function (i.e., $\mathbf{E}_a f(a)f(a+r)g(a)$ is small for any almost periodic g). Once one has this stronger weak mixing property, then one can discard all components of f which are weakly mixing relative to almost periodic functions.

One then has to figure out what is left after all these components are discarded. Because we strengthened the notion of weak mixing, we have to weaken the notion of almost periodicity to compensate. The correct notion is no longer that of almost periodicity, in which the shifts $f(\cdot+r)$ almost take values in a finite-dimensional vector space, but that of almost periodicity *relative* to almost periodic functions, in which the shifts almost take values in a finite-dimensional *module* over the algebra of almost periodic functions. A good example of such a beast is that of a *quadratic eigenfunction*, in which we have $f(a+r) = \lambda_r(a) f(a)$, where $\lambda_r(a)$ is itself an ordinary eigenfunction and is thus almost periodic in the ordinary sense; here, the relative module is the one-dimensional module formed by almost periodic multiples of f. (A typical example of a quadratic eigenfunction is $f(a) = e^{2\pi i \alpha a^2}$ for some $\alpha \in \mathbf{R}$.)

It turns out that one can "relativise" all of the previous arguments to the almost periodic "factor", and decompose an arbitrary f into a component which is weakly mixing relative to almost periodic functions, and another component which is almost periodic relative to almost periodic functions. The former type of components can be discarded. For the latter, we can once again start colouring the shifts $f(\cdot+r)$ with a finite number of colours, but with the caveat that the colour assigned is no longer independent of a but depends in an almost periodic fashion on a. Nevertheless, it is still possible to combine the van der Waerden colouring Ramsey theorem with the theory of recurrence for ordinary almost periodic functions to get a lower bound on (2.3) in this case. One can then iterate this argument to deal with arithmetic progressions of longer length, but one now needs to consider even more intricate notions of almost periodicity, e.g., almost periodicity relative to (almost periodic functions relative to almost periodic functions), etc.

It turns out that these types of ideas can be adapted (with some effort) to the density Hales-Jewett setting. It is simplest to begin with the $k = 2$

situation rather than the $k = 3$ situation. Here, we are trying to obtain nontrivial lower bounds for averages of the form

$$\mathbf{E}_\ell f(\ell(1))f(\ell(2)), \tag{2.4}$$

where ℓ ranges in some fashion over combinatorial lines in $[2]^n$, and f is some nonnegative function with large mean.

The analogues of weakly mixing and almost periodic in this setting are the 12-uniform and 12-low influence functions, respectively. Roughly speaking, a function is 12-low influence if its value usually does not change much if a 1 is flipped to a 2 or vice versa (e.g., the indicator function of a 12-insensitive set is 12-low influence); conversely, a 12-uniform function is a function g such that $\mathbf{E}_\ell f(\ell(1))g(\ell(2))$ is small for all (bounded) f. One can show that any function can be decomposed, more or less orthogonally, into a 12-uniform function and a 12-low influence function, with the upshot being that one can basically reduce the task of lower bounding (2.4) to the case when f is 12-low influence. But then $f(\ell(1))$ and $f(\ell(2))$ are approximately equal to each other, and it is straightforward to get a lower-bound in this case.

Now we turn to the $k = 3$ setting, where we are looking at lower-bounding expressions such as

$$\mathbf{E}_\ell f(\ell(1))g(\ell(2))h(\ell(3)) \tag{2.5}$$

with $f = g = h$.

It turns out that g (say) being 12-uniform is no longer enough to give a negligible contribution to the average (2.5). Instead, one needs the more complicated notion of g being 12-uniform relative to 23-low influence functions; this means that not only are the averages $\mathbf{E}_\ell f(\ell(1))g(\ell(2))$ small for all bounded f, but furthermore $\mathbf{E}_\ell f(\ell(1))g(\ell(2))h(\ell)$ is small for all bounded f and all 23-low influence h (there is a minor technical point here that h is a function of a line rather than of a point, but this should be ignored). Any component of g in (2.5) which is 12-uniform relative to 23-low influence functions are negligible and so can be removed.

One then needs to figure out what is left in g when these components are removed. The answer turns out to be functions g that are 12-almost periodic relative to 23-low influence. The precise definition of this concept is technical, but very roughly speaking it means that if one flips a digit from a 1 to a 2, then the value of g changes in a manner which is controlled by 23-low influence functions. Anyway, the upshot is that one can reduce g in (2.5) from f to the components of f which are 12-almost periodic relative to 23-low influence. Similarly, one can reduce h in (2.5) from f to the components of f which are 13-almost periodic relative to 23-low influence.

At this point, one has to use a colouring Ramsey theorem—in this case, the *Graham-Rothschild theorem*—in conjunction with the relative almost periodicity to locate lots of places in which $g(\ell(2))$ is close to $g(\ell(1))$ while $h(\ell(3))$ is simultaneously close to $h(\ell(1))$. This turns (2.5) into an expression of the form $\mathbf{E}_x f(x) g(x) h(x)$, which turns out to be relatively easy to lower bound (because g, h, being projections of f, tend to be large wherever f is large).

Notes. This article first appeared at

`terrytao.wordpress.com/2009/04/02`.

Thanks to Ben, Daniel, Kevin O'Bryant, Sune Kristian Jakobsen, and anonymous commenters for corrections.

More information about the Polymath1 project can be found at

`http://michaelnielsen.org/polymath1/index.php?title=Main_Page`.

2.2. Szemerédi's regularity lemma via random partitions

In the theory of dense graphs on n vertices, where n is large, a fundamental role is played by the *Szemerédi regularity lemma*:

Lemma 2.2.1 (Regularity lemma, standard version). *Let $G = (V, E)$ be a graph on n vertices, and let $\varepsilon > 0$ and $k_0 \geq 0$. Then there exists a partition of the vertices $V = V_1 \cup \cdots \cup V_k$, with $k_0 \leq k \leq C(k_0, \varepsilon)$ bounded below by k_0 and above by a quantity $C(k_0, \varepsilon)$ depending only on k_0, ε, obeying the following properties:*

- Equitable partition. *For any $1 \leq i, j \leq k$, the cardinalities $|V_i|, |V_j|$ of V_i and V_j differ by at most 1.*
- Regularity. *For all but at most εk^2 pairs $1 \leq i < j \leq k$, the portion of the graph G between V_i and V_j is* ε-regular *in the sense that one has*
$$|d(A, B) - d(V_i, V_j)| \leq \varepsilon$$
for any $A \subset V_i$ and $B \subset V_j$ with $|A| \geq \varepsilon |V_i|, |B| \geq \varepsilon |V_j|$, where $d(A, B) := |E \cap (A \times B)| / |A||B|$ is the density of edges between A and B.

This lemma becomes useful in the regime when n is very large compared to k_0 or $1/\varepsilon$, because all the conclusions of the lemma are uniform in n. Very roughly speaking, it says that "up to errors of size ε", a large graph can be more or less described completely by a bounded number of quantities $d(V_i, V_j)$. This can be interpreted as saying that the space of all graphs is *totally bounded* (and hence *precompact*) in a suitable metric space, thus allowing one to take formal limits of sequences (or subsequences) of graphs; see for instance [**LoSz2007**] for a discussion.

For various technical reasons it is easier to work with a slightly weaker version of the lemma, which allows for the cells $V_1, \ldots, V_k$ to have unequal sizes:

Lemma 2.2.2 (Regularity lemma, weighted version)**.** *Let $G = (V, E)$ be a graph on n vertices, and let $\varepsilon > 0$. Then there exists a partition of the vertices $V = V_1 \cup \cdots \cup V_k$, with $1 \leq k \leq C(\varepsilon)$ bounded above by a quantity $C(\varepsilon)$ depending only on ε, obeying the following properties:*

- Regularity. *One has*

$$\sum_{(V_i,V_j) \text{ not } \varepsilon\text{-regular}} |V_i||V_j| = O(\varepsilon|V|^2), \tag{2.6}$$

where the sum is over all pairs $1 \leq i \leq j \leq k$ for which G is not ε-regular between V_i and V_j.

While Lemma 2.2.2 is, strictly speaking, weaker than Lemma 2.2.1 in that it does not enforce the equitable size property between the atoms, in practice it seems that the two lemmas are roughly of equal utility; most of the combinatorial consequences of Lemma 2.2.1 can also be proven using Lemma 2.2.2. The point is that one always has to remember to weight each cell V_i by its density $|V_i|/|V|$, rather than by giving each cell an equal weight as in Lemma 2.2.1. Lemma 2.2.2 also has the advantage that one can easily generalise the result from finite vertex sets V to other probability spaces (for instance, one could weight V with something other than the uniform distribution). For applications to hypergraph regularity, it turns out to be slightly more convenient to have *two* partitions (coarse and fine) rather than just one; see for instance [**Ta2006c**]. In any event the arguments below that we give to prove Lemma 2.2.2 can be modified to give a proof of Lemma 2.2.1 also.

The proof of the regularity lemma is usually conducted by a *greedy algorithm*. Very roughly speaking, one starts with the trivial partition of V. If this partition already regularises the graph, we are done; if not, this means that there are some sets A and B in which there is a significant density fluctuation beyond what has already been detected by the original partition. One then adds these sets to the partition and iterates the argument. Every time a new density fluctuation is incorporated into the partition that models the original graph, this increases a certain "index" or "energy" of the partition. On the other hand, this energy remains bounded no matter how complex the partition, so eventually one must reach a long "energy plateau" in which no further refinement is possible, at which point one can find the regular partition.

One disadvantage of the greedy algorithm is that it is not efficient in the limit $n \to \infty$, as it requires one to search over *all* pairs of subsets A, B of a given pair V_i, V_j of cells, which is an exponentially long search. There are more algorithmically efficient ways to regularise; for instance, a polynomial time algorithm was given in [**AlDuLeRoYu1994**]. However, one can do even better, if one is willing to (a) allow cells of unequal size, (b) allow a small probability of failure, (c) have the ability to sample vertices from G at random, and (d) allow for the cells to be defined "implicitly" (via their relationships with a fixed set of reference vertices) rather than "explicitly" (as a list of vertices). In that case, one can regularise a graph in a number of operations *bounded* in n. Indeed, one has

Lemma 2.2.3 (Regularity lemma via random neighbourhoods). *Let $\varepsilon > 0$. Then there exists integers $M_1, \ldots, M_m$ with the following property: whenever $G = (V, E)$ is a graph on finitely many vertices, if one selects one of the integers M_r at random from $M_1, \ldots, M_m$ and then selects M_r vertices $v_1, \ldots, v_{M_r} \in V$ uniformly from V at random, then the 2^{M_r} vertex cells $V_1^{M_r}, \ldots, V_{2^{M_r}}^{M_r}$ (some of which can be empty) generated by the vertex neighbourhoods $A_t := \{v \in V : (v, v_t) \in E\}$ for $1 \leq t \leq M_r$ will obey the conclusions of Lemma 2.2.2 with probability at least $1 - O(\varepsilon)$.*

Thus, roughly speaking, one can regularise a graph simply by taking a large number of random vertex neighbourhoods, and using the partition (or Venn diagram) generated by these neighbourhoods as the partition. The intuition is that if there is any nonuniformity in the graph (e.g., if the graph exhibits bipartite behaviour), this will bias the random neighbourhoods to seek out the partitions that would regularise that nonuniformity (e.g., vertex neighbourhoods would begin to fill out the two vertex cells associated to the bipartite property). If one takes sufficiently many such random neighbourhoods, the probability that all detectable nonuniformity is captured by the partition should converge to 1. (It is more complicated than this, because the finer one makes the partition, the finer the types of nonuniformity one can begin to detect, but this is the basic idea.)

This fact seems to be reasonably well-known folklore, discovered independently by many authors; it is for instance quite close to the graph property testing results in [**AlSh2008**], and it also appears in [**Is2006**] and [**Au2008**] (and implicitly in [**Ta2007**]). I will present a proof of the lemma below.

2.2.1. Warmup: a weak regularity lemma.

To motivate the idea, let us first prove a weaker but simpler (and more quantitatively effective) regularity lemma, analogous to that established by Frieze and Kannan:

Lemma 2.2.4 (Weak regularity lemma via random neighbourhoods). *Let $\varepsilon > 0$. Then there exists an integer M with the following property: whenever $G = (V, E)$ is a graph on finitely many vertices, if one selects $1 \leq t \leq M$ at random and then selects t vertices $v_1, \ldots, v_t \in V$ uniformly from V at random, then the 2^t vertex cells $V_1^t, \ldots, V_{2^t}^t$ (some of which can be empty) generated by the vertex neighbourhoods $A_{t'} := \{v \in V : (v, v_{t'}) \in E\}$ for $1 \leq t' \leq t$, obey the following property with probability at least $1 - O(\varepsilon)$. For any vertex sets $A, B \subset V$, the number of edges $|E \cap (A \times B)|$ connecting A and B can be approximated by the formula*

$$(2.7) \qquad |E \cap (A \times B)| = \sum_{i=1}^{2^t} \sum_{j=1}^{2^t} d(V_i^t, V_j^t)|A \cap V_i^t||B \cap V_j^t| + O(\varepsilon|V|^2).$$

This weaker lemma only lets us count "macroscopic" edge densities $d(A, B)$, when A, B are dense subsets of V, whereas the full regularity lemma is stronger in that it also controls "microscopic" edge densities $d(A, B)$ where A, B are now dense subsets of the cells $V_i^{M_r}, V_j^{M_r}$. Nevertheless, this weaker lemma is easier to prove and already illustrates many of the ideas.

Let us now prove this lemma. Fix $\varepsilon > 0$, let M be chosen later, let $G = (V, E)$ be a graph, and select $v_1, \ldots, v_M$ at random. (There can of course be many vertices selected more than once; this will not bother us.) Let A_t and $V_1^t, \ldots, V_{2^t}^t$ be as in the above lemma. For notational purposes it is more convenient to work with the (random) σ-algebra $\mathcal{B}_t$ generated by the $A_1, \ldots, A_t$ (i.e., the collection of all sets that can be formed from $A_1, \ldots, A_t$ by boolean operations); this is an atomic σ-algebra whose atoms are precisely the (nonempty) cells $V_1^t, \ldots, V_{2^t}^t$ in the partition. Observe that these σ-algebras are nested: $\mathcal{B}_t \subset \mathcal{B}_{t+1}$.

We will use the trick of turning sets into functions, and view the graph as a function $1_E : V \times V \to \mathbf{R}$. One can then form the *conditional expectation* $\mathbf{E}(1_E|\mathcal{B}_t \times \mathcal{B}_t) : V \times V \to \mathbf{R}$ of this function to the product σ-algebra $\mathcal{B}_t \times \mathcal{B}_t$, whose value on $V_i^t \times V_j^t$ is simply the average value of 1_E on the product set $V_i^t \times V_j^t$. (When i and j are different, this is simply the edge density $d(V_i^t, V_j^t)$). One can view $\mathbf{E}(1_E|\mathcal{B}_t \times \mathcal{B}_t)$ more combinatorially, as a weighted graph on V such that all edges between two distinct cells V_i^t, V_j^t have the same constant weight of $d(V_i^t, V_j^t)$.

We give V (and $V \times V$) the uniform probability measure, and define the *energy* e_t at time t to be the (random) quantity

$$e_t := \|\mathbf{E}(1_E|\mathcal{B}_t \times \mathcal{B}_t)\|_{L^2(V \times V)}^2 = \frac{1}{|V|^2} \sum_{v,w \in V} \mathbf{E}(1_E|\mathcal{B}_t \times \mathcal{B}_t)^2.$$

One can interpret this as the mean square of the edge densities $d(V_i^t, V_j^t)$ weighted by the size of the cells V_i^t, V_j^t. From Pythagoras' theorem we have the identity

$$e_{t'} = e_t + \|\mathbf{E}(1_E|\mathcal{B}_{t'} \times \mathcal{B}_{t'}) - \mathbf{E}(1_E|\mathcal{B}_t \times \mathcal{B}_t)\|^2_{L^2(V\times V)}$$

for all $t' > t$; in particular, the e_t are increasing in t. This implies that the expectations $\mathbf{E}e_t$ are also increasing in t. On the other hand, these expectations are bounded between 0 and 1. Thus, if we select $1 \leq t \leq M$ at random, expectation of

$$\mathbf{E}(e_{t+2} - e_t)$$

telescopes to be $O(1/M)$. Thus, by Markov's inequality, with probability $1 - O(\varepsilon)$ we can freeze $v_1, \ldots, v_t$ such that we have the conditional expectation bound

$$\mathbf{E}(e_{t+2} - e_t|v_1, \ldots, v_t) = O(\frac{1}{M\varepsilon}). \tag{2.8}$$

Suppose $v_1, \ldots, v_t$ has this property. We split

$$1_E = f_{U^\perp} + f_U,$$

where

$$f_{U^\perp} := \mathbf{E}(1_E|\mathcal{B}_t \times \mathcal{B}_t)$$

and

$$f_U := 1_E - \mathbf{E}(1_E|\mathcal{B}_t \times \mathcal{B}_t).$$

We now assert that the partition $V_1^t, \ldots, V_{2^t}^t$ induced by $\mathcal{B}_t$ obeys the conclusions of Lemma 2.2.3. For this, we observe various properties on the two components of 1_E:

Lemma 2.2.5 ($f_{U^\perp}$ is structured). *$f_{U^\perp}$ is constant on each product set $V_i^t \times V_j^t$.*

Proof. This is clear from construction. □

Lemma 2.2.6 (f_U is pseudo-random). *The expression*

$$\frac{1}{|V|^4} \sum_{v,w,v',w' \in V} f_U(v,w) f_U(v,w') f_U(v',w) f_U(v',w')$$

is of size $O(\frac{1}{\sqrt{M\varepsilon}})$.

Proof. The left-hand side can be rewritten as

$$\mathbf{E}\frac{1}{|V|^2} \sum_{v,w \in V} f_U(v,w) f_U(v, v_{t+2}) f_U(v_{t+1}, w) f_U(v_{t+1}, v_{t+2}).$$

Observe that the function $(v,w) \mapsto f_U(v, v_{t+2}) f_U(v_{t+1}, w) f_U(v,w)$ is measurable with respect to $\mathcal{B}_{t+2} \times \mathcal{B}_{t+2}$, so we can rewrite this expression as

$$\mathbf{E}\frac{1}{|V|^2}\sum_{v,w\in V}\mathbf{E}(f_U|\mathcal{B}_{t+2}\times\mathcal{B}_{t+2})(v,w) f_U(v,v_{t+2}) f_U(v_{t+1},w) f_U(v_{t+1},v_{t+2}).$$

Applying Cauchy-Schwarz, one can bound this by

$$\mathbf{E}\|\mathbf{E}(f_U|\mathcal{B}_{t+2}\times\mathcal{B}_{t+2})\|_{L^2(V\times V)}.$$

But from Pythagoras we have

$$\mathbf{E}(f_U|\mathcal{B}_{t+2}\times\mathcal{B}_{t+2})^2 = e_{t+2} - e_t,$$

and so the claim follows from (2.8) and another application of Cauchy-Schwarz. □

Now we can prove Lemma 2.2.4. Observe that

$$|E\cap(A\times B)| - \sum_{i=1}^{2^t}\sum_{j=1}^{2^t} d(V_i^t, V_j^t)|A\cap V_i^t||B\cap V_j^t|$$

$$= \sum_{v,w\in V} 1_A(v) 1_B(w) f_U(v,w).$$

Applying Cauchy-Schwarz twice in v, w and using Lemma 2.2.6, we see that the right-hand side is $O((M\varepsilon)^{-1/8})$; choosing $M \gg \varepsilon^{-9}$, we obtain the claim.

2.2.2. Strong regularity via random neighbourhoods. We now prove Lemma 2.2.3, which of course implies Lemma 2.2.2.

Fix $\varepsilon > 0$ and a graph $G = (V, E)$ on n vertices. We randomly select an infinite sequence $v_1, v_2 \cdots \in V$ of vertices in V, drawn uniformly and independently at random. We define $A_t, V_i^t, \mathcal{B}_t, e_t$, as before.

Now let m be a large number depending on $\varepsilon > 0$ to be chosen later, let $F : \mathbf{Z}^+ \to \mathbf{Z}^+$ be a rapidly growing function (also to be chosen later), and set $M_1 := F(1)$ and $M_r := 2(M_{r-1} + F(M_{r-1}))$ for all $1 \le r \le m$. Thus $M_1 < M_2 < \cdots < M_{m+1}$ grows rapidly to infinity. The expected energies $\mathbf{E}e_{M_r}$ are increasing from 0 to 1, thus if we pick $1 \le r \le m$ uniformly at random, the expectation of

$$\mathbf{E}e_{M_{r+1}} - e_{M_r}$$

telescopes to be $O(1/m)$. Thus, by Markov's inequality, with probability $1 - O(\varepsilon)$, we will have

$$\mathbf{E}e_{M_{r+1}} - e_{M_r} = O(\frac{1}{m\varepsilon}).$$

Assume that r is chosen to obey this. Then, by another application of the pigeonhole principle, we can find $M_{r+1}/2 \leq t < M_{r+1}$ such that

$$\mathbf{E}(e_{t+2} - e_t) = O(\frac{1}{m\varepsilon M_{r+1}}) = O(\frac{1}{m\varepsilon F(M_r)}).$$

Fix this t. We have

$$\mathbf{E}(e_t - e_{M_r}) = O(\frac{1}{m\varepsilon}),$$

so by Markov's inequality, with probability $1 - O(\varepsilon)$, $v_1, \dots, v_t$ are such that

$$e_t - e_{M_r} = O(\frac{1}{m\varepsilon^2}), \tag{2.9}$$

and also obey the conditional expectation bound

$$\mathbf{E}(e_{t+2} - e_t | v_1, \dots, v_t) = O(\frac{1}{m\varepsilon F(M_r)}). \tag{2.10}$$

Assume that this is the case. We split

$$1_E = f_{U^\perp} + f_{err} + f_U,$$

where

$$f_{U^\perp} := \mathbf{E}(1_E | \mathcal{B}_{M_r} \times \mathcal{B}_{M_r}),$$
$$f_{err} := \mathbf{E}(1_E | \mathcal{B}_t \times \mathcal{B}_t) - \mathbf{E}(1_E | \mathcal{B}_{M_r} \times \mathcal{B}_{M_r}),$$
$$f_U := 1_E - \mathbf{E}(1_E | \mathcal{B}_t \times \mathcal{B}_t).$$

We now assert that the partition $V_1^{M_r}, \dots, V_{2^{M_r}}^{M_r}$ induced by $\mathcal{B}_{M_r}$ obeys the conclusions of Lemma 2.2.2. For this, we observe various properties on the three components of 1_E:

Lemma 2.2.7 ($f_{U^\perp}$ locally constant). *$f_{U^\perp}$ is constant on each product set $V_i^{M_r} \times V_j^{M_r}$.*

Proof. This is clear from construction. □

Lemma 2.2.8 (f_{err} small). *We have $\|f_{err}\|^2_{L^2(V \times V)} = O(\frac{1}{m\varepsilon^2})$.*

Proof. This follows from (2.9) and Pythagoras' theorem. □

Lemma 2.2.9 (f_U uniform). *The expression*

$$\frac{1}{|V|^4} \sum_{v,w,v',w' \in V} f_U(v,w) f_U(v,w') f_U(v',w) f_U(v',w')$$

is of size $O(\frac{1}{\sqrt{m\varepsilon F(M_r)}})$.

Proof. This follows by repeating the proof of Lemma 2.2.6, but using (2.10) instead of (2.8). □

Now we verify the regularity.

First, we eliminate *small atoms*: the pairs (V_i, V_j) for which $|V_i^{M_r}| \leq \varepsilon|V|/2^{M_r}$ clearly give a net contribution of at most $O(\varepsilon|V|^2)$ and are acceptable; similarly for those pairs for which $|V_j^{M_r}| \leq \varepsilon|V|/2^{M_r}$. So we may henceforth assume that

$$|V_i^{M_r}|, |V_j^{M_r}| \leq \varepsilon|V|/2^{M_r}. \tag{2.11}$$

Now, let $A \subset V_i^{M_r}$, $B \subset V_i^{M_r}$ have densities

$$\alpha := |A|/|V_i^{M_r}| \geq \varepsilon, \quad \beta := |B|/|V_j^{M_r}| \geq \varepsilon,$$

then

$$\alpha\beta d(A, B) = \frac{1}{|V_i^{M_r}||V_j^{M_r}|} \sum_{v \in V_i^{M_r}} \sum_{w \in V_i^{M_r}} 1_A(v)1_B(w)1_E(v, w).$$

We divide 1_E into the three pieces $f_{U^\perp}$, f_{err}, f_U.

The contribution of $f_{U^\perp}$ is exactly $\alpha\beta d(V_i^{M_r}, V_j^{M_r})$.

The contribution of f_{err} can be bounded using Cauchy-Schwarz as

$$O(\frac{1}{|V_i^{M_r}||V_j^{M_r}|} \sum_{v \in V_i^{M_r}} \sum_{w \in V_i^{M_r}} |f_{err}(v, w)|^2)^{1/2}.$$

Using Lemma 2.2.8 and Chebyshev's inequality, we see that the pairs (V_i, V_j) for which this quantity exceeds ε^3 will contribute at most ε^{-8}/m to (2.6), which is acceptable if we choose m so that $m \gg \varepsilon^{-9}$. Let us now discard these bad pairs.

Finally, the contribution of f_U can be bounded by two applications of Cauchy-Schwarz and (2.2.9) as

$$O(\frac{|V|^2}{|V_i^{M_r}||V_j^{M_r}|} \frac{1}{(m\varepsilon F(M_r))^{1/8}}),$$

which by (2.11) is bounded by

$$O(2^{2M_r}\varepsilon^{-2}/(m\varepsilon F(M_r))^{1/8}).$$

This can be made $O(\varepsilon^3)$ by selecting F sufficiently rapidly growing depending on ε. Putting this all together, we see that

$$\alpha\beta d(A, B) = \alpha\beta d(V_i^{M_r}, V_j^{M_r}) + O(\varepsilon^3),$$

which (since $\alpha, \beta \geq \varepsilon$) gives the desired regularity.

Remark 2.2.10. Of course, this argument gives tower-exponential bounds (as F is exponential and needs to be iterated m times), which will be familiar to any reader already acquainted with the regularity lemma.

Remark 2.2.11. One can take the partition induced by random neighbourhoods here and carve it up further to be both equitable and (mostly) regular, thus recovering a proof of Lemma 1, by following the arguments in [**Ta2006c**]. Of course, when one does so, one no longer has a partition created purely from random neighbourhoods, but it is pretty clear that one is not going to be able to make an equitable partition just from boolean operations applied to a few random neighbourhoods.

Notes. This article first appeared at

`terrytao.wordpress.com/2009/04/26.`

Thanks to Anup for corrections.

Asaf Shapira noted that in [**FiMaSh2007**] a similar (though not identical) regularisation algorithm was given which explicitly regularises a graph or hypergraph in linear time.

2.3. Szemerédi's regularity lemma via the correspondence principle

In the previous section, we discussed the *Szemerédi regularity lemma*, and how a given graph could be regularised by partitioning the vertex set into random neighbourhoods. More precisely, we gave a proof of

Lemma 2.3.1 (Regularity lemma via random neighbourhoods). *Let $\varepsilon > 0$. Then there exists integers $M_1, \ldots, M_m$ with the following property: whenever $G = (V, E)$ is a graph on finitely many vertices, if one selects one of the integers M_r at random from $M_1, \ldots, M_m$ and then selects M_r vertices $v_1, \ldots, v_{M_r} \in V$ uniformly from V at random, then the 2^{M_r} vertex cells $V_1^{M_r}, \ldots, V_{2^{M_r}}^{M_r}$ (some of which can be empty) generated by the vertex neighbourhoods $A_t := \{v \in V : (v, v_t) \in E\}$ for $1 \leq t \leq M_r$ will obey the regularity property*

$$\sum_{(V_i,V_j) \text{ not } \varepsilon\text{-regular}} |V_i||V_j| \leq \varepsilon |V|^2 \tag{2.12}$$

*with probability at least $1 - O(\varepsilon)$, where the sum is over all pairs $1 \leq i \leq j \leq k$ for which G is not ε-regular between V_i and V_j. (Recall that a pair (V_i, V_j) is ε-*regular *for G if one has*

$$|d(A, B) - d(V_i, V_j)| \leq \varepsilon$$

for any $A \subset V_i$ and $B \subset V_j$ with $|A| \geq \varepsilon|V_i|, |B| \geq \varepsilon|V_j|$, where $d(A, B) := |E \cap (A \times B)|/|A||B|$ is the density of edges between A and B.)

The proof was a combinatorial one, based on the standard energy increment argument.

In this article I would like to discuss an alternate approach to the regularity lemma, which is an infinitary approach passing through a graph-theoretic version of the Furstenberg correspondence principle. While this approach superficially looks quite different from the combinatorial approach, it in fact uses many of the same ingredients, most notably a reliance on random neighbourhoods to regularise the graph. This approach was introduced in [**Ta2007**] and used in [**Au2008, AuTa2010**] to establish some property testing results for hypergraphs. More recently, a closely related infinitary hypergraph removal lemma developed in [**Ta2007**] was also used in [**Au2009, Au2009b**] to give new proofs of the multidimensional Szemerédi theorem and of the density Hales-Jewett theorem (the latter being a spinoff of the Polymath1 project, see Section 2.1).

For various technical reasons we will not be able to use the correspondence principle to recover Lemma 2.3.1 in its full strength. Instead, we will establish the following slightly weaker variant.

Lemma 2.3.2 (Regularity lemma via random neighbourhoods, weak version). *Let $\varepsilon > 0$. Then there exists an integer M_* with the following property: whenever $G = (V, E)$ is a graph on finitely many vertices, there exists $1 \leq M \leq M_*$ such that if one selects M vertices $v_1, \ldots, v_M \in V$ uniformly from V at random, then the 2^M vertex cells $V_1^M, \ldots, V_{2^M}^M$ generated by the vertex neighbourhoods $A_t := \{v \in V : (v, v_t) \in E\}$ for $1 \leq t \leq M$ will obey the regularity property* (2.12) *with probability at least $1 - \varepsilon$.*

Roughly speaking, Lemma 2.3.1 asserts that one can regularise a large graph G with high probability by using M_r random neighbourhoods, where M_r is chosen at random from one of a number of choices $M_1, \ldots, M_m$; in contrast, the weaker Lemma 2.3.2 asserts that one can regularise a large graph G with high probability by using *some* integer M from $1, \ldots, M_*$, but the exact choice of M depends on G, and it is not guaranteed that a randomly chosen M will be likely to work. While Lemma 2.3.2 is strictly weaker than Lemma 2.3.1, it still implies the (weighted) Szemerédi regularity lemma (Lemma 2.2.2).

2.3.1. The graph correspondence principle. The first key tool in this argument is the *graph correspondence principle*, which takes a sequence of (increasingly large) graphs and uses random sampling to extract an infinitary limit object, which will turn out to be an infinite but random (and, crucially, *exchangeable*) graph. This concept of a graph limit is related to (though slightly different from) the *graphons* used as graph limits in [**LoSz2007**] or the ultraproducts used in [**ElSz2008**]. It also seems to be related to the concept of an elementary limit that I discussed in Section 1.4, though this connection is still rather tentative.

The correspondence works as follows. We start with a finite, deterministic graph $G = (V, E)$. We can then form an infinite, random graph $\hat{G} = (\mathbf{Z}, \hat{E})$ from this graph by the following recipe:

- The vertex set of $\hat{G}$ will be the integers $\mathbf{Z} = \{-2, -1, 0, 1, 2, \ldots\}$.
- For every integer n, we randomly select a vertex v_n in V, uniformly and independently at random. (Note that there will be many collisions, i.e., integers n, m for which $v_n = v_m$, but these collisions will become asymptotically negligible in the limit $|V| \to \infty$.)
- We then define the edge set $\hat{E}$ of $\hat{G}$ by declaring (n, m) to be an edge on $\hat{E}$ if and only if (v_n, v_m) is an edge in E (which in particular requires $v_n \neq v_m$).

More succinctly, $\hat{G}$ is the pullback of G under a random map from $\mathbf{Z}$ to V.

The random graph $\hat{G}$ captures all the "local" information of G, while obscuring all the "global" information. For instance, the edge density of G is essentially just the probability that a given edge, say $(1, 2)$, lies in $\hat{G}$. (There is a small error term due to the presence of collisions, but this goes away in the limit $|V| \to \infty$.) Similarly, the triangle density of G is essentially the probability that a given triangle, say $\{(1, 2), (2, 3), (3, 1)\}$, lies in $\hat{G}$. On the other hand, it is difficult to read off global properties of G, such as being connected or 4-colourable, just from $\hat{G}$.

At first glance, it may seem a poor bargain to trade in a finite deterministic graph G for an infinite random graph $\hat{G}$, which is a more complicated and less elementary object. However, there are three major advantages of working with $\hat{G}$ rather than G:

- **Exchangeability**. The probability distribution of $\hat{G}$ has a powerful symmetry or *exchangeability* property: if one takes the random graph $\hat{G}$ and interchanges any two vertices in $\mathbf{Z}$, e.g., 3 and 5, one obtains a new graph which is not equal to $\hat{G}$, but nevertheless has the same probability distribution as $\hat{G}$, basically because the v_n were selected in an iid (independent and identically distributed) manner. More generally, given any permutation $\sigma : \mathbf{Z} \to \mathbf{Z}$, the pullback $\sigma^*(\hat{G})$ of $\hat{G}$ by σ has the same probability distribution as $\hat{G}$; thus we have a measure-preserving action of the symmetric group S_∞, which places us in the general framework of ergodic theory.
- **Limits**. The space of probability measures on the space $2^{\binom{\mathbf{Z}}{2}}$ of infinite graphs is sequentially compact; given any sequence $\hat{G}_n = (\mathbf{Z}, \hat{E}_n)$ of infinite random graphs, one can find a subsequence $\hat{G}_{n_j}$

which converges in the *vague topology* to another infinite random graph. What this means is that given any event E on infinite graphs that involve only finitely many of the edges, the probability that $\hat{G}_{n_j}$ obeys E converges to the probability that $\hat{G}$ obeys E. (Thus, for instance, the probability that $\hat{G}_{n_j}$ contains the triangle $\{(1,2),(2,3),(3,1)\}$ will converge to the probability that $\hat{G}$ contains the same triangle.) Note that properties that involve infinitely many edges (e.g., connectedness) need not be preserved under vague limits.

- **Factors**. The underlying probability space for the random variable $\hat{G}$ is the space $2^{\binom{\mathbf{Z}}{2}}$ of infinite graphs, and it is natural to give this space the Borel σ-algebra $\mathcal{B}_{\mathbf{Z}}$, which is the σ-algebra generated by the *cylinder events* "$(i,j) \in \hat{G}$" for $i,j \in \mathbf{Z}$. But this σ-algebra also has a number of useful sub-σ-algebras or *factors*, representing various partial information on the graph $\hat{G}$. In particular, given any subset I of $\mathbf{Z}$, one can create the factor $\mathcal{B}_I$, defined as the σ-algebra generated by the events "$(i,j) \in \hat{G}$" for $i,j \in I$. Thus for instance, the event that $\hat{G}$ contains the triangle is measurable in $\mathcal{B}_{\{1,2,3\}}$, but not in $\mathcal{B}_{\{1,2\}}$. One can also look at compound factors such as $\mathcal{B}_I \wedge \mathcal{B}_J$, the factor generated by the union of $\mathcal{B}_I$ and $\mathcal{B}_J$. For instance, the event that $\hat{G}$ contains the edges $(1,2),(1,3)$ is measurable in $\mathcal{B}_{\{1,2\}} \vee \mathcal{B}_{\{1,3\}}$, but the event that $\hat{G}$ contains the triangle $\{(1,2),(2,3),(3,1)\}$ is not.

The connection between the infinite random graph $\hat{G}$ and partitioning through random neighbourhoods comes when contemplating the relative difference between a factor such as $\mathcal{B}_{\{-n,\dots,-1\}}$ and $\mathcal{B}_{\{-n,\dots,-1\}\cup\{1\}}$ (say). The latter factor is generated by the former factor, together with the events "$(1,-i) \in \hat{E}$" for $i = 1,\dots,n$. But observe if $\hat{G} = (\mathbf{Z},\hat{E})$ is generated from a finite deterministic graph $G = (V,E)$, then $(1,-i)$ lies in $\hat{E}$ if and only if v_1 lies in the vertex neighbourhood of v_{-i}. Thus, if one uses the vertex neighbourhoods of $v_{-1},\dots,v_{-n}$ to subdivide the original vertex set V into 2^n cells of varying sizes, the factor $\mathcal{B}_{\{-n,\dots,-1\}\cup\{1\}}$ is generated from $\mathcal{B}_{\{-n,\dots,-1\}}$, together with the random variable that computes which of these 2^n cells the random vertex v_1 falls into. We will see this connection in more detail later in this article when we use the correspondence principle to prove Lemma 2.3.2.

Combining the exchangeability and limit properties (and noting that the vague limit of exchangeable random graphs is still exchangeable), we obtain

Lemma 2.3.3 (Graph correspondence principle). *Let $G_n = (V_n, E_n)$ be a sequence of finite deterministic graphs, and let $\hat{G}_n = (\mathbf{Z}, \hat{E}_n)$ be their infinite*

random counterparts. Then there exists a subsequence n_j such that $\hat{G}_{n_j}$ converges in the vague topology to an exchangeable infinite random graph $\hat{G} = (\mathbf{Z}, \hat{E})$.

We can illustrate this principle with three main examples, two from opposing extremes of the "dichotomy between structure and randomness", and one intermediate one.

Example 2.3.4 (Random example)**.** Let $G_n = (V_n, E_n)$ be a sequence of ε_n-regular graphs of edge density p_n, where $|V_n| \to \infty$, $\varepsilon_n \to 0$, and $p_n \to p$ as $n \to \infty$. Then any graph limit $\hat{G} = (\mathbf{Z}, \hat{G})$ of this sequence will be an *Erdős-Rényi graph* $\hat{G} = G(\infty, p)$, where each edge (i, j) lies in $\hat{G}$ with an independent probability of p.

Example 2.3.5 (Structured example)**.** Let $G_n = (V_n, E_n)$ be a sequence of complete bipartite graphs, where the two cells of the bipartite graph have vertex density q_n and $1 - q_n$, respectively, with $|V_n| \to \infty$ and $q_n \to q$. Then any graph limit $\hat{G} = (\mathbf{Z}, \hat{E})$ of this sequence will be a random complete bipartite graph, constructed as follows: first, randomly colour each vertex n of $\mathbf{Z}$ red with probability q and blue with probability $1 - q$, independently for each vertex. Then define $\hat{G}$ to be the complete bipartite graph between the red vertices and the blue vertices.

Example 2.3.6 (Random+structured example)**.** Let $G_n = (V_n, E_n)$ be a sequence of *incomplete* bipartite graphs, where the two cells of the bipartite graph have vertex density p_n and $1 - p_n$, respectively, and the graph G_n is ε_n-regular between these two cells with edge density p_n, with $|V_n| \to \infty$, $\varepsilon_n \to 0$, $p_n \to p$, and $q_n \to q$. Then any graph limit $\hat{G} = (\mathbf{Z}, \hat{E})$ of this sequence will be a random bipartite graph constructed as follows: First, randomly colour each vertex n of $\mathbf{Z}$ red with probability q and blue with probability $1 - q$, independently for each vertex. Then define $\hat{G}$ to be the bipartite graph between the red vertices and the blue vertices, with each edge between red and blue having an independent probability of p of lying in $\hat{E}$.

One can use the graph correspondence principle to prove statements about finite deterministic graphs, by the usual *compactness and contradiction approach*: argue by contradiction, create a sequence of finite deterministic graph counterexamples, use the correspondence principle to pass to an infinite random exchangeable limit, and obtain the desired contradiction in the infinitary setting. This will be how we shall approach the proof of Lemma 2.3.2.

2.3.2. The infinitary regularity lemma. To prove the finitary regularity lemma via the correspondence principle, one must first develop an infinitary counterpart. We will present this infinitary regularity lemma (first introduced in this paper) shortly, but let us motivate it by a discussion based on the three model examples of infinite exchangeable graphs $\hat{G} = (\mathbf{Z}, \hat{E})$ from the previous section.

First, consider the "random" graph $\hat{G}$ from Example 2.3.4. Here, we observe that the events "$(i,j) \in \hat{E}$" are jointly independent of each other; thus, for instance

$$\mathbf{P}((1,2),(2,3),(3,1) \in \hat{E}) = \prod_{(i,j)=(1,2),(2,3),(3,1)} \mathbf{P}((i,j) \in \hat{E}).$$

More generally, we see that the factors $\mathcal{B}_{\{i,j\}}$ for all distinct $i,j \in \mathbf{Z}$ are independent, which means that

$$\mathbf{P}(E_1 \wedge \cdots \wedge E_n) = \mathbf{P}(E_1) \cdots \mathbf{P}(E_n)$$

whenever $E_1 \in \mathcal{B}_{\{i_1,j_1\}}, \ldots, E_n \in \mathcal{B}_{\{i_n,j_n\}}$ and the $\{i_1,j_1\},\ldots,\{i_n,j_n\}$ are distinct.

Next, we consider the "structured" graph $\hat{G}$ from Example 2.3.5, where we take $0 < p < 1$ to avoid degeneracies. In contrast to the preceding example, the events "$(i,j) \in \hat{E}$" are now highly dependent. For instance, if $(1,2) \in \hat{E}$ and $(1,3) \in \hat{E}$, then this forces $(2,3)$ to lie outside of $\hat{E}$, despite the fact that the events "$(i,j) \in \hat{E}$" each occur with a nonzero probability of $p(1-p)$. In particular, the factors $\mathcal{B}_{\{1,2\}}, \mathcal{B}_{\{1,3\}}, \mathcal{B}_{\{2,3\}}$ are not jointly independent.

However, one can recover a *conditional* independence by introducing some new factors. Specifically, let $\mathcal{B}_i$ be the factor generated by the event that the vertex i is coloured red. Then we see that the factors $\mathcal{B}_{\{1,2\}}, \mathcal{B}_{\{1,3\}}, \mathcal{B}_{\{2,3\}}$ now become *conditionally* jointly independent, relative to the base factor $\mathcal{B}_1 \vee \mathcal{B}_2 \vee \mathcal{B}_3$, which means that we have conditional independence identities such as

$$\begin{aligned}\mathbf{P}((1,2),(2,3),(3,1) \in \hat{E} | \mathcal{B}_1 \vee \mathcal{B}_2 \vee \mathcal{B}_3) \\ = \prod_{(i,j)=(1,2),(2,3),(3,1)} \mathbf{P}((i,j) \in \hat{E} | \mathcal{B}_1 \vee \mathcal{B}_2 \vee \mathcal{B}_3).\end{aligned}$$

Indeed, once one fixes (conditions) the information in $\mathcal{B}_1 \vee \mathcal{B}_2 \vee \mathcal{B}_3$ (i.e., once one knows what colour the vertices $1,2,3$ are), the events "$(i,j) \in \hat{E}$" for $(i,j) = (1,2),(2,3),(3,1)$ either occur with probability 1 (if i,j have distinct colours) or probability 0 (if i,j have the same colour), and so the conditional independence is trivially true.

A similar phenomenon holds for the "random+structured" graph $\hat{G}$ from Example 2.3.6, with $0 < p, q < 1$. Again, the factors $\mathcal{B}_{\{i,j\}}$ are not jointly independent in an absolute sense, but once one introduces the factors $\mathcal{B}_i$ based on the colour of the vertex i, we see once again that the $\mathcal{B}_{\{i,j\}}$ become conditionally jointly independent relative to the $\mathcal{B}_i$.

These examples suggest, more generally, that we should be able to *regularise* the graph $\hat{G}$ (or more precisely, the system of edge factors $\mathcal{B}_{\{i,j\}}$) by introducing some single-vertex factors $\mathcal{B}_i$, with respect to which the edge factors become conditionally independent; this is the infinitary analogue of a finite graph becoming ε-regular relative to a suitably chosen partition of the vertex set into cells.

Now, in Examples 2.3.5 and 2.3.6 we were able to obtain this regularisation because the vertices of the graph were conveniently coloured for us (red or blue). But for general infinite exchangeable graphs $\hat{G}$, such a vertex colouring is not provided to us, so how is one to generate the vertex factors $\mathcal{B}_i$?

The key trick—which is the infinitary analogue of using random neighbourhoods to regularise a finitary graph—is to sequester half of the infinite vertices in $\mathbf{Z}$ (e.g., the negative vertices $-1, -2, \ldots$) away as "reference" or "training" vertices, and then colorise the remaining vertices i of the graph based on how that vertex interacts with the reference vertices. More formally, we define $\mathcal{B}_i$ for $i = 0, 1, 2, \ldots$ by the formula

$$\mathcal{B}_i := \mathcal{B}_{\{-1,-2,\ldots\}\cup\{i\}}.$$

We then have

Lemma 2.3.7 (Infinitary regularity lemma)**.** *Let $\hat{G} = (\mathbf{Z}, \hat{E})$ be a infinite exchangeable random graph. Then the $\mathcal{B}_{\{i,j\}} \vee \mathcal{B}_i \vee \mathcal{B}_j$ for natural numbers i, j are conditinally jointly independent relative to the $\mathcal{B}_i$. More precisely, if I is a set of natural numbers, E is a subset of $\binom{I}{2}$, and E_e is a $\mathcal{B}_e \wedge \bigwedge_{i\in e} \mathcal{B}_i$-measurable event for all $e \in E$, then*

$$\mathbf{P}\Big(\bigwedge_{e\in E} E_e \Big| \bigwedge_{i\in I} \mathcal{B}_i\Big) = \prod_{e\in E} \mathbf{P}\Big(E_e \Big| \bigwedge_{i\in I} \mathcal{B}_i\Big).$$

Proof. By induction on E, it suffices to show that for any $e_0 \in E$, the event E_{e_0} and the event $\bigwedge_{e\in E\setminus\{e_0\}} E_e$ are independent relative to $\bigwedge_{i\in I} \mathcal{B}_i$.

By relabeling, we may take $I = \{1, \ldots, n\}$ and $e_0 = \{1, 2\}$ for some $n \geq 2$. We use the exchangeability of $\hat{G}$ (and *Hilbert's hotel trick*) to observe that the random variables

$$\mathbf{E}(1_{E_{e_0}} | \mathcal{B}_{\{-1,-2,\ldots\}\cup\{1\}} \vee \mathcal{B}_{\{-1,-2,\ldots\}\cup\{2\}})$$

and

$$\mathbf{E}(1_{E_{e_0}}|\mathcal{B}_{\{-1,-2,\ldots\}\cup\{1\}\cup\{3,\ldots,n\}} \vee \mathcal{B}_{\{-1,-2,\ldots\}\cup\{2\}\cup\{3,\ldots,n\}})$$

have the same distribution; in particular, they have the same L^2-norm. By Pythagoras' theorem, they must therefore be equal almost surely; furthermore, for any intermediate σ-algebra $\mathcal{B}$ between $\mathcal{B}_{\{-1,-2,\ldots\}\cup\{1\}} \vee \mathcal{B}_{\{-1,-2,\ldots\}\cup\{2\}}$ and $\mathcal{B}_{\{-1,-2,\ldots\}\cup\{1\}\cup\{3,\ldots,n\}} \vee \mathcal{B}_{\{-1,-2,\ldots\}\cup\{2\}\cup\{3,\ldots,n\}}$, $\mathbf{E}(1_{E_{e_0}}|\mathcal{B})$ is also equal almost surely to the above two expressions. (The astute reader will observe that we have just run the "energy increment argument"; in the infinitary world, it is somewhat slicker than in the finitary world, due to the convenience of Hilbert's hotel trick, and the fact that the existence of orthogonal projections (and in particular, conditional expectation) is itself encoding an energy increment argument.)

As a special case of the above observation, we see that

$$\mathbf{E}(1_{E_{e_0}}|\bigwedge_{i\in I}\mathcal{B}_i) = \mathbf{E}(1_{E_{e_0}}|\bigwedge_{i\in I}\mathcal{B}_i \wedge \bigwedge_{e\in E\backslash\{e_0\}} \mathcal{B}_e).$$

In particular, this implies that E_0 is conditionally independent of every event measurable in $\bigwedge_{i\in I}\mathcal{B}_i \wedge \bigwedge_{e\in E\backslash\{e_0\}}\mathcal{B}_e$, relative to $\bigwedge_{i\in I}\mathcal{B}_i$, and the claim follows. □

Remark 2.3.8. The same argument also allows one to easily regularise infinite exchangeable hypergraphs; see [**Ta2007**]. In fact one can go further and obtain a structural theorem for these hypergraphs generalising *de Finetti's theorem*, and also closely related to the graphons of Lovász and Szegedy; see [**Au2008**] for details.

2.3.3. Proof of finitary regularity lemma.

Having proven the infinitary regularity lemma, we now use the correspondence principle and the compactness and contradiction argument to recover the finitary regularity lemma, Lemma 2.3.2.

Suppose this lemma failed. Carefully negating all the quantifiers, this means that there exists $\varepsilon > 0$, a sequence M_n going to infinity, and a sequence of finite deterministic graphs $G_n = (V_n, E_n)$ such that for every $1 \leq M \leq M_n$, if one selects vertices $v_1, \ldots, v_M \in V_n$ uniformly from V_n, then the 2^M vertex cells $V_1^M, \ldots, V_{2^M}^M$ generated by the vertex neighbourhoods $A_t := \{v \in V : (v, v_t) \in E\}$ for $1 \leq t \leq M$, will obey the regularity property (2.12) with probability less than $1 - \varepsilon$.

We convert each of the finite deterministic graphs $G_n = (V_n, E_n)$ to an infinite random exchangeable graph $\hat{G}_n = (\mathbf{Z}, \hat{E}_n)$; invoking the correspondence principle and passing to a subsequence if necessary, we can assume that this graph converges in the vague topology to an exchangeable limit $\hat{G} = (\mathbf{Z}, \hat{E})$. Applying the infinitary regularity lemma to this graph, we see that the edge factors $\mathcal{B}_{\{i,j\}} \wedge \mathcal{B}_i \wedge \mathcal{B}_j$ for natural numbers i,j are conditionally jointly independent relative to the vertex factors $\mathcal{B}_i$.

Now for any distinct natural numbers i,j, let $f(i,j)$ be the indicator of the event "(i,j) lies in $\hat{E}$", thus $f=1$ when (i,j) lies in $\hat{E}$ and $f(i,j)=0$ otherwise. Clearly, $f(i,j)$ is $\mathcal{B}_{\{i,j\}}$-measurable. We can write

$$f(i,j) = f_{U^\perp}(i,j) + f_U(i,j),$$

where

$$f_{U^\perp}(i,j) := \mathbf{E}(f(i,j)|\mathcal{B}_i \wedge \mathcal{B}_j)$$

and

$$f_U(i,j) := f(i,j) - f_{U^\perp}(i,j).$$

The exchangeability of $\hat{G}$ ensures that $f, f_U, f_{U^\perp}$ are exchangeable with respect to permutations of the natural numbers, in particular $f_U(i,j) = f_U(j,i)$ and $f_{U^\perp}(i,j) = f_{U^\perp}(j,i)$.

By the infinitary regularity lemma, the $f_U(i,j)$ are jointly independent relative to the $\mathcal{B}_i$, and also have mean zero relative to these factors, so in particular they are infinitely pseudo-random in the sense that

$$\mathbf{E} f_U(1,2) f_U(3,2) f_U(1,4) f_U(3,4) = 0.$$

Meanwhile, the random variable $f_{U^\perp}(1,2)$ is measurable with respect to the factor $\mathcal{B}_1 \vee \mathcal{B}_2$, which is the limit of the factors $\mathcal{B}_{\{-1,-2,\ldots,-M\}\cup\{1\}} \vee \mathcal{B}_{\{-1,-2,\ldots,-M\}\cup\{2\}}$ as M increases. Thus, given any $\tilde{\varepsilon} > 0$ (to be chosen later), one can find an approximation $\tilde{f}_{U^\perp}(1,2)$ to $f_{U^\perp}(1,2)$, bounded between 0 and 1, which is $\mathcal{B}_{\{-1,-2,\ldots,-M\}\cup\{1\}} \vee \mathcal{B}_{\{-1,-2,\ldots,-M\}\cup\{2\}}$-measurable for some M, and such that

$$\mathbf{E}|\tilde{f}_{U^\perp}(1,2) - f_{U^\perp}(1,2)| \leq \tilde{\varepsilon}.$$

We can also impose the symmetry condition $\tilde{f}_{U^\perp}(1,2) = \tilde{f}_{U^\perp}(2,1)$. Now let $\tilde{\varepsilon}' > 0$ be an extremely small number (depending on $\tilde{\varepsilon}, n$) to be chosen later. Then one can find an approximation $\tilde{f}_U(1,2)$ to $f_U(1,2)$, bounded between -1 and 1, which is $\mathcal{B}_{\{-1,-2,\ldots,-M'\}\cup\{1\}} \vee \mathcal{B}_{\{-1,-2,\ldots,-M'\}\cup\{2\}}$-measurable for some M', and such that

$$\mathbf{E}|\tilde{f}_U(1,2) - f_U(1,2)| \leq \tilde{\varepsilon}'.$$

Again we can impose the symmetry condition $\tilde{f}_U(1,2) = \tilde{f}_U(2,1)$. We can then extend $\tilde{f}_U$ by exchangeability, so that

$$\mathbf{E}|\tilde{f}_U(i,j) - f_U(i,j)| \leq \tilde{\varepsilon}'$$

for all distinct natural numbers i,j. By the triangle inequality we then have

$$\mathbf{E}\tilde{f}_U(1,2)\tilde{f}_U(3,2)\tilde{f}_U(1,4)\tilde{f}_U(3,4) = O(\tilde{\varepsilon}'), \tag{2.13}$$

and by a separate application of the triangle inequality we have

$$\mathbf{E}|f(i,j) - \tilde{f}_{U^\perp}(i,j) - \tilde{f}_U(i,j)| = O(\tilde{\varepsilon}). \tag{2.14}$$

The bounds (2.13) and (2.14) apply to the limiting infinite random graph $\hat{G} = (\mathbf{Z}, \hat{E})$. On the other hand, all the random variables appearing in (2.13) and (2.14) involve at most finitely many of the edges of the graph. Thus, by vague convergence, the bounds (2.13) and (2.14) also apply to the graph $\hat{G}_n = (\mathbf{Z}, \hat{E}_n)$ for sufficiently large n.

Now we unwind the definitions to move back to the finite graphs $G_n = (V_n, E_n)$. Observe that, when applied to the graph $\hat{G}_n$, one has

$$\tilde{f}_{U^\perp}(1,2) = F_{U^\perp,n}(v_1, v_2),$$

where $F_{U,n} : V_n \times V_n \to [0,1]$ is a symmetric function which is constant on the pairs of cells $V_1^M, \ldots, V_{2M}^M$ generated the vertex neighbourhoods of $v_{-1}, \ldots, v_{-M}$. Similarly,

$$\tilde{f}_U(1,2) = F_{U,n}(v_1, v_2)$$

for some symmetric function $F_{U,n} : V_n \times V_n \to [-1,1]$. The estimate (2.13) can then be converted to a uniformity estimate on $F_{U,n}$

$$\mathbf{E}F_{U,n}(v_1,v_2)F_{U,n}(v_3,v_2)F_{U,n}(v_1,v_4)F_{U,n}(v_3,v_4) = O(\tilde{\varepsilon}')$$

while the estimate (2.14) can be similarly converted to

$$\mathbf{E}|1_{E_n}(v_1,v_2) - F_{U^\perp,n}(v_1,v_2) - F_{U,n}(v_1,v_2)| = O(\tilde{\varepsilon}).$$

If one then repeats the arguments in the preceding section, we conclude (if $\tilde{\varepsilon}$ is sufficiently small depending on ε, and $\tilde{\varepsilon}'$ is sufficiently small depending on ε, $\tilde{eps}$, M) that for $1-\varepsilon$ of the choices for $v_{-1}, \ldots, v_{-M}$, the partition $V_1^M, \ldots, V_{2M}^M$ induced by the corresponding vertex neighbourhoods will obey (2.12). But this contradicts the construction of the G_n, and the claim follows.

Notes. This article first appeared at

`terrytao.wordpress.com/2009/05/08.`

2.4. The two-ends reduction for the Kakeya maximal conjecture

In this article I would like to make some technical notes on a standard reduction used in the (Euclidean, maximal) Kakeya problem, known as the *two-ends reduction.* This reduction (which takes advantage of the approximate scale invariance of the Kakeya problem) was introduced by Wolff [**Wo1995**], and has since been used many times, both for the Kakeya problem and in other similar problems (e.g., in [**TaWr2003**] to study curved Radon-like transforms). I was asked about it recently, so I thought I would describe the trick here. As an application I give a proof of the $d = \frac{n+1}{2}$ case of the Kakeya maximal conjecture.

The *Kakeya maximal function conjecture* in $\mathbf{R}^n$ can be formulated as follows:

Conjecture 2.4.1 (Kakeya maximal function conjecture). *If $0 < \delta < 1$, $1 \leq d \leq n$, and $T_1, \ldots, T_N$ is a collection of $\delta \times 1$ tubes oriented in a δ-separated set of directions, then*

$$\| \sum_{i=1}^N 1_{T_i} \|_{L^{d/(d-1)}(\mathbf{R}^n)} \ll_\varepsilon (\frac{1}{\delta})^{\frac{n}{d}-1+\varepsilon} \tag{2.15}$$

for any $\varepsilon > 0$.

A standard duality argument shows that (2.15) is equivalent to the estimate

$$\sum_{i=1}^N \int_{T_i} F \ll_\varepsilon (\frac{1}{\delta})^{\frac{n}{d}-1+\varepsilon} \|F\|_{L^d(\mathbf{R}^n)}$$

for arbitrary nonnegative measurable functions F. Breaking F up into level sets via dyadic decomposition, this estimate is in turn equivalent to the estimate

$$\sum_{i=1}^N |E \cap T_i| \ll_\varepsilon (\frac{1}{\delta})^{\frac{n}{d}-1+\varepsilon} |E|^{1/d} \tag{2.16}$$

for arbitrary measurable sets E. This estimate is then equivalent to the following:

Conjecture 2.4.2 (Kakeya maximal function conjecture, second version). *If $0 < \delta, \lambda < 1$, $1 \leq d \leq n$, $T_1, \ldots, T_N$ is a collection of $\delta \times 1$ tubes oriented in a δ-separated set of directions, and E is a measurable set such that $|E \cap T_i| \geq \lambda |T_i|$ for all i, then*

$$|E| \gg_\varepsilon (N\delta^{n-1}) \lambda^d \delta^{n-d+\varepsilon}$$

for all $\varepsilon > 0$.

Indeed, to deduce (2.16) from Conjecture 2.4.2, one can perform another dyadic decomposition, this time based on the dyadic range of the densities $|E \cap T_i|/|T_i|$. Conversely, (2.16) implies Conjecture 2.4.2 in the case $N\delta^{n-1} \sim 1$, and the remaining case $N\delta^{n-1} \ll 1$ can then be deduced by the *random rotations trick* (see, e.g., [**ElObTa2009**]).

We can reformulate the conjecture again slightly:

Conjecture 2.4.3 (Kakeya maximal function conjecture, third version). *Let $0 < \delta, \lambda < 1$, $1 \leq d \leq n$, and $T_1, \ldots, T_N$ is a collection of $\delta \times 1$ tubes oriented in a δ-separated set of directions with $N \sim \delta^{1-n}$. For each $1 \leq i \leq N$, let $E_i \subset T_i$ be a set with $|E_i| \geq \lambda |T_i|$. Then*

$$|\bigcup_{i=1}^{N} E_i| \gg_\varepsilon \lambda^d \delta^{n-d+\varepsilon}$$

for all $\varepsilon > 0$.

We remark that (the Minkowski dimension version of) the Kakeya set conjecture essentially corresponds to the $\lambda = 1$ case of Conjecture 2.4.3, while the Hausdorff dimension can be shown to be implied by the case where $\lambda \gg \frac{1}{\log^2 1/\delta}$ (actually any lower bound here which is dyadically summable in δ would suffice). Thus, while the Kakeya set conjecture is concerned with how small one can make unions of tubes T_i, the Kakeya maximal function conjecture is concerned with how small one can make unions of *portions* E_i of tubes T_i, where the density λ of the tubes are fixed.

A key technical problem in the Euclidean setting (which is not present in the finite field case) is that the portions E_i of T_i may be concentrated in only a small portion of the tube, e.g., they could fill up a $\delta \times \lambda$ subtube, rather than being dispersed uniformly throughout the tube. Because of this, the set $\bigcup_{i=1}^N E_i$ could be crammed into a far tighter space than one would ideally like. Fortunately, the *two-ends reduction* allows one to eliminate this possibility, letting one only consider portions E_i which are not concentrated on just one end of the tube or another, but occupy both ends of the tube in some sense. A more precise version of this is as follows.

Definition 2.4.4 (Two-ends condition). Let E be a subset of $\mathbf{R}^n$, and let $\varepsilon > 0$. We say that E obeys the *two-ends condition* with exponent ε if one has the bound

$$|E \cap B(x,r)| \ll_\varepsilon r^\varepsilon |E|$$

for all balls $B(x,r)$ in $\mathbf{R}^n$ (note that the bound is only nontrivial when $r \ll 1$).

Informally, the two-ends condition asserts that E cannot concentrate in a small ball; it implies for instance that the diameter of E is $\gg_\varepsilon 1$.

We now have

Proposition 2.4.5 (Two-ends reduction). *To prove Conjecture* 2.4.3 *for a fixed value of d and n, it suffices to prove it under the assumption that the sets E_i all obey the two-ends condition with exponent ε for any fixed value of $\varepsilon > 0$.*

The key tool used to prove this proposition is

Lemma 2.4.6 (Every set has a large rescaled two-ends piece). *Let $E \subset \mathbf{R}^n$ be a set of positive measure and diameter $O(1)$, and let $0 < \varepsilon < n$. Then there exists a ball $B(x,r)$ of radius $r = O(1)$ such that*

$$|E \cap B(x,r)| \gg r^\varepsilon |E|$$

and

$$|E \cap B(x',r')| \ll (r'/r)^\varepsilon |E \cap B(x,r)|$$

for all other balls $B(x',r')$.

Proof. Consider the problem of maximising the quantity $|E \cap B(x,r)|/r^\varepsilon$ among al balls $B(x,r)$ of radius at most the diameter of E. On the one hand, this quantity can be at least $\gg |E|$, simply by taking $B(x,r)$ equal to the smallest ball containing E. On the other hand, using the trivial bound $|E \cap B(x,r)| \leq |B(x,r)| \ll r^n$, we see that the quantity $|E \cap B(x,r)|/r^\varepsilon$ is bounded. Thus the supremum of the $|E \cap B(x,r)|/r^\varepsilon$ is finite. If we pick a ball $B(x,r)$ which comes within a factor of 2 (say) of realising this supremum, then the claim easily follows. (Actually one can even attain the supremum exactly by a compactness argument, though this is not necessary for our applications.) □

One can view the quantity r in the above lemma as describing the "width" of the set E; this is the viewpoint taken for instance in [**TaWr2003**].

Now we prove Proposition 2.4.5.

Proof. Suppose Conjecture 2.4.3 has already been proven (assuming the two-ends condition with exponent ε) for some value of d, n, and some small value of ε. Now suppose we have the setup of Conjecture 2.4.3 without the two-ends condition.

The first observation is that the claim is easy when $\lambda \ll \delta$. Indeed, in this case we can just bound $|\bigcup_{i=1}^N E_i|$ from below the volume $\lambda |T_i| \sim \lambda \delta^{n-1}$ of a single tube. So we may assume that λ is much greater than δ.

Let $\varepsilon > 0$ be arbitrary. We apply Lemma 2.4.6 to each E_i, to find a ball $B(x_i, r_i)$ such that

$$|E_i \cap B(x_i,r_i)| \gg r_i^\varepsilon |E_i| \tag{2.17}$$

and

$$|E_i \cap B(x',r')| \ll (r'/r_i)^\varepsilon |E_i \cap B(x_i,r_i)|$$

for all $B(x',r')$. From (2.17) and the fact that $|E_i| = \lambda |T_i| \gg \lambda \delta^{n-1} \gg \delta^n$, as well as the trivial bound $|E_i \cap B(x_i,r_i)| \leq |B(x_i,r_i)| \ll r_i^n$, we obtain the lower bound $r_i \gg \delta^{1+O(\varepsilon)}$. Thus there are only about $O(\log \frac{1}{\delta})$ possible dyadic ranges $\rho \leq r_i \leq 2\rho$. Using the pigeonhole principle (refining the number N of tubes by a factor of $\log \frac{1}{\delta}$), we may assume that there is a single $\delta^{1+O(\varepsilon)} \leq \rho \ll 1$ such that *all* of the r_i lie in the same dyadic range $[\rho, 2\rho]$.

The intersection of T_i with $B(x_i, r_i)$ is then contained in a $\delta \times O(\rho)$ tube $\tilde{T}_i$, and $\tilde{E}_i := E_i \cap \tilde{T}_i$ occupies a fraction

$$|\tilde{E}_i|/|\tilde{T}_i| \gg r_i^{\varepsilon}|E_i|/|\tilde{T}_i| \gg \delta^{O(\varepsilon)}\lambda/\rho$$

of $\tilde{T}_i$. If we then rescale each of the $\tilde{E}_i$ and $\tilde{T}_i$ by $O(1/\rho)$, we can locate subsets E_i' of $O(\delta/\rho) \times 1$-tubes T_i' of density $\gg \delta^{O(\varepsilon)}\lambda/\rho$. These tubes T_i' have cardinality $\delta^{1-n+O(\varepsilon)}$ (the loss here is due to the use of the pigeonhole principle earlier), and they occupy a δ-separated set of directions, but after refining these tubes a bit we may assume that they instead occupy a δ/ρ-separated set of directions at the expense of cutting the cardinality down to $\delta^{O(\varepsilon)}(\delta/\rho)^{1-n}$ or so. Furthermore, by construction the E_i' obey the two-ends condition at exponent ε. Applying the hypothesis that Conjecture 2.4.3 holds for such sets, we conclude that

$$|\bigcup_i E_i'| \gg_\varepsilon \delta^{O(\varepsilon)}[\lambda/\rho]^d[\delta/\rho]^{n-d},$$

which on undoing the rescaling by $1/\rho$ gives

$$|\bigcup_i \tilde{E}_i| \gg_\varepsilon \delta^{O(\varepsilon)}\lambda^d\delta^{n-d}.$$

Since $\varepsilon > 0$ was arbitrary, the claim follows. □

To give an idea of how this two-ends reduction is used, we give a quick application of it:

Proposition 2.4.7. *The Kakeya maximal function conjecture is true for* $d \leq \frac{n+1}{2}$.

Proof. We use the "bush" argument of Bourgain. By the above reductions, it suffices to establish the bound

$$|\bigcup_{i=1}^N E_i| \gg_\varepsilon \lambda^{\frac{n+1}{2}}\delta^{\frac{n-1}{2}-\varepsilon}$$

whenever $N \sim \delta^{1-n}$, and $E_i \subset T_i$ are subsets of $\delta \times 1$ tubes T_i in δ-separated directions with density λ and obeying the two-ends condition with exponent ε.

Let μ be the maximum multiplicity of the E_i, i.e., $\mu := \|\sum_{i=1}^N 1_{E_i}\|_{L^\infty(\mathbf{R}^n)}$. On the one hand, we clearly have

$$|\bigcup_{i=1}^N E_i| \geq \frac{1}{\mu}\|\sum_{i=1}^N 1_{E_i}\|_{L^1(\mathbf{R}^n)} \gg \frac{1}{\mu}\lambda N\delta^{n-1} \gg \frac{\lambda}{\mu}.$$

This bound is good when μ is small. What if μ is large? Then there exists a point x_0 which is contained in μ of the E_i, and hence also contained in (at least) μ of the tubes T_i. These tubes form a "bush" centred at x_0, but the portions of that tube near the centre x_0 of the bush have high overlap. However, the two-ends condition can be used to finesse this issue. Indeed, that condition ensures that for each E_i involved in this bush, we have

$$|E_i \cap B(x_0, r)| \leq \frac{1}{2}|E_i|$$

for some $r \sim 1$, and thus

$$|E_i \backslash B(x_0, r)| \geq \frac{1}{2}|E_i| \gg \lambda\delta^{n-1}.$$

The δ-separated nature of the tubes T_i implies that the maximum overlap of the portion $T_i \backslash B(x_0, r)$ of the μ tubes in the bush away from the origin is $O(1)$, and so

$$|\bigcup_i E_i \backslash B(x_0, r)| \gg \mu\lambda\delta^{n-1}.$$

Thus we have two different lower bounds for $\bigcup_i E_i$, namely $\frac{\lambda}{\mu}$ and $\mu\lambda\delta^{n-1}$. Taking the geometric mean of these bounds to eliminate the unknown multiplicity μ, we obtain

$$|\bigcup_i E_i| \gg \lambda\delta^{(n-1)/2},$$

which certainly implies the desired bound since $\lambda \leq 1$. □

Remark 2.4.8. Note that the two-ends condition actually proved a *better* bound than what was actually needed for the Kakeya conjecture, in that the power of λ was more favourable than necessary. However, this gain disappears under the rescaling argument used in the proof of Proposition 2.4.5. Nevertheless, this does illustrate one of the advantages of employing the two-ends reduction; the bounds one gets upon doing so tend to be better (especially for small values of λ) than what one would have had without it, and so getting the right bound tends to be a bit easier in such cases. Note though that for the Kakeya set problem, where λ is essentially 1, the two-ends reduction is basically redundant.

Remark 2.4.9. One technical drawback to using the two-ends reduction is that if at some later stage one needs to refine the sets E_i to smaller sets, then one may lose the two-ends property. However, one could invoke the arguments used in Proposition 2.4.5 to recover this property again by refining E_i further. One may then lose some other property by this further refinement, but one convenient trick that allows one to take advantage of multiple refinements simultaneously is to iteratively refine the various sets involved and use the pigeonhole principle to find some place along this iteration where all relevant statistics of the system (e.g., the "width" r of the

E_i) stabilise (here one needs some sort of monotonicity property to obtain this stabilisation). This type of trick was introduced in [**Wo1998**] and has been used in several subsequent papers, for instance in [**LaTa2001**].

Notes. This article first appeared at

`terrytao.wordpress.com/2009/05/15.`

Thanks to Arie Israel, Josh Zahl, Shuanglin Shao, and an anonymous commenter for corrections.

2.5. The least quadratic nonresidue, and the square root barrier

A large portion of analytic number theory is concerned with the distribution of number-theoretic sets such as the primes, or *quadratic residues* in a certain modulus. At a local level (e.g., on a short interval $[x, x + y]$), the behaviour of these sets may be quite irregular. However, in many cases one can understand the *global* behaviour of such sets on very large intervals, (e.g., $[1, x]$), with reasonable accuracy (particularly if one assumes powerful additional conjectures, such as the Riemann hypothesis and its generalisations). For instance, in the case of the primes, we have the *prime number theorem*, which asserts that the number of primes in a large interval $[1, x]$ is asymptotically equal to $x/\log x$; in the case of quadratic residues modulo a prime p, it is clear that there are exactly $(p - 1)/2$ such residues in $[1, p]$. With elementary arguments, one can also count statistics such as the number of pairs of consecutive quadratic residues; and with the aid of deeper tools such as the Weil sum estimates, one can count more complex patterns in these residues also (e.g., k-point correlations).

One is often interested in converting this sort of "global" information on long intervals into "local" information on short intervals. If one is interested in the behaviour on a *generic* or *average* short interval, then the question is still essentially a global one, basically because one can view a long interval as an average of a long sequence of short intervals. (This does not mean that the problem is automatically easy, because not every global statistic about, say, the primes is understood. For instance, we do not know how to rigorously establish the conjectured asymptotic for the number of *twin primes* $n, n + 2$ in a long interval $[1, N]$, and so we do not fully understand the local distribution of the primes in a typical short interval $[n, n + 2]$.)

However, suppose that instead of understanding the *average-case* behaviour of short intervals, one wants to control the *worst-case* behaviour of such intervals (i.e., to establish bounds that hold for *all* short intervals, rather than *most* short intervals). Then it becomes substantially harder to convert global information to local information. In many cases one encounters a "square root barrier", in which global information at scale x (e.g., statistics on $[1, x]$) cannot be used to say anything nontrivial about a fixed (and possibly worst-case) short interval at scales $x^{1/2}$ or below. (Here we ignore factors of $\log x$ for simplicity.) The basic reason for this is that even randomly distributed sets in $[1, x]$ (which are basically the most uniform type of global distribution one could hope for) exhibit random fluctuations of size $x^{1/2}$ or so in their global statistics (as can be seen for instance from the *central limit theorem*). Because of this, one could take a random (or pseudo-random) subset of $[1, x]$ and delete all the elements in a short interval of length $o(x^{1/2})$, without anything suspicious showing up on the global statistics level; the edited set still has essentially the same global statistics as the original set. On the other hand, the worst-case behaviour of this set on a short interval has been drastically altered.

One stark example of this arises when trying to control the largest gap between consecutive prime numbers in a large interval $[x, 2x]$. There are convincing heuristics that suggest that this largest gap is of size $O(\log^2 x)$ (*Cramér's conjecture*). But even assuming the Riemann hypothesis, the best upper bound on this gap is only of size $O(x^{1/2} \log x)$, basically because of this square root barrier.

On the other hand, in some cases one can use additional tricks to get past the square root barrier. The key point is that many number-theoretic sequences have special structure that distinguish them from being exactly like random sets. For instance, quadratic residues have the basic but fundamental property that the product of two quadratic residues is again a quadratic residue. One way to use this sort of structure is to amplify bad behaviour in a single short interval into bad behaviour across many short intervals (cf. Section 1.9 of *Structure and Randomness*). Because of this amplification, one can sometimes get new worst-case bounds by tapping the average-case bounds.

In this article I would like to indicate a classical example of this type of amplification trick, namely Burgess's bound on short character sums. To narrow the discussion, I would like to focus primarily on the following classical problem:

Problem 2.5.1. What are the best bounds one can place on the first quadratic nonresidue n_p in the interval $[1, p-1]$ for a large prime p?

(The first quadratic residue is, of course, 1; the more interesting problem is the first quadratic *non*residue.)

Probabilistic heuristics (presuming that each nonsquare integer has a 50-50 chance of being a quadratic residue) suggests that n_p should have size $O(\log p)$, and indeed Vinogradov conjectured that $n_p = O_\varepsilon(p^\varepsilon)$ for any $\varepsilon > 0$. Using the *Pólya-Vinogradov inequality*, one can get the bound $n_p = O(\sqrt{p}\log p)$ (and can improve it to $n_p = O(\sqrt{p})$ using *smoothed sums*); combining this with a sieve theory argument (exploiting the multiplicative nature of quadratic residues) one can boost this to $n_p = O(p^{\frac{1}{2\sqrt{e}}}\log^2 p)$. Inserting Burgess's amplification trick one can boost this to $n_p = O_\varepsilon(p^{\frac{1}{4\sqrt{e}}+\varepsilon})$ for any $\varepsilon > 0$. Apart from refinements to the ε factor, this bound has stood for five decades as the "world record" for this problem, which is a testament to the difficulty in breaching the square root barrier.

Note: in order not to obscure the presentation with technical details, I will be using asymptotic notation $O()$ in a somewhat informal manner.

2.5.1. Character sums. To approach the problem, we begin by fixing the large prime p and introducing the *Legendre symbol* $\chi(n) = \left(\frac{n}{p}\right)$, defined to equal 0 when n is divisible by p, $+1$ when n is an invertible quadratic residue modulo p, and -1 when n is an invertible quadratic nonresidue modulo p. Thus, for instance, $\chi(n) = +1$ for all $1 \leq n < n_p$. One of the main reasons one wants to work with the function χ is that it enjoys two easily verified properties:

- χ is periodic with period p.
- One has the total multiplicativity property $\chi(nm) = \chi(n)\chi(m)$ for all integers n, m.

In the jargon of number theory, χ is a *Dirichlet character* with conductor p. Another important property of this character is of course the law of *quadratic reciprocity*, but this law is more important for the *average-case* behaviour in p, whereas we are concerned here with the *worst-case* behaviour in p, and so we will not actually use this law here.

An obvious way to control n_p is via the character sum

$$\sum_{1\leq n\leq x} \chi(n). \tag{2.18}$$

From the triangle inequality, we see that this sum has magnitude at most x. If we can then obtain a nontrivial bound of the form

$$\sum_{1\leq n\leq x} \chi(n) = o(x) \tag{2.19}$$

for some x, this forces the existence of a quadratic residue less than or equal to x, thus $n_p \leq x$. So one approach to the problem is to bound the character sum (2.18).

As there are just as many residues as nonresidues, the sum (2.18) is periodic with period p and we obtain a trivial bound of p for the magnitude of the sum. One can achieve a nontrivial bound by Fourier analysis. One can expand

$$\chi(n) = \sum_{a=0}^{p-1} \hat{\chi}(a) e^{2\pi i a n/p},$$

where $\hat{\chi}(a)$ are the Fourier coefficients of χ,

$$\hat{\chi}(a) := \frac{1}{p} \sum_{n=0}^{p-1} \chi(n) e^{-2\pi i a n/p}.$$

As there are just as many quadratic residues as nonresidues, $\hat{\chi}(0) = 0$, so we may drop the $a = 0$ term. From summing the geometric series we see that

$$\sum_{1 \le n \le x} e^{2\pi i a n/p} = O(1/\|a/p\|), \tag{2.20}$$

where $\|a/p\|$ is the distance from a/p to the nearest integer (0 or 1); inserting these bounds into (2.18) and summing what is essentially a harmonic series in a, we obtain

$$\sum_{1 \le n \le x} \chi(n) = O(p \log p \sup_{a \neq 0} |\hat{\chi}(a)|).$$

Now, how big is $\hat{\chi}(a)$? Taking absolute values, we get a bound of 1, but this gives us something worse than the trivial bound. To do better, we use the Plancherel identity

$$\sum_{a=0}^{p-1} |\hat{\chi}(a)|^2 = \frac{1}{p} \sum_{n=0}^{p-1} |\chi(n)|^2,$$

which tells us that

$$\sum_{a=0}^{p-1} |\hat{\chi}(a)|^2 = O(1).$$

This tells us that $\hat{\chi}$ is small *on the average*, but does not immediately tell us anything new about the *worst-case* behaviour of χ, which is what we need here. But now we use the multiplicative structure of χ to relate average-case and worst-case behaviour. Note that if b is coprime to p, then $\chi(bn)$ is a scalar multiple of $\chi(n)$ by a quantity $\chi(b)$ of magnitude 1; taking Fourier transforms, this implies that $\hat{\chi}(a/b)$ and $\hat{\chi}(a)$ also differ by this factor. In particular, $|\hat{\chi}(a/b)| = |\hat{\chi}(a)|$. As b was arbitrary, we thus see that $|\hat{\chi}(a)|$ is constant for all a coprime to p; in other words, the worst case is the *same* as the average case. Combining this with the Plancherel bound one obtains $|\hat{\chi}(a)| = O(1/\sqrt{p})$, leading to the *Pólya-Vinogradov inequality*

$$\sum_{1 \le n \le x} \chi(n) = O(\sqrt{p} \log p).$$

(In fact, a more careful computation reveals the slightly sharper bound $|\sum_{1\leq n\leq x}\chi(n)| \leq \sqrt{p}\log p$; this is nontrivial for $x > \sqrt{p}\log p$.)

Remark 2.5.2. Up to logarithmic factors, this is consistent with what one would expect if χ fluctuated like a random sign pattern (at least for x comparable to p; for smaller values of x, one expects instead a bound of the form $O(\sqrt{x})$, up to logarithmic factors). It is conjectured that the $\log p$ factor can be replaced with a $O(\log\log p)$ factor, which would be consistent with the random fluctuation model and is best possible; this is known for GRH, but unconditionally the Pólya-Vinogradov inequality is still the best known. (See however `http://arxiv.org/abs/math/0503113` this paper of Granville and Soundararajan for an improvement for nonquadratic characters χ.)

A direct application of the Pólya-Vinogradov inequality gives the bound $n_p \leq \sqrt{p}\log p$. One can get rid of the logarithmic factor (which comes from the harmonic series arising from (2.20)) by replacing the sharp cutoff $1_{1\leq n\leq x}$ by a smoother sum, which has a better behaved Fourier transform. But one can do better still by exploiting the multiplicativity of χ again, by the following trick of Vinogradov. Observe that not only does one have $\chi(n) = +1$ for all $n \leq n_p$, but also $\chi(n) = +1$ for any n which is $n_p - 1$-smooth, i.e., is the product of integers less than n_p. So even if n_p is significantly less than x, one can show that the sum (2.18) is large if the majority of integers less than x are $n_p - 1$-smooth.

Since every integer n less than x is either n_p-smooth (in which case $\chi(n) = +1$) or divisible by a prime q between n_p and x (in which case $\chi(n)$ is at least -1), we obtain the lower bound

$$\sum_{1\leq n\leq x}\chi(n) \geq \sum_{1\leq n\leq x} 1 - \sum_{n_p<q\leq x}\ \sum_{1\leq n\leq x:q|n} 2.$$

Clearly, $\sum_{1\leq n\leq x} 1 = x + O(1)$ and $\sum_{1\leq n\leq x:q|n} 2 = 2\frac{x}{q} + O(1)$. The total number of primes less than x is $O(\frac{x}{\log x}) = o(x)$ by the prime number theorem, thus

$$\sum_{1\leq n\leq x}\chi(n) \geq x - \sum_{n_p<q\leq x} 2\frac{x}{q} + o(x).$$

Using the classical asymptotic $\sum_{q\leq y}\frac{1}{q} = \log\log y + C + o(1)$ for some absolute constant C (which basically follows from the prime number theorem, but also has an elementary proof), we conclude that

$$\sum_{1\leq n\leq x}\chi(n) \geq x[1 - 2\log\frac{\log x}{\log n_p} + o(1)].$$

If $n_p \geq x^{\frac{1}{\sqrt{e}}+\varepsilon}$ for some fixed $\varepsilon > 0$, then the expression in brackets is bounded away from zero for x large; in particular, this is incompatible with (2.19) for x large enough. As a consequence, we see that if we have a bound of the form (2.19), then we can conclude $n_p = O_\varepsilon(x^{\frac{1}{\sqrt{e}}+\varepsilon})$ for all $\varepsilon > 0$; in particular, from the Pólya-Vinogradov inequality one has

$$n_p = O_\varepsilon(p^{\frac{1}{2\sqrt{e}}+\varepsilon})$$

for all $\varepsilon > 0$, or equivalently that $n_p \leq p^{\frac{1}{2\sqrt{e}}+o(1)}$. (By being a bit more careful, one can refine this to $n_p = O(p^{\frac{1}{2\sqrt{e}}} \log^{2/\sqrt{e}} p)$.)

Remark 2.5.3. The estimates on the Gauss-type sums

$$\hat{\chi}(a) := \frac{1}{p}\sum_{n=0}^{p-1} \chi(n) e^{-2\pi i a n/p}$$

are sharp; nevertheless, they fail to penetrate the square root barrier in the sense that no nontrivial estimates are provided below the scale $\sqrt{p}$. One can also see this barrier using the *Poisson summation formula* (Exercise 1.12.41 of *Volume I*), which basically gives a formula that (very roughly) takes the form

$$\sum_{n=O(x)} \chi(n) \sim \frac{x}{\sqrt{p}} \sum_{n=O(p/x)} \chi(n)$$

for any $1 < x < p$, and is basically the limit of what one can say about character sums using Fourier analysis alone. In particular, we see that the Pólya-Vinogradov bound is basically the Poisson dual of the trivial bound. The scale $x = \sqrt{p}$ is the crossing point where Poisson summation does not achieve any nontrivial modification of the scale parameter.

2.5.2. Average-case bounds.

The Pólya-Vinogradov bound establishes a nontrivial estimate (2.18) for x significantly larger than $\sqrt{p}\log p$. We are interested in extending (2.18) to shorter intervals.

Before we address this issue for a fixed interval $[1, x]$, we first study the *average-case* bound on short character sums. Fix a short length y, and consider the shifted sum

$$\sum_{a \leq n \leq a+y} \chi(n), \tag{2.21}$$

where a is a parameter. The analogue of (2.18) for such intervals would be

$$\sum_{a \leq n \leq a+y} \chi(n) = o(y). \tag{2.22}$$

For y very small (e.g., $y = p^\varepsilon$ for some small $\varepsilon > 0$), we do not know how to establish (2.22) for *all* a; but we can at least establish (2.22) for *almost all*

a, with only about $O(\sqrt{p})$ exceptions (here we see the square root barrier again!).

More precisely, we will establish the moment estimates

$$\frac{1}{p}\sum_{a=0}^{p-1}|\sum_{a\leq n\leq a+y}\chi(n)|^k = O_k(y^{k/2}+y^k p^{-1/2}) \tag{2.23}$$

for any positive even integer $k = 2, 4, \ldots$. If y is not too tiny, say $y \geq p^\varepsilon$ for some $\varepsilon > 0$, then by applying (2.23) for a sufficiently large k and using Chebyshev's inequality (or Markov's inequality), we see (for any given $\delta > 0$) that one has the nontrivial bound

$$|\sum_{a\leq n\leq a+y}\chi(n)| \leq \delta y$$

for all but at most $O_{\delta,\varepsilon}(\sqrt{p})$ values of $a \in [1, p]$.

To see why (2.23) is true, let us just consider the easiest case $k = 2$. Squaring both sides, we expand (2.23) as

$$\frac{1}{p}\sum_{a=0}^{p-1}\sum_{a\leq n,m\leq a+y}\chi(n)\chi(m) = O(y) + O(y^2 p^{-1/2}).$$

We can write $\chi(n)\chi(m)$ as $\chi(nm)$. Writing $m = n + h$, and using the periodicity of χ, we can rewrite the left-hand side as

$$\sum_{h=-y}^{y}(y-|h|)[\frac{1}{p}\sum_{n\in F_p}\chi(n(n+h))],$$

where we have abused notation and identified the finite field F_p with $\{0, 1, \ldots, p-1\}$.

For $h = 0$, the inner average is $O(1)$. For h nonzero, we claim the bound

$$\sum_{n\in F_p}\chi(n(n+h)) = O(\sqrt{p}), \tag{2.24}$$

which is consistent with (and is in fact slightly stronger than) what one would get if χ was a random sign pattern; assuming this bound gives (2.23) for $k = 2$ as required.

The bound (2.24) can be established by quite elementary means (as it comes down to counting points on the hyperbola $y^2 = x(x + h)$, which can be done by transforming the hyperbola to be rectangular), but for larger values of k we will need the more general estimate

$$\sum_{n\in F_p}\chi(P(n)) = O_k(\sqrt{p}) \tag{2.25}$$

whenever P is a polynomial over F of degree k which is not a constant multiple of a perfect square; this can be easily seen to give (2.23) for general k.

An equivalent form of (2.25) is that the *hyperelliptic curve*

$$\{(x, y) \in F_p \times F_p : y^2 = P(x)\} \tag{2.26}$$

contains $p+O_k(\sqrt{p})$ points. This fact follows from a general theorem of Weil establishing the Riemann hypothesis for curves over function fields, but can also be deduced by a more elementary argument of Stepanov [**St1969**], using the polynomial method, which we now give here. (This arrangement of the argument is based on the exposition in [**IwKo2004**].)

By translating the x variable, we may assume that $P(0)$ is nonzero. The key lemma is the following. Assume p large, and take l to be an integer comparable to $\sqrt{p}$ (other values of this parameter are possible, but this is the optimal choice). All polynomials $Q(x)$ are understood to be over the field F_p (i.e., they lie in the polynomial ring $F_p[X]$), although indeterminate variables x need not lie in this field.

Lemma 2.5.4. *There exists a nonzero polynomial $Q(x)$ of one indeterminate variable x over F_p of degree at most $lp/2 + O_k(p)$ which vanishes to order at least l at every point $x \in F_p$ for which $P(x)$ is a quadratic residue.*

Note from the factor theorem that Q can vanish to order at least l on at most $\deg(Q)/l \leq p/2 + O_k(\sqrt{p})$ points, and so we see that $P(x)$ is an invertible quadratic residue for at most $p/2 + O_k(\sqrt{p})$ values of F_p. Multiplying P by a quadratic nonresidue and running the same argument, we also see that $P(x)$ is an invertible quadratic nonresidue for at most $p/2 + O_k(\sqrt{p})$ values of F_p, and (2.25) (or the asymptotic for the number of points in (2.26)) follows.

We now prove the lemma. The polynomial Q will be chosen to be of the form

$$Q(x) = P^l(x)(R(x, x^p) + P^{\frac{p-1}{2}}(x)S(x, x^p)),$$

where $R(x, z), S(x, z)$ are polynomials of degree at most $\frac{p-k-1}{2}$ in x, and degree at most $\frac{l}{2} + C$ in z, where C is a large constant (depending on k) to be chosen later (these parameters have been optimised for the argument that follows). Since P has degree at most k, such a Q will have degree

$$\leq kl + \frac{p-k-1}{2} + \frac{p-1}{2}k + p(\frac{l}{2} + C') = \frac{lp}{2} + O_k(p)$$

as required. We claim (for suitable choices of C, C') that

(a) The degrees are small enough that $Q(x)$ is a nonzero polynomial whenever $R(x, z), S(x, z)$ are nonzero polynomials; and

(b) The degrees are large enough that there exists a nontrivial choice of $R(x,z)$ and $S(x,z)$ that $Q(x)$ vanishes to order at least l whenever $x \in F_p$ is such that $P(x)$ is a quadratic residue.

Claims (a) and (b) together establish the lemma.

We first verify (a). We can cancel off the initial P^l factor, so that we need to show that $R(x,x^p) + P^{\frac{p-1}{2}}(x)S(x,x^p)$ does not vanish when at least one of $R(x,z), Q(x,z)$ is not vanishing. We may assume that R, Q are not both divisible by z, since we could cancel out a common factor of x^p otherwise.

Suppose for contradiction that the polynomial $R(x,x^p) + P^{\frac{p-1}{2}}S(x,x^p)$ vanished, which implies that $R(x,0) = -P^{\frac{p-1}{2}}(x)S(x,0)$ modulo x^p. Squaring and multiplying by P, we see that

$$R(x,0)^2P(x) = P(x)^pS(x,0)^2 \bmod x^p,$$

but over F_p and modulo x^p, $P(x)^p = P(0)$ by Fermat's little theorem. Observe that $R(x,0)^2P(x)$ and $P(0)S(x,0)^2$ both have degree at most $p-1$, and so we can remove the x^p modulus and conclude that $R(x,0)^2P(x) = P(0)S(x,0)^2$ over F_p. But this implies (by the fundamental theorem of arithmetic for $F_p[X]$) that P is a constant multiple of a square, a contradiction. (Recall that $P(0)$ is nonzero, and that $R(x,0)$ and $S(x,0)$ are not both zero.)

Now we prove (b). Let $x \in F_p$ be such that $P(x)$ is a quadratic residue, thus $P(x)^{\frac{p-1}{2}} = +1$ by Fermat's little theorem. To get vanishing to order l, we need

$$\frac{d^j}{dx^j}[P^l(x)(R(x,x^p) + P^{\frac{p-1}{2}}(x)S(x,x^p))] = 0 \tag{2.27}$$

for all $0 \le j < l$. (Of course, we cannot define derivatives using limits and Newton quotients in this finite characteristic setting, but we can still define derivatives of polynomials formally, thus for instance $\frac{d}{dx}x^n := nx^{n-1}$, and enjoy all the usual rules of calculus, such as the product rule and chain rule.)

Over F_p, the polynomial x^p has derivative zero. If we then compute the derivative in (2.27) using many applications of the product and chain rule, we see that the left-hand side of (2.27) can be expressed in the form

$$P^{l-j}(x)[R_j(x,x^p) + P^{\frac{p-1}{2}}(x)S_j(x,x^p))],$$

where $R_j(x,z), S_j(x,z)$ are polynomials of degree at most $\frac{p-k-1}{2} + O_k(j)$ in x and at most $\frac{l}{2} + C$ in z, whose coefficients depend in some linear fashion on the coefficients of R and S. (The exact nature of this linear relationship will depend on k, p, P, but this will not concern us.) Since we only need to evaluate this expression when $P(x)^{\frac{p-1}{2}} = +1$ and $x^p = p$ (by Fermat's little theorem), we thus see that we can verify (2.27) provided that the polynomial

$$P^{l-j}(x)[R_j(x,x) + S_j(x,x))]$$

vanishes identically. This is a polynomial of degree at most

$$O(l-j) + \frac{p-k-1}{2} + O_k(j) + \frac{l}{2} + C = \frac{p}{2} + O_k(p^{1/2}) + C,$$

and there are $l+1$ possible values of j, so this leads to

$$\frac{lp}{2} + O_k(p) + O(C\sqrt{p})$$

linear constraints on the coefficients of R and S to be satisfied. On the other hand, the total number of these coefficients is

$$2 \times (\frac{p-k-1}{2} + O(1)) \times (\frac{l}{2} + C + O(1)) = \frac{lp}{2} + Cp + O_k(p).$$

For C large enough, there are more coefficients than constraints, and so one can find a nontrivial choice of coefficients obeying the constraints (2.27), and (b) follows.

Remark 2.5.5. If one optimises all the constants here, one gets an upper bound of basically $8k\sqrt{p}$ for the deviation in the number of points in (2.26). This is only a little worse than the sharp bound of $2g\sqrt{p}$ given from Weil's theorem, where $g = \lfloor \frac{k-1}{2} \rfloor$ is the genus; however, it is possible to boost the former bound to the latter by using a version of the tensor power trick (generalising F_p to F_{p^m} and then letting $m \to \infty$) combined with the theory of *Artin L-functions* and the *Riemann-Roch theorem*. This is (very briefly!) sketched in Section 1.9 of *Structure and Randomness*.

Remark 2.5.6. Once again, the global estimate (2.25) is very sharp, but cannot penetrate below the square root barrier, in that one is allowed to have about $O(\sqrt{p})$ exceptional values of a for which no cancellation exists. One expects that these exceptional values of a in fact do not exist, but we do not know how to do this unless y is larger than $x^{1/4}$ (so that the Burgess bounds apply).

2.5.3. The Burgess bound. The average case bounds in the previous section give an alternate demonstration of a nontrivial estimate (2.18) for $x > p^{1/2+\varepsilon}$, which is just a bit weaker than what the Pólya-Vinogradov inequality gives. Indeed, if (2.18) failed for such an x, thus

$$|\sum_{n \in [1,x]} \chi(n)| \gg x,$$

then by taking a small y (e.g., $y = p^{\varepsilon/2}$) and covering $[1,x]$ by intervals of length y, we see (from a first moment method argument) that

$$|\sum_{a \leq n \leq a+y} \chi(n)| \gg y$$

for a positive fraction of the a in $[1, x]$. But this contradicts the results of the previous section.

Burgess observed that by exploiting the multiplicativity of χ one last time to amplify the above argument, one can extend the range for which (2.18) can be proved from $x > p^{1/2+\varepsilon}$ to also cover the range $p^{1/4+\varepsilon} < x < p^{1/2}$. The idea is not to cover $[1, x]$ by intervals of length y, but rather by *arithmetic progressions* $\{a, a+r, \ldots, a+yr\}$ of length y, where $a = O(x)$ and $r = O(x/y)$. Another application of the first moment method then shows that

$$|\sum_{0 \leq j \leq y} \chi(a + jr)| \gg y$$

for a positive fraction of the a in $[1, x]$ and r in $[1, x/y]$ (i.e., $\gg x^2/y$ such pairs (a, r)).

For technical reasons, it will be inconvenient if a and r have too large of a common factor, so we pause to remove this possibility. Observe that for any $d \geq 1$, the number of pairs (a, r) which have d as a common factor is $O(\frac{1}{d^2}x^2/y)$. As $\sum_{d=1}^{\infty} \frac{1}{d^2}$ is convergent, we may thus remove those pairs which have too large of a common factor, and assume that all pairs (a, r) have common factor $O(1)$ at most (so are "almost coprime").

Now we exploit the multiplicativity of χ to write $\chi(a+jr)$ as $\chi(r)\chi(b+j)$, where b is a residue which is equal to a/r mod q. Dividing out by $\chi(r)$, we conclude that

$$|\sum_{0 \leq j \leq y} \chi(b + j)| \gg y \tag{2.28}$$

for $\gg x^2/y$ pairs (a, r).

Now for a key observation: the $\gg x^2/y$ values of b arising from the pairs (a, r) are mostly disjoint. Indeed, suppose that two pairs $(a, r), (a', r')$ generated the same value of b, thus $a/r = a'/r'$ mod p. This implies that $ar' = a'r$ mod p. Since $x < p^{1/2}$, we see that $ar', a'r$ do not exceed p, so we may remove the modulus and conclude that $ar' = a'r$. But since we are assuming that a, r and a', r' are almost coprime, we see that for each (a, r) there are at most $O(1)$ values of a', r' for which $ar' = a'r$. So the b's here only overlap with multiplicity $O(1)$, and we conclude that (2.28) holds for $\gg x^2/y$ values of b. But comparing this with the previous section, we obtain a contradiction unless $x^2/y \ll \sqrt{p}$. Setting y to be a sufficiently small power of p, we obtain Burgess's result that (2.18) holds for $x > p^{1/4+\varepsilon}$ for any fixed $\varepsilon > 0$.

Combining Burgess's estimate with Vinogradov's sieving trick, we conclude the bound $n_p = O_\varepsilon(p^{1/4\sqrt{e}+\varepsilon})$ for all $\varepsilon > 0$, which is the best bound known today on the least quadratic nonresidue except for refinements of the p^ε error term.

Remark 2.5.7. There are many generalisations of this result, for instance to more general characters (with possibly composite conductor), or to shifted sums (2.21). However, the $p^{1/4}$ type exponent has not been improved except with the assistance of powerful conjectures such as the *generalised Riemann hypothesis*.

Notes. This article first appeared at

`terrytao.wordpress.com/2009/08/18.`

Thanks to Efthymios Sofos, Joshua Zelinsky, K, and Seva Lev for corrections.

Boris noted the similarity between the use of the Frobenius map $x \mapsto x^p$ in Stepanov's argument and Thue's trick from the proof of his famous result on the Diophantine approximations to algebraic numbers, where instead of the exact equality $x = x^p$ that is used here, he used two very good approximations to the same algebraic number.

2.6. Determinantal processes

Given a set S, a (simple) *point process* is a random subset A of S. (A nonsimple point process would allow multiplicity; more formally, A is no longer a subset of S, but is a Radon measure on S, where we give S the structure of a locally compact Polish space, but I do not wish to dwell on these sorts of technical issues here.) Typically, A will be finite or countable, even when S is uncountable. Basic examples of point processes include the following.

- *Bernoulli point process.* S is an at most countable set, $0 \leq p \leq 1$ is a parameter, and A a random set such that the events $x \in A$ for each $x \in S$ are jointly independent and occur with a probability of p each. This process is automatically simple.
- *Discrete Poisson point process.* S is an at most countable space, λ is a measure on S (i.e., an assignment of a nonnegative number $\lambda(\{x\})$ to each $x \in S$), and A is a *multiset* where the multiplicity of x in A is a *Poisson random variable* with intensity $\lambda(\{x\})$, and the multiplicities of $x \in A$ as x varies in S are jointly independent. This process is usually not simple.
- *Continuous Poisson point process.* S is a locally compact Polish space with a Radon measure λ, and for each $\Omega \subset S$ of finite measure, the number of points $|A \cap \Omega|$ that A contains inside Ω is a *Poisson random variable* with intensity $\lambda(\Omega)$. Furthermore, if $\Omega_1, \ldots, \Omega_n$ are disjoint sets, then the random variables

$|A \cap \Omega_1|, \ldots, |A \cap \Omega_n|$ are jointly independent. (The fact that Poisson processes exist at all requires a nontrivial amount of measure theory, and will not be discussed here.) This process is almost surely simple iff all points in S have measure zero.

- *Spectral point processes.* The spectrum of a random matrix is a point process in $\mathbf{C}$ (or in $\mathbf{R}$, if the random matrix is Hermitian). If the spectrum is almost surely simple, then the point process is almost surely simple. In a similar spirit, the zeroes of a random polynomial are also a point process.

A remarkable fact is that many natural (simple) point processes are *determinantal processes*. Very roughly speaking, this means that there exists a positive semidefinite kernel $K : S \times S \to \mathbf{R}$ such that, for any $x_1, \ldots, x_n \in S$, the probability that $x_1, \ldots, x_n$ all lie in the random set A is proportional to the determinant $\det((K(x_i, x_j))_{1 \leq i,j \leq n})$. Examples of processes known to be determinantal include nonintersecting random walks, spectra of random matrix ensembles such as GUE, and zeroes of polynomials with Gaussian coefficients.

I would be interested in finding a good explanation (even at the heuristic level) as to why determinantal processes are so prevalent in practice. I do have a very weak explanation, namely that determinantal processes obey a large number of rather pretty algebraic identities, and so it is plausible that any other process which has a very algebraic structure (in particular, any process involving Gaussians, characteristic polynomials, etc.) would be connected in some way with determinantal processes. I am not particularly satisfied with this explanation, but I thought I would at least describe some of these identities below to support this case. (This is partly for my own benefit, as I am trying to learn about these processes, particularly in connection with the spectral distribution of random matrices.) The material here is partly based on [**HoKrPeVi2006**].

2.6.1. Discrete determinantal processes. In order to ignore all measure-theoretic distractions and focus on the algebraic structure of determinantal processes, we will first consider the discrete case when the space S is just a finite set $S = \{1, \ldots, N\}$ of cardinality $|S| = N$. We say that a process $A \subset S$ is a *determinantal process* with kernel K, where K is a $k \times k$ symmetric real matrix, if one has

$$\mathbf{P}(\{i_1, \ldots, i_k\} \subset A) = \det(K(i_a, i_b))_{1 \leq a,b \leq k} \tag{2.29}$$

for all distinct $i_1, \ldots, i_k \in S$.

To build determinantal processes, let us first consider point processes of a fixed cardinality n, thus $0 \leq n \leq N$ and A is a random subset of

S of size n, or in other words a random variable taking values in the set $\binom{S}{n} := \{B \subset S : |B| = n\}$.

In this simple model, an n-element point process is basically just a collection of $\binom{N}{n}$ probabilities $p_B = \mathbf{P}(A = B)$, one for each $B \in \binom{S}{n}$, which are nonnegative numbers that add up to 1. For instance, in the uniform point process where A is drawn uniformly at random from $\binom{S}{n}$, each of these probabilities p_B would equal $1/\binom{N}{n}$. How would one generate other interesting examples of n-element point processes?

For this, we can borrow the idea from quantum mechanics that probabilities can arise as the square of coefficients of unit vectors, though unlike quantum mechanics it will be slightly more convenient here to work with real vectors rather than complex ones. To formalise this, we work with the nth *exterior power* $\bigwedge^n \mathbf{R}^N$ of the Euclidean space $\mathbf{R}^N$; this space is sort of a "quantisation" of $\binom{S}{n}$ and is analogous to the space of quantum states of n identical *fermions*, if each fermion can exist classically in one of N states (or "spins"). (The requirement that the process be simple is then analogous to the *Pauli exclusion principle*.)

This space of n-vectors in $\mathbf{R}^N$ is spanned by the wedge products $e_{i_1} \wedge \cdots \wedge e_{i_n}$ with $1 \leq i_1 < \cdots < i_n \leq N$, where $e_1, \ldots, e_N$ is the standard basis of $\mathbf{R}^N$. There is a natural inner product to place on $\bigwedge^n \mathbf{R}^N$ by declaring all the $e_{i_1} \wedge \cdots \wedge e_{i_n}$ to be orthonormal.

Lemma 2.6.1. *If $f_1, \ldots, f_N$ is any orthonormal basis of $\mathbf{R}^N$, then the $f_{i_1} \wedge \cdots \wedge f_{i_n}$ for $1 \leq i_1 < \cdots < i_n \leq N$ are an orthonormal basis for $\bigwedge^n \mathbf{R}^N$.*

Proof. By definition, this is true when $(f_1, \ldots, f_N) = (e_1, \ldots, e_N)$. If the claim is true for some orthonormal basis $f_1, \ldots, f_N$, it is not hard to see that the claim also holds if one rotates f_i and f_j in the plane that they span by some angle θ, where $1 \leq i < j \leq n$ are arbitrary. But any orthonormal basis can be rotated into any other by a sequence of such rotations (e.g., by using *Euler angles*), and the claim follows. □

Corollary 2.6.2. *If $v_1, \ldots, v_n$ are vectors in $\mathbf{R}^N$, then the magnitude of $v_1 \wedge \cdots \wedge v_n$ is equal to the n-dimensional volume of the parallelopiped spanned by $v_1, \ldots, v_n$.*

Proof. Observe that applying row operations to v_i (i.e., modifying one v_i by a scalar multiple of another v_j) does not affect either the wedge product or the volume of the parallelopiped. Thus by using the *Gram-Schmidt process*, we may assume that the v_i are orthogonal; by normalising, we may assume they are orthonormal. The claim now follows from the preceding lemma. □

From this and the ordinary Pythagorean theorem in the inner product space $\bigwedge^n \mathbf{R}^N$, we conclude the *multidimensional Pythagorean theorem*: the square of the n-dimensional volume of a parallelopiped in $\mathbf{R}^N$ is the sum of squares of the n-dimensional volumes of the projection of that parallelopiped to each of the $\binom{N}{n}$ coordinate subspaces $\mathrm{span}(e_{i_1}, \ldots, e_{i_n})$. (I believe this theorem was first observed in this generality by Donchian and Coxeter.) We also note another related fact:

Lemma 2.6.3 (Gram identity). *If $v_1, \ldots, v_n$ are vectors in $\mathbf{R}^N$, then the square of the magnitude of $v_1 \wedge \cdots \wedge v_n$ is equal to the determinant of the* Gram matrix $(v_i \cdot v_j)_{1 \le i,j \le n}$.

Proof. Again, the statement is invariant under row operations, and one can reduce as before to the case of an orthonormal set, in which case the claim is clear. (Alternatively, one can proceed via the *Cauchy-Binet formula.*) $\square$

If we define $e_{\{i_1,\ldots,i_n\}} := e_{i_1} \wedge \cdots \wedge e_{i_n}$, then we have identified the standard basis of $\bigwedge^n \mathbf{R}^N$ with $\binom{S}{n}$ by identifying e_B with B. As a consequence of this and the multidimensional Pythagorean theorem, every unit n-vector ω in $\bigwedge^n \mathbf{R}^N$ determines an n-element point process A on S, by declaring the probability p_B of A taking the value B to equal $|\omega \cdot e_B|^2$ for each $B \in \binom{S}{n}$. Note that multiple n-vectors can generate the same point process, because only the magnitude of the coefficients $\omega \cdot e_B$ are of interest; in particular, ω and $-\omega$ generate the same point process. (This is analogous to how multiplying the wave function in quantum mechanics by a complex phase has no effect on any physical observable.)

Now we can introduce determinantal processes. If V is an n-dimensional subspace of $\mathbf{R}^N$, we can define the (projection) *determinantal process* $A = A_V$ associated to V to be the point process associated to the *volume form* of V, i.e., to the wedge product of an orthonormal basis of V. (This volume form is only determined up to sign, because the orientation of V has not been fixed, but as observed previously, the sign of the form has no impact on the resulting point process.)

By construction, the probability that the point process A is equal to a set $\{i_1, \ldots, i_n\}$ is equal to the square of the determinant of the $n \times n$ matrix consisting of the $i_1, \ldots, i_n$ coordinates of an arbitrary orthonormal basis of V. By extending such an orthonormal basis to the rest of $\mathbf{R}^N$, and representing $e_{i_1}, \ldots, e_{i_n}$ in this basis, it is not hard to see that $\mathbf{P}(A = \{i_1, \ldots, i_n\})$ can be interpreted geometrically as the square of the volume of the parallelopiped generated by $Pe_{i_1}, \ldots, Pe_{i_n}$, where P is the orthogonal projection onto V.

In fact we have the more general fact:

Lemma 2.6.4. *If $k \ge 1$ and $i_1, \ldots, i_k$ are distinct elements of S, then $\mathbf{P}(\{i_1, \ldots, i_k\} \subset A)$ is equal to the square of the k-dimensional volume of* *the parallelopiped generated by the orthogonal projections of* $Pe_{i_1}, \ldots, Pe_{i_k}$ *to* V.

Proof. We can assume that $k \leq n$, since both expressions in the lemma vanish otherwise.

By (anti-)symmetry we may assume that $\{i_1, \ldots, i_k\} = \{1, \ldots, k\}$. By the Gram-Schmidt process we can find an orthonormal basis $v_1, \ldots, v_n$ of V such that each v_i is orthogonal to $e_1, \ldots, e_{i-1}$.

Now consider the $n \times N$ matrix M with rows $v_1, \ldots, v_n$, thus M vanishes below the diagonal. The probability $\mathbf{P}(\{1, \ldots, k\} \in A)$ is equal to the sum of squares of the determinants of all the $n \times n$ minors of M that contain the first k rows. As M vanishes below the diagonal, we see from cofactor expansion that this is equal to the product of the squares of the first k diagonal entries, times the sum of squares of the determinants of all the $n - k \times n - k$ minors of the bottom $n - k$ rows. But by the generalised Pythagorean theorem, this latter factor is the square of the volume of the parallelopiped generated by $v_{k+1}, \ldots, v_n$, which is 1. Meanwhile, by the base times height formula, we see that the product of the first k diagonal entries of M is equal in magnitude to the k-dimensional volume of the orthogonal projections of $e_1, \ldots, e_k$ to V. The claim follows. □

As a special case of Lemma 2.6.4, we have $\mathbf{P}(i \in A) = \|Pe_i\|^2$ for any i. In particular, if e_i lies in V, then i almost surely lies in A, and when e_i is orthogonal to V, i almost surely is disjoint from A.

Let $K(i, j) = Pe_i \cdot e_j = Pe_i \cdot Pe_j$ denote the matrix coefficients of the orthogonal projection P. From Lemma 2.6.4 and the Gram identity, we conclude that A is a determinantal process (see (2.29)) with kernel K. Also, by combining Lemma 2.6.4 with the generalised Pythagorean theorem, we conclude a monotonicity property:

Lemma 2.6.5 (Monotonicity property). *If* $V \subset W$ *are nested subspaces of* $\mathbf{R}^N$, *then* $\mathbf{P}(B \subset A_V) \leq \mathbf{P}(B \subset A_W)$ *for every* $B \subset S$.

This seems to suggest that there is some way of representing A_W as the union of A_V with another process coupled with A_V, but I was not able to build a nonartificial example of such a representation. On the other hand, if $V \subset \mathbf{R}^N$ and $V' \subset \mathbf{R}^{N'}$, then the process $A_{V \oplus V'}$ associated with the direct sum $V \oplus V' \subset \mathbf{R}^{N+N'}$ has the same distribution of the disjoint union of A_V with an independent copy of $A_{V'}$.

The determinantal process interacts nicely with complements:

Lemma 2.6.6 (Hodge duality). *Let* V *be an* n*-dimensional subspace of* $\mathbf{R}^N$. *The* $N - n$*-element determinantal process* $A_{V^\perp}$ *associated to the orthogonal* *complement* $V^\perp$ *of* V *has the same distribution as the complement* $S\backslash A_V$ *of the* n*-element determinantal process* A_V *associated to* V.

Proof. We need to show that $\mathbf{P}(A_V = B) = \mathbf{P}(A_{V^\perp} = S\backslash B)$ for all $B \in \binom{N}{n}$. By symmetry we can take $B = \{1, \ldots, n\}$. Let $v_1, \ldots, v_n$ and $v_{n+1}, \ldots, v_N$ be an orthonormal basis for V and $V^\perp$, respectively, and let M be the resulting $N \times N$ orthogonal matrix; then the task is to show that the top $n \times n$ minor X of M has the same determinant squared as the bottom $N-n \times N-n$ minor Y. But if one splits $M = \begin{pmatrix} X & Z \\ W & Y \end{pmatrix}$, we see from the orthogonality property that $XX^* = I_n - ZZ^*$ and $Y^*Y = I_{N-n} - Z^*Z$, where I_n is the $n \times n$ identity matrix. But from the *singular value decomposition*, we see that $I_n - ZZ^*$ and $I_{N-n} - Z^*Z$ have the same determinant, and the claim follows. (One can also establish this lemma using the *Hodge star operation*.) □

From this lemma we see that $S\backslash A$ is a determinantal process with kernel $I_N - K$. In particular, we have

$$\mathbf{P}(\{i_1, \ldots, i_k\} \cap A = \emptyset) = \det(I_k - (K(i_a, i_b))_{1 \le a,b \le k}). \tag{2.30}$$

The construction of the determinantal process given above is somewhat indirect. A more direct way to build the process exploits the following lemma:

Lemma 2.6.7. *Let* V *be an* n*-dimensional subspace of* $\mathbf{R}^N$, *let* A_V *be the corresponding* n*-element determinantal process, and let* $1 \le i_1 < \cdots < i_k \le N$ *for some* $1 \le k \le n$. *Then if one conditions on the event that* $\{i_1, \ldots, i_k\} \in A_V$ *(assuming this event has nonzero probability), the resulting* $n-k$*-element process* $A_V \backslash \{i_1, \ldots, i_k\}$ *has the same distribution as the* $n-k$*-element determinantal process* A_W *associated to the* $n-k$*-dimensional subspace* W *of* V *that is orthogonal to* $e_{i_1}, \ldots, e_{i_k}$.

Proof. By symmetry it suffices to consider the case $\{i_1, \ldots, i_k\} = \{1, \ldots, k\}$. By a further application of symmetry it suffices to show that

$$\mathbf{P}(A_V = \{1, \ldots, n\}) = \mathbf{P}(\{1, \ldots, k\} \subset A_V)\mathbf{P}(A_W = \{k+1, \ldots, n\}).$$

By the Gram-Schmidt process, we can find an orthonormal basis $v_1, \ldots, v_n$ of V whose $n \times N$ matrix of coefficients vanishes below the diagonal. One then easily verifies (using Lemma 2.6.4) that $\mathbf{P}(A_V = \{1, \ldots, n\})$ is the product of the n diagonal entries, $\mathbf{P}(\{1, \ldots, k\} \subset A_V)$ is the product of the first k, and $\mathbf{P}(A_W = \{k+1, \ldots, n\})$ is the product of the last $n-k$, and the claim follows. □

Remark 2.6.8. There is a dual version of this lemma: if one conditions on the event that $\{i_1, \ldots, i_k\}$ is disjoint from A_V, then the resulting process is the determinantal process associated to the orthogonal projection of V to the orthogonal complement of $e_{i_1}, \ldots, e_{i_k}$.

From this lemma, it is not difficult to see that one can build A_V recursively as $A_V = \{a\} \cup A_{V_a}$, where a is a random variable drawn from S with a $\mathbf{P}(a = i) = \|Pe_i\|^2/\dim(V)$ for each i, and V_a is the subspace of V orthogonal to e_a. Another consequence of this lemma and the monotonicity property is the negative dependence inequality

$$\mathbf{P}(B_1 \cup B_2 \subset A) \leq \mathbf{P}(B_1 \subset A)\mathbf{P}(B_2 \subset A)$$

for any disjoint $B_1, B_2 \subset S$. Thus the presence of A on one set B_1 reduces the chance of A being present on a disjoint set B_2 (not surprising, since A has fixed size).

Thus far, we have only considered point processes with a fixed number n of points. As a consequence, the determinantal kernel K involved here is of a special form, namely the coefficients of an orthogonal projection matrix to an n-dimensional space (or equivalently, a symmetric matrix whose eigenvalues consist of n ones and $N - n$ zeroes). But one can create more general point processes by taking a *mixture* of the fixed-number processes, e.g., first picking a projection kernel K (or a subspace V) by some random process, and then sampling A from the point process associated to that kernel or subspace.

For instance, let $\phi_1, \ldots, \phi_N$ be an orthonormal basis of $\mathbf{R}^N$, and let $0 \leq \lambda_1, \ldots, \lambda_N \leq 1$ be weights. Then we can create a random subspace V of $\mathbf{R}^N$ by setting V equal to the span V_J of some random subset $\{\phi_j : j \in J\}$ of the basis $v_1, \ldots, v_N$, where each j lies in J with an independent probability of λ_j, and then sampling A from A_V. Then A will be a point process whose cardinality can range from 0 to N. Given any set $\{i_1, \ldots, i_k\} \subset S$, we can then compute the probability $\mathbf{P}(\{i_1, \ldots, i_k\} \subset A)$ as

$$\mathbf{P}(\{i_1, \ldots, i_k\} \subset A) = \mathbf{E}_J \mathbf{P}(\{i_1, \ldots, i_k\} \subset A_{V_J}),$$

where J is selected as above. Using (2.29), we have

$$\mathbf{P}(\{i_1, \ldots, i_k\} \subset A_{V_J}) = \det(K_{V_J}(i_a, i_b))_{1 \leq a,b \leq k}.$$

But $K_{V_J}(i_a, i_b) = \sum_{j \in J} \phi_j(i_a)\phi_j(i_b)$, where $\phi_j(i)$ is the ith coordinate of ϕ_j. Thus we can write

$$(K_{V_J}(i_a, i_b))_{1 \leq a,b \leq k} = \sum_{j=1}^{N} \mathbf{I}(j \in J) R_j,$$

where $\mathbf{I}(j \in J)$ is the indicator of the event $j \in J$, and R_j is the rank one matrix $(\phi_j(i_a)\phi_j(i_b))_{1 \leq a,b \leq k}$. Using multilinearity of the determinant and the fact that any determinant involving two or more rows of the same rank one matrix automatically vanishes, we see that we can express

$$\det((K_{V_J}(i_a,i_b))_{1\leq a,b\leq k}) = \sum_{1\leq j_1,\ldots,j_k\leq N,\text{distinct}} \mathbf{I}(j_1,\ldots,j_k \in J)\det(R_{j_1,\ldots,j_k}),$$

where $R_{j_1,\ldots,j_k}$ is the matrix whose first row is the same as that of R_{j_1}, the second row is the same as that of R_{j_2}, and so forth. Taking expectations in J, the quantity $\mathbf{I}(j_1,\ldots,j_k \in J)$ becomes $\lambda_{j_1}\ldots\lambda_{j_k}$. Undoing the multilinearity step, we conclude that

$$\mathbf{E}_J \det(K_{V_J}(i_a,i_b))_{1\leq a,b\leq k} = \det(\sum_{j=1}^N \lambda_j R_j),$$

and thus A is a determinantal process with kernel

$$K(x,y) := \sum_{j=1}^N \lambda_j \phi_j(x)\phi_j(y).$$

To summarise, we have created a determinantal process A whose kernel K is now an arbitrary symmetric matrix with eigenvalues $\lambda_1,\ldots,\lambda_n \in [0,1]$, and it is a mixture of constant-size processes A_{V_J}. In particular, the cardinality $|A|$ of this process has the same distribution as the cardinality $|J|$ of the random subset of $\{1,\ldots,N\}$, or in other words $|A| \equiv I_{\lambda_1} + \cdots + I_{\lambda_k}$, where $I_{\lambda_1},\ldots,I_{\lambda_k}$ are independent Bernoulli variables with expectation $\lambda_1,\ldots,\lambda_k$, respectively.

Observe that if one takes a determinantal process $A \subset S$ with kernel K, and restricts it to a subset S' of S, then the resulting process $A \cap S' \subset S'$ is a determinantal process whose kernel K' is simply the restriction of K to the $S' \times S'$ block of $S \times S$. Applying the previous observation, we conclude that the random variable $|A \cap S'|$ has the same distribution as the sum of $|S'|$ independent Bernoulli variables, whose expectations are the eigenvalues of the restriction of K to S'. (Compare this to the Poisson point process A with some intensity measure λ, where the distribution of $|A\cap\Omega|$ is a Poisson process with intensity $\lambda(\Omega)$.) Note that most point processes do not obey this property (e.g., the uniform distribution on $\binom{S}{n}$ does not unless $n = 0, 1$ or $n = N, N-1$), and so most point processes are not determinantal.

It is known that increasing a positive semidefinite matrix by another positive semidefinite matrix does not decrease the determinant (indeed, it does not decrease any eigenvalue, by the minimax characterisation of those eigenvalues). As a consequence, if the kernel K' of a determinantal process A' is larger than the kernel K of another determinantal process A in the sense that $K - K'$ is positive semidefinite, then A' is "larger" than A in the sense that $\mathbf{P}(B \subset A') \geq \mathbf{P}(B \subset A)$ for all $B \subset S$. A particularly nice special case is when $K = cK'$ for some $0 \leq c \leq 1$, then $\mathbf{P}(B \subset A) = c^{|B|}\mathbf{P}(B \subset A')$ for all B, and one can interpret A as the process obtained from A' by deleting each element of A' independently at random with probability $1 - c$ (i.e., keeping that element independently at random with probability c).

As a consequence of this, one can obtain a converse to our previous construction of determinantal processes, and conclude that a determinantal process can be associated to a symmetric kernel K only if the eigenvalues of K lie between zero and one. The fact that K is positive semidefinite follows from the fact that all symmetric minors of K have nonnegative determinant (thanks to (2.29)). Now suppose for contradiction that K has an eigenvalue larger than 1, then one can find $0 \leq c < 1$ such that the largest eigenvalue of cK is exactly 1. By our previous discussion, the process A_{cK} associated to cK is then formed from the process A_K by deleting each element of A with nonzero probability; in particular, A_K is empty with nonzero probability. On the other hand, we know that $|A_K|$ has the distribution of the sum of independent Bernoulli variables, at least one of which is 1 with probability one, a contradiction. (This proof is due to [**HoKrPeVi2006**], though the result is originally due to Soshnikov [**So2000**]. An alternate proof is to extend the identity (2.30) to all determinantal processes and conclude that $I - K$ is necessarily positive definite.)

2.6.2. Continuous determinantal processes. One can extend the theory of discrete determinantal processes to the continuous setting. For simplicity we restrict our attention to (simple) point processes $A \subset \mathbf{R}$ on the real line. A process A is said to have *correlation functions* $\rho_k : \mathbf{R}^k \to \mathbf{R}$ for $k \geq 1$ if the ρ_k are symmetric, nonnegative, and locally integrable, and one has the formula

$$\mathbf{E} \sum_{x_1,\ldots,x_k \in A, \text{distinct}} f(x_1, \ldots, x_k) = \int_{\mathbf{R}^k} f(x_1, \ldots, x_k)\rho_k(x_1, \ldots, x_k)\, dx_1 \cdots dx_k$$

for any bounded measurable symmetric f with compact support, where the left-hand side is summed over all k-tuples of distinct points in A (this sum is of course empty if $|A| \leq k$). Intuitively, the probability that A contains an element in the infinitesimal interval $[x_i, x_i + dx_i]$ for all $1 \leq i \leq k$ and distinct $x_1, \ldots, x_k$ is equal to $\rho_k(x_1, \ldots, x_k)dx_1 \cdots dx_k$. The ρ_k are not quite probability distributions; instead, the integral $\int_{\mathbf{R}^k} \rho_k$ is equal to $k!\mathbf{E}\binom{|A|}{k}$. Thus, for instance, if A is a constant-size process of cardinality n, then ρ_k has integral $\frac{n!}{(n-k)!}$ on $\mathbf{R}^n$ for $1 \leq k \leq n$ and vanishes for $k > n$.

If the correlation functions exist, it is easy to see that they are unique (up to almost everywhere equivalence), and can be used to compute various statistics of the process. For instance, an application of the inclusion-exclusion principle shows that for any bounded measurable set Ω, the probability that

$A \cap \Omega = \emptyset$ is (formally) equal to

$$\sum_{k=0}^{\infty} \frac{(-1)^k}{k!} \int_{(\mathbf{R}\backslash\Omega)^k} \rho_k(x_1, \ldots, x_k) \, dx_1 \cdots dx_k.$$

A process is *determinantal* with some symmetric measurable kernel $K : \mathbf{R} \times \mathbf{R} \to \mathbf{R}$ if it has correlation functions ρ_k given by the formula

$$\rho_k(x_1, \ldots, x_k) = \det(K(x_i, x_j))_{1 \leq i,j \leq k}. \tag{2.31}$$

Informally, the probability that A intersects the infinitesimal intervals $[x_i, x_i + dx_i]$ for distinct $x_1, \ldots, x_k$ is $\det(K(x_i, x_j) dx_i^{1/2} dx_j^{1/2})_{1 \leq i,j \leq k}$. (Thus, K is most naturally interpreted as a half-density, or as an integral operator from $L^2(\mathbf{R})$ to $L^2(\mathbf{R})$.)

There are analogues of the discrete theory in this continuous setting. For instance, one can show that a symmetric measurable kernel K generates a determinantal process if and only if the associated integral operator $\mathcal{K}$ has spectrum lies in the interval $[0, 1]$. The analogue of (2.30) is the formula

$$\mathbf{P}(A \cap \Omega = \emptyset) = \det(I - \mathcal{K}|_\Omega);$$

more generally, the distribution of $|A \cap \Omega|$ is the sum of independent Bernoulli variables, whose expectations are the eigenvalues of $\mathcal{K}|_\Omega$. Finally, if $\mathcal{K}$ is an orthogonal projection onto an n-dimensional space, then the process has a constant size of n. Conversely, if A is a process of constant size n, whose nth correlation function $\rho_n(x_1, \ldots, x_n)$ is given by (2.31), where $\mathcal{K}$ is an orthogonal projection onto an n-dimensional space, then (2.31) holds for all other values of k as well, and so A is a determinantal process with kernel K. (This is roughly the analogue of Lemma 2.6.4.)

These facts can be established either by approximating a continuous process as the limit of discrete ones or by obtaining alternate proofs of several of the facts in the previous section which do not rely as heavily on the discrete hypotheses; see [**HoKrPeVi2006**] for details.

A Poisson process can be viewed as the limiting case of a determinantal process in which $\mathcal{K}$ degenerates to a (normalisation of) a multiplication operator $f \mapsto \lambda f$, where λ is the intensity function.

2.6.3. The spectrum of GUE.

Now we turn to a specific example of a continuous point process, namely the spectrum $A = \{\lambda_1, \ldots, \lambda_n\} \subset \mathbf{R}$ of the *Gaussian unitary ensemble* $M_n = (\zeta_{ij})_{1 \leq i,j \leq n}$, where the ζ_{ij} are independent for $1 \leq i \leq j \leq n$ with mean zero and variance 1, with ζ_{ij} being the standard complex Gaussian for $i < j$ and the standard real Gaussian $N(0, 1)$ for $i = j$. The probability distribution of M_n can be expressed as

$$c_n \exp(-\frac{1}{2} \operatorname{tr}(M_n^2)) \, dM_n,$$

where dM_n is Lebesgue measure on the space of Hermitian $n \times n$ matrices, and $c_n > 0$ is some explicit normalising constant.

The n-point correlation function of A can be computed explicitly:

Lemma 2.6.9 (Ginibre formula). *The n-point correlation function $\rho_n(x_1, \ldots, x_n)$ of the GUE spectrum A is given by*

$$\rho_n(x_1, \ldots, x_n) = c'_n\Big(\prod_{1 \le i < j \le n} |x_i - x_j|^2\Big) \exp\Big(-\sum_{i=1}^n x_i^2/2\Big), \tag{2.32}$$

where the normalising constant c'_n is chosen so that ρ_n has integral 1.

The constant $c'_n > 0$ is essentially the reciprocal of the *partition function* for this ensemble and can be computed explicitly, but we will not do so here.

Proof. Let D be a diagonal random matrix $D = \operatorname{diag}(x_1, \ldots, x_n)$ whose entries are drawn using the distribution $\rho_n(x_1, \ldots, x_n)$ defined by (2.32), and let $U \in U(n)$ be a unitary matrix drawn uniformly at random (with respect to Haar measure on $U(n)$) and independently of D. It will suffice to show that the GUE M_n has the same probability distribution as U^*DU. Since probability distributions have total mass one, it suffices to show that their distributions differ up to multiplicative constants.

The distributions of M_n and U^*DU are easily seen to be continuous and invariant under unitary rotations. Thus, it will suffice to show that their probability density at a given diagonal matrix $D_0 = \operatorname{diag}(x_1^0, \ldots, x_n^0)$ are the same up to multiplicative constants. We may assume that the x_i^0 are distinct, since this occurs for almost every choice of D_0.

On the one hand, the probability density of M_n at D_0 is proportional to $\exp(-\sum_{i=1}^n (x_i^0)^2/2)$. On the other hand, a short computation shows that if U^*DU is within a distance $O(\varepsilon)$ of D_0 for some infinitesimal $\varepsilon > 0$, then (up to permutations) D must be a distance $O(\varepsilon)$ from D_0, and the ij entry of U must be a complex number of size $O(\varepsilon/|x_i^0 - x_j^0|)$ for $1 \le i < j \le n$, while the diagonal entries of U can be arbitrary phases. Pursuing this computation more rigorously (e.g., using the Harish-Chandra formula) and sending $\varepsilon \to 0$, one can show that the probability density of U^*DU at D_0 is a constant multiple of

$$\rho_n(x_1, \ldots, x_n) \prod_{1 \le i < j \le n} \frac{1}{|x_i^0 - x_j^0|^2}$$

(the square here arising because of the complex nature of the ij coefficient of U), and the claim follows. □

One can also represent the k-point correlation functions as a determinant:

Lemma 2.6.10 (Gaudin-Mehta formula). *The k-point correlation function $\rho_k(x_1,\dots,x_n)$ of the GUE spectrum A is given by*

$$\rho_k(x_1,\dots,x_k) = \det(K_n(x_i,x_j))_{1\le i<j\le k}, \tag{2.33}$$

where $K_n(x,y)$ is the kernel of the orthogonal projection $\mathcal{K}$ in $L^2(\mathbf{R})$ to the space spanned by the polynomials $x^i e^{-x^2/4}$ for $i = 0,\dots,n-1$. In other words, A is the n-point determinantal process with kernel K_n.

Proof. By the material in the preceding section, it suffices to establish this for $k = n$. As K is the kernel of an orthogonal projection to an n-dimensional space, it generates an n-point determinantal process and so $\det(K_n(x_i,x_j))_{1\le i<j\le n}$ has integral $\binom{n}{n} = 1$. Thus it will suffice to show that ρ_n and $\det(K_n(x_i,x_j))_{1\le i<j\le n}$ agree up to multiplicative constants.

By Gram-Schmidt, one can find an orthonormal basis $\phi_i(x)e^{-x^2/4}$, $i = 0,\dots,n-1$ for the range of $\mathcal{K}$, with each ϕ_i a polynomial of degree i (these are essentially the *Hermite polynomials*). Then we can write

$$K_n(x_i,x_j) = \sum_{k=0}^{n-1} \phi_k(x_i)\phi_k(x_j)e^{-(x_i^2+x_j^2)/4}.$$

Cofactor expansion then shows that $\det(K_n(x_i,x_j))_{1\le i<j\le n}$ is equal to

$$\exp(-\sum_{i=1}^n x_i^2/2)$$

times a polynomial $P(x_1,\dots,x_n)$ in $x_1,\dots,x_n$ of degree at most $2\sum_{k=0}^{n-1} k = n(n-1)$. On the other hand, this determinant is always nonnegative, and vanishes whenever $x_i = x_j$ for any $1 \le i < j \le n$, and so must contain $(x_i - x_j)^2$ as a factor for all $1 \le i < j \le n$. As the total degree of all these (relatively prime) factors is $n(n-1)$, the claim follows. □

This formula can be used to obtain asymptotics for the (renormalised) GUE eigenvalue spacings in the limit $n \to \infty$, by using asymptotics for (renormalised) Hermite polynomials; this was first established by Dyson [**Dy1970**].

Notes. This article first appeared at

`terrytao.wordpress.com/2009/08/23.`

Thanks to anonymous commenters for corrections.

Craig Tracy noted that some nondeterminantal processes, such as TASEP, still enjoy many of the spacing distributions as their determinantal counterparts.

Manju Krishnapur raised the relevant question of how one could determine quickly whether a given process is determinantal.

Russell Lyons noted the open problem on coupling determinantal processes together was also raised in Question 10.1 of [**Ly2003**] (which also covers most of the other material in this article).

2.7. The Cohen-Lenstra distribution

At a conference recently, I learned of the recent work of Ellenberg, Venkatesh, and Westerland [**ElVeWe2009**], which concerned the conjectural behaviour of *class groups* of quadratic fields, and in particular it explained the numerically observed phenomenon that about 75.4% of all quadratic fields $\mathbf{Q}[\sqrt{d}]$ (with d prime) enjoy unique factorisation (i.e., have trivial class group). (Class groups, as I learned at this conference, are arithmetic analogues of the (abelianised) fundamental groups in topology, with Galois groups serving as the analogue of the full fundamental group.) One thing I learned here was that there was a canonical way to randomly generate a (profinite) abelian group, by taking the product of randomly generated finite abelian *p-groups* for each prime p. The way to canonically randomly generate a finite abelian p-group is to take large integers n, d, and look at the cokernel of a random homomorphism from $(\mathbf{Z}/p^n\mathbf{Z})^d$ to $(\mathbf{Z}/p^n\mathbf{Z})^d$. In the limit $n, d \to \infty$ (or by replacing $\mathbf{Z}/p^n\mathbf{Z}$ with the p-adics and just sending $d \to \infty$), this stabilises and generates any given p-group G with probability

$$\frac{1}{|\operatorname{Aut}(G)|} \prod_{j=1}^{\infty} (1 - \frac{1}{p^j}), \tag{2.34}$$

where $\operatorname{Aut}(G)$ is the group of automorphisms of G. In particular this leads to the strange identity

$$\sum_G \frac{1}{|\operatorname{Aut}(G)|} = \prod_{j=1}^{\infty} (1 - \frac{1}{p^j})^{-1}, \tag{2.35}$$

where G ranges over all p-groups; I do not know how to prove this identity other than via the above probability computation, the proof of which I give below.

Based on the heuristic that the class group should behave "randomly" subject to some "obvious" constraints, it is expected that a randomly chosen real quadratic field $\mathbf{Q}[\sqrt{d}]$ has unique factorisation (i.e., the class group has trivial p-group component for every p) with probability

$$\prod_{p \text{ odd}} \prod_{j=2}^{\infty} (1 - \frac{1}{p^j}) \approx 0.754,$$

whereas a randomly chosen imaginary quadratic field $\mathbf{Q}[\sqrt{-d}]$ has unique factorisation with probability

$$\prod_{p \text{ odd}} \prod_{j=1}^{\infty} (1 - \frac{1}{p^j}) = 0.$$

The former claim is conjectural, whereas the latter claim follows from (for instance) Siegel's theorem on the size of the class group, as discussed in Section 1.12.4. The work in [**ElVeWe2009**] establishes some partial results towards the function field analogues of these heuristics.

2.7.1. p-groups. Henceforth the prime p will be fixed. We will abbreviate "finite abelian p-group" as "p-group" for brevity. Thanks to the *classification of finite abelian groups*, the p-groups are all isomorphic to the products

$$(\mathbf{Z}/p^{n_1}\mathbf{Z}) \times \cdots \times (\mathbf{Z}/p^{n_d}\mathbf{Z})$$

of cyclic p-groups.

The cokernel of a random homomorphism from $(\mathbf{Z}/p^n\mathbf{Z})^d$ to $(\mathbf{Z}/p^n\mathbf{Z})^d$ can be written as the quotient of the p-group $(\mathbf{Z}/p^n\mathbf{Z})^d$ by the subgroup generated by d randomly chosen elements $x_1, \ldots, x_d$ from that p-group. One can view this quotient as a d-fold iterative process, in which one starts with the p-group $(\mathbf{Z}/p^n\mathbf{Z})^d$, and then one iterates d times the process of starting with a p-group G, and quotienting out by a randomly chosen element x of that group G. From induction, one sees that at the jth stage of this process $(0 \leq j \leq d)$, one ends up with a p-group isomorphic to $(\mathbf{Z}/p^n\mathbf{Z})^{d-j} \times G_j$ for some p-group G_j.

Let us see how the group $(\mathbf{Z}/p^n\mathbf{Z})^{d-j} \times G_j$ transforms to the next group $(\mathbf{Z}/p^n\mathbf{Z})^{d-j-1} \times G_{j+1}$. We write a random element of $(\mathbf{Z}/p^n\mathbf{Z})^{d-j} \times G_j$ as (x, y), where $x \in (\mathbf{Z}/p^n\mathbf{Z})^{d-j}$ and $y \in G_j$. Observe that for any $0 \leq i < n$, x is a multiple of p^i (but not p^{i+1}) with probability $(1 - p^{-(d-j)})p^{-i(d-j)}$. (The remaining possibility is that x is zero, but this event will have negligible probability in the limit $n \to \infty$.) If x is indeed divisible by p^i but not p^{i+1}, and i is not too close to n, a little thought will then reveal that $|G_{j+1}| = p^i|G_j|$. Thus the size of the p-groups G_j only grow as j increases. (Things go wrong when i gets close to n, e.g., $p^i \geq p^n/|G_j|$, but the total size of this event as j ranges from 0 to d sums to be $o(1)$ as $n \to \infty$ (uniformly in d), by using the tightness bounds on $|G_j|$ mentioned below. Alternatively, one can avoid a lot of technicalities by taking the limit $n \to \infty$ before taking the limit $d \to \infty$ (instead of studying the double limit $n, d \to \infty$), or equivalently by replacing the cyclic group $\mathbf{Z}/p^n\mathbf{Z}$ with the p-adics $\mathbf{Z}_p$).

Furthermore, the exponentially decreasing nature of the probability $(1 - p^{-(d-j)})p^{-i(d-j)}$ in i (and in $d - j$) implies that the distribution of $|G_j|$ forms a *tight sequence* in n, j, d: for every $\varepsilon > 0$, one has an $R > 0$ such that the probability that $|G_j| \geq R$ is less than ε for all choices of n, j, d. (This tightness is necessary to prove the equality in (2.35) rather than just an inequality (from Fatou's lemma).) Indeed, the probability that $|G_j| = p^m$ converges as $n, d \to \infty$ to the t^m coefficient in the generating function

$$\prod_{k=1}^{\infty} \sum_{i=0}^{\infty} t^i (1 - p^{-k}) p^{-ik} = \prod_{k=1}^{\infty} \frac{1 - p^{-k}}{1 - tp^{-k}}. \tag{2.36}$$

In particular, this claim is true for the final cokernel G_d. Note that this (and the geometric series formula) already yields (2.34) in the case of the trivial group $G = \{0\}$ and the order p group $G = \mathbf{Z}/p\mathbf{Z}$ (note that $\mathrm{Aut}(G)$ has order 1 and p in these respective cases). But it is not enough to deal with higher groups. For instance, up to isomorphism there are two p-groups of order p^2, namely $\mathbf{Z}/p^2\mathbf{Z}$ and $(\mathbf{Z}/p\mathbf{Z})^2$, whose automorphism group has order $p^2 - p$ and $(p^2 - 1)(p^2 - p)$, respectively. Summing up the corresponding two expressions (2.34) one can observe that this matches the t^2 coefficient of (2.36) (after some applications of the geometric series formula). Thus we see that (2.36) is consistent with the claim (2.34), but does not fully imply that claim.

To get the full asymptotic (2.34), we try a slightly different tack. Fix a p-group G, and consider the event that the cokernel of a random map $T : (\mathbf{Z}/p^n\mathbf{Z})^d \to (\mathbf{Z}/p^n\mathbf{Z})^d$ is isomorphic to G. We assume n so large that all elements in G have order at most p^n. If this is the case, then there must be a surjective homomorphism $\phi : (\mathbf{Z}/p^n\mathbf{Z})^d \to G$ such that the range of T is equal to the kernel of ϕ. The number of homomorphisms from $(\mathbf{Z}/p^n\mathbf{Z})^d$ to G is $|G|^d$ (one has to pick d generators in G). If d is large, it is easy to see that most of these homomorphisms are surjective (the proportion of such homomorphisms is $1 - o(1)$ as $d \to \infty$). On the other hand, there is some multiplicity; the range of T can emerge as the kernel of ϕ in $|\mathrm{Aut}(G)|$ different ways (since any two surjective homomorphisms $\phi, \phi' : (\mathbf{Z}/p^n\mathbf{Z})^d \to G$ with the same kernel arise from an automorphism of G). So to prove (2.34), it suffices to show that for any surjective homomorphism $\phi : (\mathbf{Z}/p^n\mathbf{Z})^d \to G$, the probability that the range of T equals the kernel of ϕ is

$$(1 + o(1))|G|^{-d} \prod_{j=1}^{\infty} (1 - \frac{1}{p^j}).$$

The range of T is the same thing as the subgroup of $(\mathbf{Z}/p^n\mathbf{Z})^d$ generated by d random elements $x_1, \ldots, x_d$ of that group. The kernel of ϕ has index $|G|$ inside $(\mathbf{Z}/p^n\mathbf{Z})^d$, so the probability that all of those random elements lie in the kernel of ϕ is $|G|^{-d}$. So it suffices to prove the following claim: if ϕ is a fixed surjective homomorphism from $(\mathbf{Z}/p^n\mathbf{Z})^d$ to G and $x_1, \ldots, x_d$ are chosen randomly from the kernel of ϕ, then $x_1, \ldots, x_d$ will generate that kernel with probability

$$(1+o(1))\prod_{j=1}^{\infty}(1-\frac{1}{p^j}). \tag{2.37}$$

But from the classification of p-groups, the kernel of ϕ (which has bounded index inside $(\mathbf{Z}/p^n\mathbf{Z})^d$) is isomorphic to

$$(\mathbf{Z}/p^{n-O(1)}\mathbf{Z}) \times \cdots \times (\mathbf{Z}/p^{n-O(1)}\mathbf{Z}), \tag{2.38}$$

where $O(1)$ means "bounded uniformly in n", and there are d factors here. As in the previous argument, one can now imagine starting with the group (2.38), and then iterating d times the operation of quotienting out by the group generated by a randomly chosen element; our task is to compute the probability that one ends up with the trivial group by applying this process.

As before, at the jth stage of the iteration, one ends up with a group of the form

$$(\mathbf{Z}/p^{n-O(1)}\mathbf{Z}) \times \cdots \times (\mathbf{Z}/p^{n-O(1)}\mathbf{Z}) \times G_j, \tag{2.39}$$

where there are $d-j$ factors of $(\mathbf{Z}/p^{n-O(1)}\mathbf{Z})$. The group G_j is increasing in size, so the only way in which one ends up with the trivial group is if all the G_j are trivial. But if G_j is trivial, the only way that G_{j+1} is trivial is if the randomly chosen element from (2.39) has a $(\mathbf{Z}/p^{n-O(1)}\mathbf{Z}) \times \cdots \times (\mathbf{Z}/p^{n-O(1)}\mathbf{Z})$ component which is invertible (i.e., not a multiple of p), which occurs with probability $1-p^{-(d-j)}$ (assuming n is large enough). Multiplying all these probabilities together gives (2.37).

Notes. This article first appeared at

`terrytao.wordpress.com/2009/10/02.`

Thanks to David Speyer and an anonymous commenter for corrections.

2.8. An entropy Plünnecke-Ruzsa inequality

A handy inequality in additive combinatorics is the *Plünnecke-Ruzsa inequality* [**Ru1989**]:

Theorem 2.8.1 (Plünnecke-Ruzsa inequality). *Let $A, B_1, \ldots, B_m$ be finite nonempty subsets of an additive group G such that $|A+B_i| \leq K_i|A|$ for all $1 \leq i \leq m$ and some scalars $K_1, \ldots, K_m \geq 1$. Then there exists a subset A' of A such that $|A'+B_1+\cdots+B_m| \leq K_1 \cdots K_m|A'|$.*

The proof uses graph-theoretic techniques. Setting $A = B_1 = \cdots = B_m$, we obtain a useful corollary: if A has small doubling in the sense that $|A+A| \leq K|A|$, then we have $|mA| \leq K^m|A|$ for all $m \geq 1$, where $mA = A+\cdots+A$ is the sum of m copies of A.

In a recent paper [**Ta2010c**], I adapted a number of sum set estimates to the entropy setting, in which finite sets such as A in G are replaced with discrete random variables X taking values in G, and (the logarithm of) cardinality $|A|$ of a set A is replaced by *Shannon entropy* $\mathbf{H}(X)$ of a random variable X. (Throughout this note I assume all entropies to be finite.) However, at the time, I was unable to find an entropy analogue of the Plünnecke-Ruzsa inequality, because I did not know how to adapt the graph theory argument to the entropy setting.

I recently discovered, however, that buried in a classic paper of Kaimonovich and Vershik [**KaVe1983**] (implicitly in Proposition 1.3, to be precise) there was the following analogue of Theorem 2.8.1:

Theorem 2.8.2 (Entropy Plünnecke-Ruzsa inequality). *Let $X, Y_1, \ldots, Y_m$ be independent random variables of finite entropy taking values in an additive group G such that $\mathbf{H}(X + Y_i) \leq \mathbf{H}(X) + \log K_i$ for all $1 \leq i \leq m$ and some scalars $K_1, \ldots, K_m \geq 1$. Then $\mathbf{H}(X+Y_1+\cdots+Y_m) \leq \mathbf{H}(X)+\log K_1 \cdots K_m$.*

In fact Theorem 2.8.2 is a bit "better" than Theorem 2.8.1 in the sense that Theorem 2.8.1 needed to refine the original set A to a subset A', but no such refinement is needed in Theorem 2.8.2. One corollary of Theorem 2.8.2 is that if $\mathbf{H}(X_1 + X_2) \leq \mathbf{H}(X) + \log K$, then $\mathbf{H}(X_1 + \cdots + X_m) \leq \mathbf{H}(X)+(m-1)\log K$ for all $m \geq 1$, where $X_1, \ldots, X_m$ are independent copies of X; this improves slightly over the analogous combinatorial inequality. Indeed, the function $m \mapsto \mathbf{H}(X_1 + \cdots + X_m)$ is concave (this can be seen by using the $m = 2$ version of Theorem 2.8.2 (or (2.41) below) to show that the quantity $\mathbf{H}(X_1 + \cdots + X_{m+1}) - \mathbf{H}(X_1 + \cdots + X_m)$ is decreasing in m).

Theorem 2.8.2 is actually a quick consequence of the *submodularity inequality*

$$\mathbf{H}(W) + \mathbf{H}(X) \leq \mathbf{H}(Y) + \mathbf{H}(Z) \tag{2.40}$$

in information theory, which is valid whenever X, Y, Z, W are discrete random variables such that Y and Z each determine X (i.e., X is a function of Y, and also a function of Z), and Y and Z jointly determine W (i.e., W is a function of Y and Z). To apply this, let X, Y, Z be independent discrete random variables taking values in G. Observe that the pairs $(X, Y + Z)$ and $(X + Y, Z)$ each determine $X + Y + Z$, and jointly determine (X, Y, Z). Applying (2.40) we conclude that

$$\mathbf{H}(X, Y, Z) + \mathbf{H}(X + Y + Z) \leq \mathbf{H}(X, Y + Z) + \mathbf{H}(X + Y, Z),$$

which after using the independence of X, Y, Z simplifies to the *sumset submodularity inequality*

$$\mathbf{H}(X + Y + Z) + \mathbf{H}(Y) \leq \mathbf{H}(X + Y) + \mathbf{H}(Y + Z) \tag{2.41}$$

(this inequality was also recently observed

`http://www.stat.yale.edu/~mm888/Pubs/2008/ITW-sums08.pdf.`

by Madiman; it is the $m = 2$ case of Theorem 2.8.2). As a corollary of this inequality, we see that if $\mathbf{H}(X + Y_i) \leq \mathbf{H}(X) + \log K_i$, then

$$\mathbf{H}(X + Y_1 + \cdots + Y_i) \leq \mathbf{H}(X + Y_1 + \cdots + Y_{i-1}) + \log K_i,$$

and Theorem 2.8.2 follows by telescoping series.

The proof of Theorem 2.8.2 seems to be genuinely different from the graph-theoretic proof of Theorem 2.8.1. It would be interesting to see if the above argument can be somehow adapted to give a stronger version of Theorem 2.8.1. Note also that both Theorem 2.8.1 and Theorem 2.8.2 have extensions to more general combinations of $X, Y_1, \ldots, Y_m$ than $X + Y_i$; see [**GyMaRu2008**] and [**Ma2008**], respectively.

It is also worth remarking that the above inequalities largely carry over to the non-abelian setting. For instance, if $X_1, X_2, \ldots$ are iid copies of a discrete random variable in a multiplicative group G, the above arguments show that the function $m \mapsto \mathbf{H}(X_1 \cdots X_m)$ is concave. In particular, the expression $\frac{1}{m}\mathbf{H}(X_1 \cdots X_m)$ decreases monotonically to a limit, the *asymptotic entropy* $\mathbf{H}(G, X)$. This quantity plays an important role in the theory of bounded harmonic functions on G, as observed by [**KaVe1983**]:

Proposition 2.8.3. *Let G be a discrete group, and let X be a discrete random variable in G with finite entropy whose support generates G. Then there exists a nonconstant bounded function $f : G \to \mathbf{R}$ which is harmonic with respect to X (which means that $\mathbf{E}f(Xx) = f(x)$ for all $x \in G$) if and only if $\mathbf{H}(G, X) \neq 0$.*

Proof (Sketch). Suppose first that $\mathbf{H}(G, X) = 0$, then we see from concavity that the successive differences $\mathbf{H}(X_1 \cdots X_m) - \mathbf{H}(X_1 \cdots X_{m-1})$ converge to zero. From this it is not hard to see that the *mutual information*

$$\mathbf{I}(X_m, X_1 \cdots X_m) := \mathbf{H}(X_m) + \mathbf{H}(X_1 \cdots X_m) - \mathbf{H}(X_m | X_1 \cdots X_m)$$

goes to zero as $m \to \infty$. Informally, knowing the value of X_m reveals very little about the value of $X_1 \cdots X_m$ when m is large.

Now let $f : G \to \mathbf{R}$ be a bounded harmonic function, and let m be large. For any $x \in G$ and any value s in the support of X_m, we observe from harmonicity that

$$f(sx) = \mathbf{E}(f(X_1 \cdots X_m x) | X_m = s).$$

But from the asymptotic vanishing of mutual information and the boundedness of f, one can show that the right-hand side will converge to

$\mathbf{E}(f(X_1 \cdots X_m x))$, which by harmonicity is equal to $f(x)$. Thus f is invariant with respect to the support of X, and is thus constant since this support generates G.

Conversely, if $\mathbf{H}(G, X)$ is nonzero, then the above arguments show that $\mathbf{I}(X_m, X_1 \cdots X_m)$ stays bounded away from zero as $m \to \infty$, thus $X_1 \cdots X_m$ reveals a nontrivial amount of information about X_m. This turns out to be true even if m is not deterministic, but is itself random, varying over some medium-sized range. From this, one can find a bounded function F such that the conditional expectation $\mathbf{E}(F(X_1 \cdots X_m)|X_m = s)$ varies nontrivially with s. On the other hand, the bounded function $x \mapsto \mathbf{E}F(X_1 \cdots X_{m-1}x)$ is approximately harmonic (because we are varying m) and has some nontrivial fluctuation near the identity (by the preceding sentence). Taking a limit as $m \to \infty$ (using Arzelá-Ascoli) we obtain a nonconstant bounded harmonic function as desired. □

Notes. This article first appeared at

`terrytao.wordpress.com/2009/10/27.`

Thanks to Seva Lev and an anonymous commenter for corrections.

2.9. An elementary noncommutative Freiman theorem

Let X be a finite subset of a noncommutative group G. As mentioned in Section 3.2 of *Structure and Randomness*, there is some interest in classifying those X which obey *small doubling* conditions such as $|X \cdot X| = O(|X|)$ or $|X \cdot X^{-1}| = O(|X|)$. A full classification here has still not been established. However, I wanted to record here an elementary argument of Freiman [**Fr1973**] (see also [**TaVu2006b**, Exercise 2.6.5], which in turn is based on an argument in [**La2001**]) that handles the case when $|X \cdot X|$ is very close to $|X|$:

Proposition 2.9.1. *If $|X^{-1} \cdot X| < \frac{3}{2}|X|$, then $X \cdot X^{-1}$ and $X^{-1} \cdot X$ are both finite groups, which are conjugate to each other. In particular, X is contained in the right-coset (or left-coset) of a group of order less than $\frac{3}{2}|X|$.*

Remark 2.9.2. The constant $\frac{3}{2}$ is completely sharp; consider the case when $X = \{e, x\}$ where e is the identity and x is an element of order larger than 2. This is a small example, but one can make it as large as one pleases by taking the direct product of X and G with any finite group. In the converse direction, we see that whenever X is contained in the right-coset $S \cdot x$ (resp. left-coset $x \cdot S$) of a group of order less than $2|X|$, then $X \cdot X^{-1}$ (resp. $X^{-1} \cdot X$) is necessarily equal to all of S, by the inclusion-exclusion principle (see the proof below for a related argument).

Proof. We begin by showing that $S := X \cdot X^{-1}$ is a group. As S is symmetric and contains the identity, it suffices to show that this set is closed under addition.

Let $a, b \in S$. Then we can write $a = xy^{-1}$ and $b = zw^{-1}$ for $x, y, z, w \in X$. If y were equal to z, then $ab = xw^{-1} \in X \cdot X^{-1}$ and we would be done. Of course, there is no reason why y should equal z; but we can use the hypothesis $|X^{-1} \cdot X| < \frac{3}{2}|X|$ to boost this as follows. Observe that $x^{-1} \cdot X$ and $y^{-1} \cdot X$ both have cardinality $|X|$ and lie inside $X^{-1} \cdot X$, which has cardinality strictly less than $\frac{3}{2}|X|$. By the inclusion-exclusion principle, this forces $x^{-1} \cdot X \cap y^{-1} \cdot X$ to have cardinality greater than $\frac{1}{2}|X|$. In other words, there exist more than $\frac{1}{2}|X|$ pairs $x', y' \in X$ such that $x^{-1}x' = y^{-1}y'$, which implies that $a = x'(y')^{-1}$. Thus there are more than $\frac{1}{2}|X|$ elements $y' \in X$ such that $a = x'(y')^{-1}$ for some $x' \in X$ (since x' is uniquely determined by y'); similarly, there exists more than $\frac{1}{2}|X|$ elements $z' \in X$ such that $b = z'(w')^{-1}$ for some $w' \in X$. Again by inclusion-exclusion, we can thus find $y' = z'$ in X for which one has simultaneous representations $a = x'(y')^{-1}$ and $b = y'(z')^{-1}$, and so $ab = x'(z')^{-1} \in X \cdot X^{-1}$, and the claim follows.

In the course of the above argument, we showed that every element of the group S has more than $\frac{1}{2}|X|$ representations of the form xy^{-1} for $x, y \in X$. But there are only $|X|^2$ pairs (x, y) available, and thus $|S| < 2|X|$.

Now let x be any element of X. Since $X \cdot x^{-1} \subset S$, we have $X \subset S \cdot x$, and so $X^{-1} \cdot X \subset x^{-1} \cdot S \cdot x$. Conversely, every element of $x^{-1} \cdot S \cdot x$ has exactly $|S|$ representations of the form $z^{-1}w$ where $z, w \in S \cdot x$. Since X occupies more than half of $S \cdot x$, we thus see from the inclusion-exclusion principle, there is thus at least one representation $z^{-1}w$ for which z, w both lie in X. In other words, $x^{-1} \cdot S \cdot x = X^{-1} \cdot X$, and the claim follows. □

To relate this to the classical doubling constants $|X \cdot X|/|X|$, we first make an easy observation:

Lemma 2.9.3. *If $|X \cdot X| < 2|X|$, then $X \cdot X^{-1} = X^{-1} \cdot X$.*

Again, this is sharp; consider X equal to $\{x, y\}$ where x, y generate a free group.

Proof. Suppose that xy^{-1} is an element of $X \cdot X^{-1}$ for some $x, y \in X$. Then the sets $X \cdot x$ and $X \cdot y$ have cardinality $|X|$ and lie in $X \cdot X$, so by the inclusion-exclusion principle, the two sets intersect. Thus there exist $z, w \in X$ such that $zx = wy$, thus $xy^{-1} = z^{-1}w \in X^{-1} \cdot X$. This shows that $X \cdot X^{-1}$ is contained in $X^{-1} \cdot X$. The converse inclusion is proven similarly. □

Proposition 2.9.4. *If $|X \cdot X| < \frac{3}{2}|X|$, then $S := X \cdot X^{-1}$ is a finite group of order $|X \cdot X|$, and $X \subset S \cdot x = x \cdot S$ for some x in the normaliser of S.*

The factor $\frac{3}{2}$ is sharp, by the same example used to show sharpness of Proposition 2.9.1. However, there seems to be some room for further improvement if one weakens the conclusion a bit; see below.

Proof. Let $S = X^{-1} \cdot X = X \cdot X^{-1}$ (the two sets being equal by Lemma 2.9.3). By the argument used to prove Lemma 2.9.3, every element of S has more than $\frac{1}{2}|X|$ representations of the form xy^{-1} for $x, y \in X$. By the argument used to prove Proposition 2.9.1, this shows that S is a group; also, since there are only $|X|^2$ pairs (x, y), we also see that $|S| < 2|X|$.

Pick any $x \in X$; then $x^{-1} \cdot X, X \cdot x^{-1} \subset S$, and so $X \subset x \cdot S, S \cdot x$. Because every element of $x \cdot S \cdot x$ has $|S|$ representations of the form yz with $y \in x \cdot S$, $z \in S \cdot x$, and X occupies more than half of $x \cdot S$ and of $S \cdot x$, we conclude that each element of $x \cdot S \cdot x$ lies in $X \cdot X$, and so $X \cdot X = x \cdot S \cdot x$ and $|S| = |X \cdot X|$.

The intersection of the groups S and $x \cdot S \cdot x^{-1}$ contains $X \cdot x^{-1}$, which is more than half the size of S, and so we must have $S = x \cdot S \cdot x^{-1}$, i.e., x normalises S, and the proposition follows. □

Because the arguments here are so elementary, they extend easily to the infinitary setting in which X is now an infinite set, but has finite measure with respect to some translation-invariant Kiesler measure μ. We omit the details. (I am hoping that this observation may help simplify some of the theory in that setting.)

2.9.1. Beyond the 3/2 barrier.

It appears that one can push the arguments a bit beyond the 3/2 barrier, though of course one has to weaken the conclusion in view of the counterexample in Remark 2.9.2. Here I give a result that increases $3/2 = 1.5$ to the golden ratio $\phi := (1+\sqrt{5})/2 = 1.618\cdots$:

Proposition 2.9.5 (Weak noncommutative Kneser theorem)**.** *If $|X^{-1} \cdot X|, |X \cdot X^{-1}| \leq K|X|$ for some $1 < K < \phi$, then $X \cdot X^{-1} = H \cdot Z$ for some finite subgroup H, and some finite set Z with $|Z| \leq C(K)$ for some $C(K)$ depending only on K.*

Proof. Write $S := X \cdot X^{-1}$. Let us say that h *symmetrises* S if $h \cdot S = S$, and let H be the set of all h that symmetrise S. It is clear that H is a finite group with $H \cdot S = S$ and thus $S \cdot H = S$ also.

For each $z \in S$, let $r(z)$ be the number of representations of z of the form $z = xy^{-1}$ with $x, y \in X$. Double counting shows that $\sum_{z \in S} r(z) = |X|^2$, while by hypothesis $|S| \leq K|X|$; thus the average value of $r(z)$ is at least $|X|/K$. Since $1 < K < \phi$, $1/K > K - 1$. Since $r(z) \leq |X|$ for all z, we conclude that $r(z) > (K-1)|X|$ for at least $c(K)|X|$ values of $z \in S$, for some explicitly computable $c(K) > 0$.

Suppose $z, w \in S$ is such that $r(z) > (K-1)|X|$, thus z has more than $(K-1)|X|$ representations of the form xy^{-1} with $x, y \in X$. On the other hand, the argument used to prove Proposition 2.9.1 shows that w has at least $(2-K)|X|$ representations of the form $x'(y')^{-1}$ with $x', y' \in X$. By the inclusion-exclusion formula, we can thus find representations for which $y = x'$, which implies that $zw \in S$. Since $w \in S$ was arbitrary, this implies that $z \in H$. Thus $|H| \geq c(K)|X|$. Since $S = H \cdot S$ and $|S| \leq K|X|$, this implies that S can be covered by at most $C(K)$ right-cosets of S for some $C(K)$ depending only on K, and the claim follows. $\square$

This result appears in [**Fr1973**], and a related argument also appears in [**Le2000**].

It looks like one should be able to get a bit more structural information on X than is given by the above conclusion, and I doubt the golden ratio is sharp either (the correct threshold should be 2, in analogy with the commutative Kneser theorem; after that, the conclusion will fail, as can be seen by taking X to be a long geometric progression). Readers here are welcome to look for improvements to these results, of course.

Notes. This article first appeared at

`terrytao.wordpress.com/2009/11/10.`

Thanks to Miguel Lacruz for corrections, and Ben Green and Seva Lev for references.

2.10. Nonstandard analogues of energy and density increment arguments

This article assumes some familiarity with nonstandard analysis (see, e.g., Section 1.5 of *Structure and Randomness*).

Let us call a model M of a language L *weakly countably saturated*[2] if, every countable sequence $P_1(x), P_2(x), \ldots$ of formulae in L (involving countably many constants in M) which is finitely satisfiable in M (i.e., any finite collection $P_1(x), \ldots, P_n(x)$ in the sequence has a solution x in M) is automatically satisfiable in M (i.e., there is a solution x to all $P_n(x)$ simultaneously). Equivalently, a model is weakly countably saturated if the topology generated by the definable sets is *countably compact*.

[2] The stronger property of being *countably saturated* asserts that if an *arbitrary* sequence of formulae involving countably many constants is finitely satisfiable, then it is satisfiable; the relation between the two concepts is thus analogous to compactness and countable compactness. If one chooses a special type of ultrafilter, namely a "countably incomplete" ultrafilter, one can recover the full strength of countable saturation, though it is not needed for the remarks here.

Most models are not (weakly) countably saturated. Consider for instance the standard natural numbers $\mathbf{N}$ as a model for arithmetic. Then the sequence of formulae "$x > n$" for $n = 1, 2, 3, \ldots$ is finitely satisfiable in $\mathbf{N}$, but not satisfiable.

However, if one takes a model M of L and passes to an *ultrapower* $*M$, whose elements x consist of sequences $(x_n)_{n \in \mathbf{N}}$ in M, modulo equivalence with respect to some fixed nonprincipal ultrafilter p, then it turns out that such models are automatically weakly countably saturated. Indeed, if $P_1(x), P_2(x), \ldots$ are finitely satisfiable in $*M$, then they are also finitely satisfiable in M (either by inspection, or by appeal to *Łos's theorem* and/or the *transfer principle* in nonstandard analysis), so for each n there exists $x_n \in M$ which satisfies $P_1, \ldots, P_n$. Letting $x = (x_n)_{n \in \mathbf{N}} \in *M$ be the ultralimit of the x_n, we see that x satisfies all of the P_n at once.

In particular, nonstandard models of mathematics, such as the nonstandard model $*\mathbf{N}$ of the natural numbers, are automatically countably saturated. (This fact is closely related to the *idealisation axiom* in internal set theory.)

This has some cute consequences. For instance, suppose one has a nonstandard metric space $*X$ (an ultralimit of standard metric spaces), and suppose one has a standard sequence $(x_n)_{n \in \mathbf{N}}$ of elements of $*X$ which are standard-Cauchy, in the sense that for any standard $\varepsilon > 0$ one has $d(x_n, x_m) < \varepsilon$ for all sufficiently large n, m. Then there exists a nonstandard element $x \in *X$ such that x_n standard-converges to x in the sense that for every standard $\varepsilon > 0$ one has $d(x_n, x) < \varepsilon$ for all sufficiently large n. Indeed, from the standard-Cauchy hypothesis, one can find a standard $\varepsilon(n) > 0$ for each standard n that goes to zero (in the standard sense), such that the formulae "$d(x_n, x) < \varepsilon(n)$" are finitely satisfiable, and hence satisfiable by countable saturation. Thus we see that nonstandard metric spaces are automatically "standardly complete" in some sense.

This leads to a nonstandard structure theorem for Hilbert spaces, analogous to the orthogonal decomposition in Hilbert spaces:

Theorem 2.10.1 (Nonstandard structure theorem for Hilbert spaces). *Let $*H$ be a nonstandard Hilbert space, let S be a bounded (external) subset of $*H$, and let $x \in H$. Then there exists a decomposition $x = x_S + x_{S^\perp}$, where $x_S \in *H$ is "almost standard-generated by S" in the sense that for every standard $\varepsilon > 0$, there exists a standard finite linear combination of elements of S which is within ε of S, and $x_{S^\perp} \in *H$ is "standard-orthogonal to S" in the sense that $\langle x_{S^\perp}, s \rangle = o(1)$ for all $s \in S$.*

Proof. Let d be the infimum of all the (standard) distances from x to a standard linear combination of elements of S, then for every standard n one can find a standard linear combination x_n of elements of S which lie within $d+1/n$ of x. From the parallelogram law we see that x_n is standard-Cauchy, and thus it standard-converges to some limit $x_S \in *H$, which is then almost standard-generated by S by construction. An application of Pythagoras then shows that $x_{S^\perp} := x - x_S$ is standard-orthogonal to every element of S. □

This is the nonstandard analogue of a combinatorial structure theorem for Hilbert spaces (see, e.g., [**Ta2007b**, Theorem 2.6]). There is an analogous nonstandard structure theorem for σ-algebras (the counterpart of [**Ta2007b**, Theorem 3.6]) which I will not discuss here, but I will give just one sample corollary:

Theorem 2.10.2 (Nonstandard arithmetic regularity lemma). *Let $*G$ be a nonstandardly finite abelian group, and let $f : *G \to [0,1]$ be a function. Then one can split $f = f_{U^\perp} + f_U$, where $f_U : *G \to [-1,1]$ is standard-uniform in the sense that all Fourier coefficients are (uniformly) $o(1)$, and $f_{U^\perp} : *G \to [0,1]$ is standard-almost periodic in the sense that for every standard $\varepsilon > 0$, one can approximate $f_{U^\perp}$ to error ε in $L^1(*G)$ norm by a standard linear combination of characters (which is also bounded).*

This can be used for instance to give a nonstandard proof of Roth's theorem (which is not much different from the "finitary ergodic" proof of Roth's theorem, given for instance in [**TaVu2006b**, Section 10.5]). There is also a nonstandard version of the Szemerédi regularity lemma which can be used, among other things, to prove the hypergraph removal lemma (the proof then becomes rather close to the infinitary proof of this lemma in [**Ta2007**]). More generally, the above structure theorem can be used as a substitute for various "energy increment arguments" in the combinatorial literature, though it does not seem that there is a significant saving in complexity in doing so unless one is performing quite a large number of these arguments.

One can also cast density increment arguments in a nonstandard framework. Here is a typical example. Call a nonstandard subset X of a nonstandard finite set Y *dense* if one has $|X| \geq \varepsilon|Y|$ for some standard $\varepsilon > 0$.

Theorem 2.10.3. *Suppose Szemerédi's theorem (every set of integers of positive upper density contains an arithmetic progression of length k) fails for some k. Then there exists an unbounded nonstandard integer N, a dense subset A of $[N] := \{1, \dots, N\}$ with no progressions of length k, and with the additional property that*

$$\frac{|A \cap P|}{|P|} \leq \frac{|A \cap [N]|}{N} + o(1)$$

for any subprogression P of $[N]$ of unbounded size (thus there is no sizeable density increment on any large progression).

Proof. Let $B \subset \mathbf{N}$ be a (standard) set of positive upper density which contains no progression of length k. Let $\delta := \limsup_{|P| \to \infty} |B \cap P|/|P|$ be the asymptotic maximal density of B inside a long progression, thus $\delta > 0$. For any $n > 0$, one can then find a standard integer $N_n \geq n$ and a standard subset A_n of $[N_n]$ of density at least $\delta - 1/n$ such that A_n can be embedded (after a linear transformation) inside B, so in particular A_n has no progressions of length k. Applying the saturation property, one can then find an unbounded N and a set A of $[N]$ of density at least $\delta - 1/n$ for every standard n (i.e., of density at least $\delta - o(1)$) with no progressions of length k. By construction, we also see that for any subprogression P of $[N]$ of unbounded size, A has density at most $\delta + 1/n$ for any standard n, thus has density at most $\delta + o(1)$, and the claim follows. □

This can be used as the starting point for any density-increment based proof of Szemerédi's theorem for a fixed k, e.g., Roth's proof for $k = 3$, Gowers' proof for arbitrary k, or Szemerédi's proof for arbitrary k. (It is likely that Szemerédi's proof, in particular, simplifies a little bit when translated to the nonstandard setting, though the savings are likely to be modest.)

I am also hoping that the recent results of Hrushovski [**Hr2009**] on the noncommutative Freiman problem require only countable saturation, as this makes it more likely that they can be translated to a nonstandard setting and thence to a purely finitary framework.

Notes. This article first appeared at

`terrytao.wordpress.com/2009/11/10.`

Balazs Szegedy noted the connection to his recent work [**Sz2009**] on higher order Fourier analysis from a nonstandard perspective.

2.11. Approximate bases, sunflowers, and nonstandard analysis

One of the most basic theorems in linear algebra is that every finite-dimensional vector space has a finite basis. Let us give a statement of this theorem in the case when the underlying field is the rationals:

Theorem 2.11.1 (Finite generation implies finite basis, infinitary version). *Let V be a vector space over the rationals $\mathbf{Q}$, and let $v_1, \ldots, v_n$ be a finite collection of vectors in V. Then there exists a collection $w_1, \ldots, w_k$ of vectors in V, with $1 \leq k \leq n$, such that*

- w generates v. *Every v_j can be expressed as a rational linear combination of the $w_1, \ldots, w_k$.*
- w independent. *There is no nontrivial linear relation $a_1w_1 + \cdots + a_kw_k = 0$, $a_1, \ldots, a_k \in \mathbf{Q}$ among the $w_1, \ldots, w_m$ (where nontrivial means that the a_i are not all zero).*

In fact, one can take $w_1, \ldots, w_m$ to be a subset of the $v_1, \ldots, v_n$.

Proof. We perform the following "rank reduction argument". Start with $w_1, \ldots, w_k$ initialised to $v_1, \ldots, v_n$ (so initially we have $k = n$). Clearly, w generates v. If the w_i are linearly independent, then we are done. Otherwise, there is a nontrivial linear relation between them. After shuffling things around, we see that one of the w_i, say w_k, is a rational linear combination of the $w_1, \ldots, w_{k-1}$. In such a case, w_k becomes redundant, and we may delete it (reducing the rank k by one). We repeat this procedure; it can only run for at most n steps and so terminates with $w_1, \ldots, w_m$ obeying both of the desired properties. □

In additive combinatorics, one often wants to use results like this in finitary settings, such as that of a cyclic group $\mathbf{Z}/p\mathbf{Z}$ where p is a large prime. Now, technically speaking, $\mathbf{Z}/p\mathbf{Z}$ is not a vector space over $\mathbf{Q}$, because one only multiplies an element of $\mathbf{Z}/p\mathbf{Z}$ by a rational number if the denominator of that rational does not divide p. But for p very large, $\mathbf{Z}/p\mathbf{Z}$ "behaves" like a vector space over $\mathbf{Q}$, at least if we restrict our attention to the rationals of "bounded height"—where the numerator and denominator of the rationals are bounded. Thus we shall refer to elements of $\mathbf{Z}/p\mathbf{Z}$ as "vectors" over $\mathbf{Q}$, even though strictly speaking this is not quite the case.

On the other hand, saying that one element of $\mathbf{Z}/p\mathbf{Z}$ is a rational linear combination of another set of elements is not a very interesting statement: any nonzero element of $\mathbf{Z}/p\mathbf{Z}$ already generates the entire space! However, if one again restricts attention to rational linear combinations *of bounded height*, then things become interesting again. For instance, the vector 1 can generate elements such as 37 or $\frac{p-1}{2}$ using rational linear combinations of bounded height, but will not be able to generate such elements of $\mathbf{Z}/p\mathbf{Z}$ as $\lfloor\sqrt{p}\rfloor$ without using rational numbers of unbounded height.

For similar reasons, the notion of linear independence over the rationals does not initially look very interesting over $\mathbf{Z}/p\mathbf{Z}$: any two nonzero elements of $\mathbf{Z}/p\mathbf{Z}$ are of course rationally dependent. But again, if we restrict our attention to rational numbers of bounded height, then independence begins to emerge: for instance, 1 and $\lfloor\sqrt{p}\rfloor$ are independent in this sense.

Thus, it becomes natural to ask whether there is a "quantitative" analogue of Theorem 2.11.1, with nontrivial content in the case of "vector spaces over the bounded height rationals" such as $\mathbf{Z}/p\mathbf{Z}$, which asserts that given any bounded collection $v_1, \ldots, v_n$ of elements, one can find another set $w_1, \ldots, w_k$ which is linearly independent "over the rationals up to some height", such that the $v_1, \ldots, v_n$ can be generated by the $w_1, \ldots, w_k$ "over the rationals up to some height". Of course to make this rigorous, one needs to quantify the two heights here, one giving the independence, and the other giving the generation. In order to be useful for applications, it turns out that one often needs the former height to be much larger than the latter; exponentially larger, for instance, is not an uncommon request. Fortunately, one can accomplish this, at the cost of making the height somewhat large:

Theorem 2.11.2 (Finite generation implies finite basis, finitary version). *Let $n \geq 1$ be an integer, and let $F : \mathbf{N} \to \mathbf{N}$ be a function. Let V be an abelian group which admits a well-defined division operation by any natural number of size at most $C(F,n)$ for some constant $C(F,n)$ depending only on F, n; for instance one can take $V = \mathbf{Z}/p\mathbf{Z}$ for p a prime larger than $C(F,n)$. Let $v_1, \ldots, v_n$ be a finite collection of "vectors" in V. Then there exists a collection $w_1, \ldots, w_k$ of vectors in V, with $1 \leq k \leq n$, as well an integer $M \geq 1$, such that*

- Complexity bound. *$M \leq C(F,n)$ for some $C(F,n)$ depending only on F, n.*
- w generates v. *Every v_j can be expressed as a rational linear combination of the $w_1, \ldots, w_k$ of height at most M (i.e., the numerator and denominator of the coefficients are at most M).*
- w independent. *There is no nontrivial linear relation $a_1 w_1 + \cdots + a_k w_k = 0$ among the $w_1, \ldots, w_k$ in which the $a_1, \ldots, a_k$ are rational numbers of height at most $F(M)$.*

In fact, one can take $w_1, \ldots, w_k$ to be a subset of the $v_1, \ldots, v_n$.

Proof. We perform the same "rank reduction argument" as before, but translated to the finitary setting. Start with $w_1, \ldots, w_k$ initialised to $v_1, \ldots, v_n$ (so initially we have $k = n$), and initialise $M = 1$. Clearly, w generates v at this height. If the w_i are linearly independent up to rationals of height $F(M)$, then we are done. Otherwise, there is a nontrivial linear relation between them. After shuffling things around, we see that one of the w_i, say w_k, is a rational linear combination of the $w_1, \ldots, w_{k-1}$, whose height is bounded by some function depending on $F(M)$ and k. In such a case, w_k becomes redundant, and we may delete it (reducing the rank k by one), but note that in order for the remaining $w_1, \ldots, w_{k-1}$ to generate $v_1, \ldots, v_n$, we need to raise the height upper bound for the rationals involved from M to some quantity M' depending on $M, F(M), k$. We then replace M by M' and continue the process. We repeat this procedure; it can only run for at most

n steps and so terminates with $w_1, \ldots, w_m$ and M obeying all of the desired properties. (Note that the bound on M is quite poor, being essentially an n-fold iteration of F! Thus, for instance, if F is exponential, then the bound on M is *tower-exponential* in nature.) □

Remark 2.11.3. A variant of this type of approximate basis lemma was used in [**TaVu2007**].

Looking at the statements and proofs of these two theorems, it is clear that the two results are in some sense the "same" result, except that the latter has been made sufficiently quantitative that it is meaningful in such finitary settings as $\mathbf{Z}/p\mathbf{Z}$. In this article I will show how this equivalence can be made formal using the language of nonstandard analysis (see Section 1.9 of *Structure and Randomness*). This is not a particularly deep (or new) observation, but it is perhaps the simplest example I know of that illustrates how nonstandard analysis can be used to transfer a quantifier-heavy finitary statement, such as Theorem 2.11.2, into a quantifier-light infinitary statement, such as Theorem 2.11.1, thus lessening the need to perform "epsilon management" duties, such as keeping track of unspecified growth functions such as F. This type of transference is discussed at length in Section 1.3 of *Structure and Randomness.*

In this particular case, the amount of effort needed to set up the nonstandard machinery in order to reduce Theorem 2.11.2 from Theorem 2.11.1 is too great for this transference to be particularly worthwhile, especially given that Theorem 2.11.2 has such a short proof. However, when performing a particularly intricate argument in additive combinatorics, in which one is performing a number of "rank reduction arguments", "energy increment arguments", "regularity lemmas", "structure theorems", and so forth, the purely finitary approach can become bogged down with all the epsilon management one needs to do. The nonstandard approach can efficiently hide a large number of these parameters from view, and it can then become worthwhile to invest in the nonstandard framework in order to clean up the rest of a lengthy argument. Furthermore, an advantage of moving up to the infinitary setting is that one can then deploy all the firepower of an existing well-developed infinitary theory of mathematics (in this particular case, this would be the theory of linear algebra) *out of the box*, whereas in the finitary setting one would have to painstakingly finitise each aspect of such a theory that one wished to use (imagine for instance trying to finitise the *rank-nullity theorem* for rationals of bounded height).

The nonstandard approach is very closely related to the use of compactness arguments, or of the technique of taking ultralimits and ultraproducts; indeed we will use an ultrafilter in order to create the nonstandard model in the first place.

I will also discuss two variants of both Theorem 2.11.1 and Theorem 2.11.2 which have actually shown up in my research. The first is that of the *regularity lemma* for polynomials over finite fields, which came up when studying the equidistribution of such polynomials in [**GrTa2007**]. The second comes up when one is dealing not with a single finite collection $v_1, \ldots, v_n$ of vectors, but rather with a *family* $(v_{h,1}, \ldots, v_{h,n})_{h \in H}$ of such vectors, where H ranges over a large set; this gives rise to what we call the *sunflower lemma*, and it came up in [**GrTaZi2009**].

This article is mostly concerned with nonstandard translations of the "rank reduction argument". Nonstandard translations of the "energy increment argument" and "density increment argument" were briefly discussed in Section 2.10.

2.11.1. Equivalence of Theorems 2.11.1 and 2.11.2. Both Theorem 2.11.1 and Theorem 2.11.2 are easy enough to prove. But we will now spend a certain amount of effort in showing that one can deduce each theorem from the other without actually going through the proof of either. This may not seem particularly worthwhile (or to be serious overkill) in the case of these two particular theorems, but the method of deduction is extremely general, and can be used to relate much more deep and difficult infinitary and finitary theorems to each other without a significant increase in effort.[3]

Let us first show why the finitary theorem, Theorem 2.11.2, implies Theorem 2.11.1. We argue by contradiction. If Theorem 2.11.1 failed, then we could find a vector space V over the rationals, and a finite collection $v_1, \ldots, v_n$ of vectors, for which *no* finite subcollection $w_1, \ldots, w_k$ of the $v_1, \ldots, v_n$ obeyed both the generation property and the linear independence property. In other words, whenever a subcollection $w_1, \ldots, w_k$ happened to generate $v_1, \ldots, v_n$ by rationals, then it must necessarily contain a linear dependence.

We use this to create a function $F : \mathbf{N} \to \mathbf{N}$ as follows. Given any natural number M, consider all the finite subcollections $w_1, \ldots, w_k$ of $v_1, \ldots, v_n$ which can generate the $v_1, \ldots, v_n$ using rationals of height at most M. By the above hypothesis, all such subcollections contain a linear dependence involving rationals of some finite height. There may be many such dependences; we pick one arbitrarily. We then choose $F(M)$ to be any natural number larger than the heights of all the rationals involved in all the linear dependencies thus chosen. (Here we implicitly use the fact that there are only finitely many subcollections of the $v_1, \ldots, v_n$ to search through.)

[3]This is closely related to various *correspondence principles* between combinatorics and parts of infinitary mathematics, such as ergodic theory; see also Section 1.3 of *Structure and Randomness* for a closely related equivalence.

Having chosen this function F, we then apply Theorem 2.11.2 to the vectors $v_1, \ldots, v_n$ and this choice of function F, to obtain a subcollection $w_1, \ldots, w_k$ which generates the $v_1, \ldots, v_n$ using rationals of height at most M, and which has no linear dependence involving rationals of height at most $F(M)$. But this contradicts the construction of F, and gives the claim.

Remark 2.11.4. Note how important it is here that the growth function F in Theorem 2.11.2 is not specified in advance, but is instead a parameter that can be set to be as "large" as needed. Indeed, for Theorem 2.11.2 for any fixed F (e.g., exponential, tower-exponential, Ackermann, etc.) gives a statement which is strictly "weaker" than Theorem 2.11.1 in a sense that I will not try to make precise here; it is only the union of *all* these statements for *all* conceivable F that gives the full strength of Theorem 2.11.1. A similar phenomenon occurs with the *finite convergence principle* (Section 1.3 of *Structure and Randomness*). It is this "second order" nature of infinitary statements (they quantify not just over numerical parameters such as N or ε, but also over functional parameters such as F) that make such statements appear deeper than finitary ones, but the distinction largely disappears if one is willing to perform such second-order quantifications.

Now we turn to the more interesting deduction, which is to obtain Theorem 2.11.2 from Theorem 2.11.1. Again, one argues by contradiction. Suppose that Theorem 2.11.2 failed. Carefully negating all the quantifiers (and using the axiom of choice), we conclude that there exists a function $F : \mathbf{N} \to \mathbf{N}$ and a natural number n with the following property: given any natural number K, there exists an abelian group V_K which is divisible up to height K, and elements $v_{1,K}, \ldots, v_{n,K}$ in V_K such that there is *no* subcollection $w_{1,K}, \ldots, w_{k,K}$ of the $v_{1,K}, \ldots, v_{n,K}$, together with an integer $M \leq K$, such that $w_{1,K}, \ldots, w_{k,K}$ generate $v_{1,K}, \ldots, v_{n,K}$ using rationals of height at most M, and such that the $w_{1,K}, \ldots, w_{k,K}$ have no linear dependence using rationals of height at most $F(M)$.

We now perform an *ultralimit* as $K \to \infty$. We will not pause here to recall the machinery of ultrafilters, ultralimits, and ultraproducts, but refer the reader instead to Section 1.5 of *Structure and Randomness* for discussion.

We pick a nonprincipal ultrafilter p of the natural numbers. Starting with the "standard" abelian groups V_K, we then form their *ultraproduct* $V = \prod_K V_K/p$, defined as the space of sequences $v = (v_K)_{K \in \mathbf{N}}$ with $v_K \in V_K$ for each K, modulo equivalence by p. Thus two sequences $v = (v_K)_{K \in \mathbf{N}}$ and $v' = (v'_K)_{K \in \mathbf{N}}$ are considered equal if $v_K = v'_K$ for a p-large set of K (i.e., for a set of K that lies in p).

Now that nonstandard objects are in play, we will need to take some care to distinguish between standard objects (e.g., standard natural numbers) and their nonstandard counterparts.

Since each of the V_K are an abelian group, V is also an abelian group (an easy special case of the *transfer principle*). Since each V_K is divisible up to height K, V is divisible up to all (standard) heights; in other words, V is actually a vector space over the (standard) rational numbers $\mathbf{Q}$. The point is that while none of the V_K are, strictly speaking, vector spaces over $\mathbf{Q}$, they increasingly behave as if they were such spaces, and in the limit one recovers genuine vector space structure.

For each $1 \leq i \leq n$, one can take an ultralimit of the elements $v_{i,K} \in V_K$ to generate an element $v_i := (v_{i,K})_{K \in \mathbf{N}}$ of the ultraproduct V. So now we have n vectors $v_1, \ldots, v_n$ of a vector space V over $\mathbf{Q}$—precisely the setting of Theorem 2.11.1! So we apply that theorem and obtain a subcollection $w_1, \ldots, w_k \in V$ of the $v_1, \ldots, v_n$, such that each v_i can be generated from the $w_1, \ldots, w_k$ using (standard) rationals, and such that the $w_1, \ldots, w_k$ are linearly independent over the (standard) rationals.

Since all (standard) rationals have a finite height, one can find a (standard) natural number M such that each of the v_i can be generated from the $w_1, \ldots, w_k$ using (standard) rationals of height at most M. Undoing the ultralimit, we conclude that for a p-large set of K's, all of the $v_{i,K}$ can be generated from the $w_{1,K}, \ldots, w_{k,K}$ using rationals of height at most M. But by hypothesis, this implies for all sufficiently large K in this p-large set, the $w_{1,K}, \ldots, w_{k,K}$ contain a nontrivial rational dependence of height at most $F(M)$, thus

$$\frac{a_{1,K}}{q_{1,K}} w_{1,K} + \cdots + \frac{a_{k,K}}{q_{k,K}} w_{k,K} = 0$$

for some integers $a_{i,K}, q_{i,K}$ of magnitude at most $F(M)$, with the $a_{k,K}$ not all zero.

By the pigeonhole principle (and the finiteness of $F(M)$), each of the $a_{i,K}, q_{i,K}$ is constant in K on a p-large set of K. So if we take an ultralimit again to go back to the nonstandard world, the quantities $a_i := (a_{i,K})_{K \in \mathbf{N}}$, $q_i := (q_{i,K})_{K \in \mathbf{N}}$ are standard integers (rather than merely nonstandard integers). Thus we have

$$\frac{a_1}{q_1} w_1 + \cdots + \frac{a_k}{q_k} w_k = 0$$

with the a_i not all zero, i.e., we have a linear dependence amongst the $w_1, \ldots, w_k$. But this contradicts Theorem 2.11.1.

2.11.2. Polynomials over finite fields. Let $\mathbf{F}$ a fixed finite field (e.g., the field $\mathbf{F}_2$ of two elements), and consider a high-dimensional finite vector space V over $\mathbf{F}$. A polynomial $P : \mathbf{F}^n \to \mathbf{F}$ of degree $\leq d$ can then be defined as a combination of monomials each of degree at most d, or alternatively as a function whose $d+1$-th derivative vanishes; see Section 1.12 of *Poincaré's Legacies, Vol. I* for some discussion of this equivalence.

We define the *rank* $\text{rank}_{\leq d-1}(P)$ of a degree $\leq d$ polynomial P to be the least number k of degree $\leq d-1$ polynomials $Q_1, \ldots, Q_k$, such that P is completely determined by $Q_1, \ldots, Q_k$, i.e., $P = f(Q_1, \ldots, Q_k)$ for some function $f : \mathbf{F}^k \to \mathbf{F}$. In the case when P has degree ≤ 2, this concept is very close to the familiar *rank* of a quadratic form or matrix.

A generalisation of the notion of linear independence is that of linear independence modulo low rank. Let us call a collection $P_1, \ldots, P_n$ of degree $\leq d$ polynomials *M-linearly independent* if every nontrivial linear combination $a_1P_1 + \cdots + a_nP_n$ with $a_1, \ldots, a_n \in \mathbf{F}$ not all zero, has rank at least M:

$$\text{rank}_{\leq d-1}(a_1P_1 + \cdots + a_nP_n) \leq M.$$

There is then the following analogue of Theorem 2.11.2:

Theorem 2.11.5 (Polynomial regularity lemma at one degree, finitary version). *Let $n, d \geq 1$ be integers, let $\mathbf{F}$ be a finite field, and let $F : \mathbf{N} \to \mathbf{N}$ be a function. Let V be a vector space over $\mathbf{F}$, and let $P_1, \ldots, P_n : V \to F$ be polynomials of degree $\leq d$. Then there exists a collection $Q_1, \ldots, Q_k : V \to F$ of polynomials of degree $\leq d$, with $1 \leq k \leq n$, as well an integer $M \geq 1$, such that*

- Complexity bound. *$M \leq C(F, n, d, \mathbf{F})$ for some $C(F, n, d, \mathbf{F})$ depending only on $F, n, d, \mathbf{F}$.*
- Q generates P. *Every P_j can be expressed as a $\mathbf{F}$-linear combination of the $Q_1, \ldots, Q_k$, plus an error E which has rank $\text{rank}_{\leq d-1}(E)$ at most M.*
- P independent. *There is no nontrivial linear relation $a_1Q_1 + \cdots + a_kQ_k = E$ among the $w_1, \ldots, w_m$ in which E has rank $\text{rank}_{\leq d-1}(E)$ at most $F(M)$.*

In fact, one can take $Q_1, \ldots, Q_k$ to be a subset of the $P_1, \ldots, P_n$.

This theorem can be proven in much the same way as Theorem 2.11.2, and the reader is invited to do so as an exercise. The constant $C(F, n, d, \mathbf{F})$ can in fact be taken to be independent of d and $\mathbf{F}$, but this is not important to us here.

Roughly speaking, Theorem 2.11.5 asserts that a finite family of degree $\leq d$ polynomials can be expressed as a linear combination of degree $\leq d$ polynomials which are "linearly independent modulo low rank errors", plus some lower rank objects. One can think of this as *regularising* the degree $\leq d$

polynomials, modulo combinations of lower degree polynomials. For applications (and in particular, for understanding the equidistribution) one also needs to regularise the degree $\leq d-1$ polynomials that arise this way, and so forth for increasingly lower degrees until all polynomials are regularised. (A similar phenomenon occurs for the hypergraph regularity lemma.)

When working with theorems like this, it is helpful to think conceptually of "quotienting out" by all polynomials of low rank. Unfortunately, in the finitary setting, the polynomials of low rank do not form a group, and so the quotient is ill defined. However, this can be rectified by passing to the infinitary setting. Indeed, once one does so, one can quotient out the low rank polynomials, and Theorem 2.11.5 follows directly from Theorem 2.11.1 (or more precisely, the analogue of that theorem in which the field of rationals $\mathbf{Q}$ is replaced by the finite field $\mathbf{F}$).

Let us see how this works. To prove Theorem 2.11.5, suppose for contradiction that the theorem failed. Then one can find $F, n, d, \mathbf{F}$, such that for every natural K, one can find a vector space V_K and polynomials $P_{1,K}, \ldots, P_{n,K} : V_K \to \mathbf{F}$ of degree $\leq d$, for which there do not exist polynomials $Q_{1,K}, \ldots, Q_{k,K}$ with $k \leq n$ and an integer $M \leq K$ such that each $P_{j,K}$ can be expressed as a linear combination of the $Q_{i,K}$ modulo an error of rank at most M, and such that there are no nontrivial linear relations amongst the $Q_{i,K}$ modulo errors of rank at most $F(M)$.

Taking an ultralimit as before, we end up with a (nonstandard) vector space V over $\mathbf{F}$ (which is likely to be infinite) and (nonstandard) polynomials $P_1, \ldots, P_n : V \to \mathbf{F}$ of degree $\leq d$ (here it is best to use the "local" definition of a polynomial of degree $\leq d$, as a (nonstandard) function whose $d+1$-th derivative vanishes, but one can also view this as a (nonstandard) sum of monomials if one is careful).

The space $\mathrm{Poly}_{\leq d}(V)$ of (nonstandard) degree $\leq d$ polynomials on V is a (nonstandard) vector space over $\mathbf{F}$. Inside this vector space, one has the subspace $\mathrm{Lowrank}_{\leq d}(V)$ consisting of all polynomials $P \in \mathrm{Poly}_{\leq d}(V)$ whose rank $\mathrm{rank}_{\leq d-1}(V)$ is a standard integer (as opposed to a nonstandard integer); call these the *bounded rank* polynomials. This is easily seen to be a subspace of $\mathrm{Poly}_{\leq d}(V)$ (although it is not a *nonstandard* or *internal* subspace, i.e., the ultralimit of subspaces of the $\mathrm{Poly}_{\leq d}(V_K)$). As such, one can rigorously form the quotient space $\mathrm{Poly}_{\leq d}(V)/\mathrm{Lowrank}_{\leq d}(V)$ of degree $\leq d$ polynomials, modulo bounded rank $\leq d$ polynomials.

The polynomials $P_1, \ldots, P_n$ then have representatives

$$P_1, \ldots, P_n \bmod \mathrm{Lowrank}_{\leq d}(V)$$

in this quotient space. Applying Theorem 2.11.1 (for the field $\mathbf{F}$), one can then find a subcollection $Q_1, \ldots, Q_k \bmod \mathrm{Lowrank}_{\leq d}(V)$ which are linearly independent in this space, which generate $P_1, \ldots, P_n$. Undoing the quotient, we see that the $P_1, \ldots, P_n$ are linear combinations of the $Q_1, \ldots, Q_k$ plus a bounded rank error, while no nontrivial linear combination of $Q_1, \ldots, Q_k$ has bounded rank. Undoing the ultralimit as in the previous section, we obtain the desired contradiction.

We thus see that in the nonstandard world, the somewhat nonrigorous concepts of "low rank" and "high rank" can be formalised as that of "bounded rank" and "unbounded rank". Furthermore, the former space forms a subspace, so in the nonstandard world one can rigorously talk about "quotienting out by bounded rank errors". Thus we see that the algebraic machinery of quotient spaces can be applied in the nonstandard world directly, whereas in the finitary world it can only be applied heuristically. In principle, one could also start deploying more advanced tools of abstract algebra (e.g., exact sequences, cohomology, etc.) in the nonstandard setting, although this has not yet seriously begun to happen in additive combinatorics (although there are strong hints of some sort of "additive cohomology" emerging in the body of work surrounding the inverse conjecture for the Gowers norm, especially on the ergodic theory side of things).

2.11.3. Sunflowers. Now we return to vector spaces (or approximate vector spaces) V over the rationals, such as $V = \mathbf{Z}/p\mathbf{Z}$ for a large prime p. Instead of working with a single (small) tuple $v_1, \ldots, v_n$ of vectors in V, we now consider a *family* $(v_{1,h}, \ldots, v_{n,h})_{h \in H}$ of such vectors in V, where H ranges over a large set, for instance a dense subset of the interval $X := [-N, N] = \{-N, \ldots, N\}$ for some large N. This situation happens to show up in our recent work on the inverse conjecture for the Gowers norm, where the $v_{1,h}, \ldots, v_{n,h}$ represent the various "frequencies" that arise in a derivative $\Delta_h f$ of a function f with respect to the shift h. (This need to consider families is an issue that also comes up in the finite field ergodic theory analogue [**BeTaZi2009**] of the inverse conjectures, due to the unbounded number of generators in that case, but interestingly can be avoided in the ergodic theory over $\mathbf{Z}$.)

In Theorem 2.11.2, the main distinction was between linear dependence and linear independence of the tuple $v_1, \ldots, v_n$ (or some reduction of this tuple, such as $w_1, \ldots, w_k$). We will continue to be interested in the linear dependence or independence of the tuples $v_{1,h}, \ldots, v_{n,h}$ for various h. But we also wish to understand how the $v_{i,h}$ vary with h as well. At one extreme (the "structured" case), there is no dependence on h: $v_{i,h} = v_i$ for all i and all h. At the other extreme (the "pseudo-random" case), the $v_{i,h}$ are basically independent as h varies; in particular, for (almost) all of the pairs $h, h' \in H$, the tuples $v_{1,h}, \ldots, v_{n,h}$ and $v_{1,h'}, \ldots, v_{n,h'}$ are not just separately independent, but are *jointly* independent. One can think of $v_{1,h}, \ldots, v_{n,h}$ and $v_{1,h'}, \ldots, v_{n,h'}$ as being in "general position" relative to each other.

The *sunflower lemma* asserts that any family $(v_{1,h}, \ldots, v_{n,h})_{h \in H}$ is basically a combination of the above scenarios, thus one can divide the family into a linearly independent *core* collection of vectors $(w_1, \ldots, w_m)$ that do not depend on h, together with *petals* $(v'_{1,h}, \ldots, v'_{k,h})_{h \in H'}$, which are in "general position" in the above sense, relative to the core. However, as a price one pays for this, one has to refine H to a dense subset H' of H. This lemma, which significantly generalises Theorem 2.11.2, is formalised as follows:

Theorem 2.11.6 (Sunflower lemma, finitary version). *Let $n \geq 1$ be an integer, and let $F : \mathbf{N} \to \mathbf{N}$ be a function. Let V be an abelian group which admits a well-defined division operation by any natural number of size at most $C(F,n)$ for some constant $C(F,n)$ depending only on F, n. Let H be a finite set, and let $(v_{1,h}, \ldots, v_{n,h})_{h \in H}$ be a collection of n-tuples of vectors in V indexed by H. Then there exists a subset H' of H, integers $k, m \geq 0$ with $m + k \leq n$, a collection $w_1, \ldots, w_m$ of "core" vectors in V for some m, a collection of "petal" vectors $(v'_{1,h}, \ldots, v'_{k,h})_{h \in H'}$ for each $h \in H'$, as well an integer $M \geq 1$, such that*

- Complexity bound. *$M \leq C(F,n)$ for some $C(F,n)$ depending only on F, n.*
- H' dense. *One has $|H'| \geq c(F,n)|H|$ for some $c(F,n) > 0$ depending only on F, n.*
- w, v' generates v. *Every $v_{j,h}$ with $1 \leq j \leq n$ and $h \in H'$ can be expressed as a rational linear combination of the $w_1, \ldots, w_m$ and $v'_{1,h}, \ldots, v'_{k,h}$ of height at most M.*
- w independent. *There is no nontrivial rational linear relation among the $w_1, \ldots, w_m$ of height at most $F(M)$.*
- v' in general position relative to w. *More generally, for $1 - \frac{1}{F(M)}$ of the pairs $(h, h') \in H' \times H'$, there is no nontrivial linear relation among $w_1, \ldots, w_m, v'_{1,h}, \ldots, v'_{k,h}, v'_{1,h'}, \ldots, v'_{k,h'}$ of height at most $F(M)$.*

One can take the $v'_{1,h}, \ldots, v'_{k,h}$ to be a subcollection of the $v_{1,h}, \ldots, v_{n,h}$, though this is not particularly useful in applications.

Proof. We perform a two-parameter "rank reduction argument", where the rank is indexed by the pair (k, m) (ordered lexicographically). We initially set $m = 0$, $k = n$, $H' = H$, $M = 1$, and $v'_{i,h} = v_{i,h}$ for $h \in H$.

At each stage of the iteration, w, v' will generate v (at height M), and we will have some complexity bound on M, m and some density bound on

H'. So one needs to check the independence of w and the general position of v' relative to w.

If there is a linear relation of w at height $F(M)$, then one can use this to reduce the size m of the core by one, leaving the petal size k unchanged, just as in the proof of Theorem 2.11.2. So let us move on, and suppose that there is no linear relation of w at height $F(M)$, but instead there is a failure of the general position hypothesis. In other words, for at least $|H'|^2/F(M)$ pairs $(h,h') \in H' \times H'$, one can find a relation of the form

$$a_{1,h,h'}w_1 + \cdots + a_{m,h,h'}w_m + b_{1,h,h'}v'_{1,h} \\ + \cdots + b_{k,h,h'}v'_{k,h} + c_{1,h,h'}v'_{1,h'} + \cdots + c_{k,h,h'}v'_{k,h'} = 0,$$

where the $a_{i,h,h'}, b_{i,h,h'}, c_{i,h,h'}$ are rationals of height at most $F(M)$, not all zero. The number of possible values for such rationals is bounded by some quantity depending on $m, k, F(M)$. Thus, by the pigeonhole principle, we can find $\gg_{F(M),m,k} |H'|^2$ pairs (i.e., at least $c(F(M),m,k)|H'|^2$ pairs for some $c(F(M),m,k) > 0$ depending only on $F(M), m, k$) such that

$$a_1w_1 + \cdots + a_mw_m + b_1v'_{1,h} + \cdots + b_kv'_{k,h} + c_1v'_{1,h'} + \cdots + c_kv'_{k,h'} = 0$$

for some fixed rationals a_i, b_i, c_i of height at most $F(M)$. By the pigeonhole principle again, we can then find a fixed $h_0 \in H'$ such that

$$a_1w_1 + \cdots + a_mw_m + b_1v'_{1,h} + \cdots + b_kv'_{k,h} = u_{h_0}$$

for all h in some subset H'' of H' with $|H''| \gg_{F(M),m,k} |H'|$, where

$$u_{h_0} := -c_1v'_{1,h_0} - \cdots - c_kv'_{k,h_0}.$$

If the b_i and c_i all vanished, then we would have a linear dependence amongst the core vectors, which we already know how to deal with. So suppose that we have at least one active petal coefficient, say b_k. Then upon rearranging, we can express $v'_{k,h}$ as some rational linear combination of the original core vectors $w_1, \ldots, w_m$, a new core vector u_{h_0}, and the other petals $v'_{1,h}, \ldots, v'_{k-1,h}$, with heights bounded by $\ll_{F(M),k,m} 1$. We may thus refine H' to H'', delete the petal vector $v'_{k,h}$, and add the vector u to the core, thus decreasing k by one and increasing m by one. One still has the generation property so long as one replaces M with a larger M' depending on $M, F(M), k, m$.

Since each iteration of this process either reduces m by one keeping k fixed or reduces k by one increasing m, we see that after at most $2n$ steps, the process must terminate, when we have both the linear independence of the w property and the general position of the v' property. (Note here that we are basically performing a *proof by infinite descent.*) At that stage, one easily verifies that we have obtained all the required conclusions of the theorem. □

As one can see, this result is a little bit trickier to prove than Theorem 2.11.2. Let us now see how it will translate to the nonstandard setting, and see what the nonstandard analogue of Theorem 2.11.6 is. We will skip some details and get to the point where we can motivate and prove this nonstandard analogue; this analogue does in fact imply Theorem 2.11.6 by repeating the arguments from previous sections, but we will leave this as an exercise for the interested reader.

As before, the starting point is to introduce a parameter K, so that the approximate vector space V_K now depends on K (and becomes an actual vector space in the ultralimit V), and the parameter set H_K now also depends on K. We will think of $|H_K|$ as going to infinity as $K \to \infty$, as this is the most interesting case (for bounded H_K, the result basically collapses back to Theorem 2.11.2). In that case, the ultralimit H of the H_K is a nonstandard finite set (i.e., an ultralimit of finite sets) whose (nonstandard) cardinality $|H|$ is an *unbounded* nonstandard integer: it is a nonstandard integer (indeed, it is the ultralimit of the $|H_K|$) which is larger than any standard integer. On the other hand, n and F remain standard (i.e., they do not involve K).

For each K, one starts with a family $(v_{1,h,K}, \ldots, v_{n,h,K})_{h \in H_K}$ of n-tuples of vectors in V_K. Taking ultralimits, one ends up with a family $(v_{1,h}, \ldots, v_{n,h})_{h \in H}$ of n-tuples of vectors in V. Furthermore, for each $1 \leq i \leq n$, the maps $h \mapsto v_{i,h}$ are *nonstandard* (or *internal*) functions from H to V, i.e., they are ultralimits of maps from H_K to V_K. The internal nature of these maps (which is a kind of "measurability" condition on these functions) will be important later. Of course, H and V are also internal (being ultralimits of H_K and V_K, respectively).

We say that a subset H' of H is *dense* if it is an internal subset (i.e., it is the ultralimit of some subsets H'_K of H_K), and if $|H'| \geq \varepsilon |H|$ for some standard $\varepsilon > 0$ (recall that $|H'|, |H|$ are nonstandard integers). If an internal subset is not dense, we say that it is *sparse*, which in nonstandard asymptotic notation (see Section 1.3 of *Structure and Randomness*) is equivalent to $|H'| = o(|H|)$. If a statement $P(h)$ holds on all h in dense set of H, we say that it holds for *many* h; if it holds for all h outside of a sparse set, we say it holds for *almost all* h. These are analogous to the more familiar concepts of "holding with positive probability" and "holding almost surely" in probability theory. For instance, if $P(h)$ holds for many h in H, and $Q(h)$ holds for almost all h in H, then $P(h)$ and $Q(h)$ jointly hold for many h in H. Note how all the epsilons have been neatly hidden away in this nonstandard framework.

Now we state the nonstandard analogue of Theorem 2.11.6.

Theorem 2.11.7 (Sunflower lemma, nonstandard version). *Let $n \geq 1$ be a (standard) integer, let V be a (nonstandard) vector space over the standard rationals $\mathbf{Q}$, and let H be a (nonstandard) set. Let $(v_{1,h}, \ldots, v_{n,h})_{h \in H}$ be a collection of n-tuples of vectors in V indexed by H, such that all the maps $h \mapsto v_{i,h}$ for $1 \leq i \leq n$ are internal. Then there exists a dense subset H' of H, a bounded-dimensional subspace W of V, a (standard) integer $k \geq 0$ with $\dim(W) + k \leq n$, and a collection of "petal" vectors $(v'_{1,h}, \ldots, v'_{k,h})_{h \in H'}$ for each $h \in H'$, with the maps $h \mapsto v'_{i,h}$ being internal for all $1 \leq i \leq k$, such that*

- W, v' generates v. *Every $v_{j,h}$ with $1 \leq j \leq n$ and $h \in H'$ lies in the span of W and the $v'_{1,h}, \ldots, v'_{k,h}$.*
- v' in general position relative to W. *For almost all of the pairs $(h, h') \in H' \times H'$, the vectors $v'_{1,h}, \ldots, v'_{k,h}, v'_{1,h'}, \ldots, v'_{k,h'}$ are linearly independent modulo W over $\mathbf{Q}$.*

Of course, using Theorem 2.11.1 one could obtain a basis $w_1, \ldots, w_m$ for W with $m = \dim(W)$, at which point the theorem more closely resembles Theorem 2.11.6.

Proof. Define a *partial representation* of the family $(v_{1,h}, \ldots, v_{n,h})$ to be a dense subset H' of H, a bounded dimensional space W, a standard integer k with $\dim(W) + k \leq n$, and a collection of $(v'_{1,h}, \ldots, v'_{k,h})_{h \in H'}$ depending internally on h that obeys the generation property (but not necessarily the general position property). Clearly, we have at least one partial representation, namely the trivial one where W is empty, $k = n$, $H' := H$, and $v'_{i,h} := v_{i,h}$. Now, among all such partial representations, let us take a representation with the minimal value of k. (Here we are of course using the *well-ordering property* of the standard natural numbers.) We claim that this representation enjoys the general position property, which will give the claim.

Indeed, suppose this was not the case. Then, for many pairs $(h, h') \in H' \times H'$, the vectors $v'_{1,h}, \ldots, v'_{k,h}, v'_{1,h'}, \ldots, v'_{k,h'}$ have a linear dependence modulo W over $\mathbf{Q}$. (Actually, there is a technical "measurability" issue to address here, which I will return to later.) By symmetry and pigeonholing, we may assume that the $v'_{k,h}$ coefficient of (say) this dependence is nonzero. (Again, there is a measurability issue here.) Applying the pigeonhole principle, one can find $h_0 \in H'$ such that

$$v'_{1,h}, \ldots, v'_{k,h}, v'_{1,h_0}, \ldots, v'_{k,h_0}$$

have a linear dependence over $\mathbf{Q}$ modulo W for many h. (Again, there is a measurability issue here.)

Fix h_0. The number of possible linear combinations of $v'_{1,h_0}, \ldots, v'_{k,h_0}$ is countable. Because of this (and using a "countable pigeonhole principle") that I will address below, we can find a *fixed* rational linear combination u_{h_0} of the $v'_{1,h_0}, \ldots, v'_{k,h_0}$ such that

$$v'_{1,h}, \ldots, v'_{k,h}, u_{h_0}$$

have a linear dependence over $\mathbf{Q}$ modulo W for all h in some dense subset H'' of H'. But now one can pass from H' to the dense subset H'', delete the petal $v'_{k,h}$, and add the vector u_{h_0} to the core space W, thus creating a partial representation with a smaller value of k, contradicting minimality, and we are done. □

We remark here that whereas the finitary analogue of this result was proven using the method of infinite descent, the nonstandard version could instead be proven using the (equivalent) well-ordering principle. One could easily recast the nonstandard version in descent form also, but it is somewhat more difficult to cast the finitary argument using well-ordering due to the extra parameters and quantifiers in play.

Let us now address the measurability issues. The main problem here is that the property of having a linear dependence over the standard rationals $\mathbf{Q}$ is not an internal property, because it requires knowledge of what the standard rationals are, which is not an internal concept in the language of vector spaces. However, for each fixed choice of rational coefficients, the property of having a specific linear dependence with those selected coefficients *is* an internal concept (here we crucially rely on the hypothesis that the maps $h \mapsto v_{i,h}$ were internal), so really what we have here is a sort of "σ-internal" property (a countable union of internal properties). But this is good enough for many purposes. In particular, we have

Lemma 2.11.8 (Countable pigeonhole principle). *Let H be a nonstandardly finite set (i.e., the ultralimit of finite sets H_K), and for each standard natural number n, let E_n be an internal subset of H. Then one of the following holds:*

- Positive density. *There exists a natural number n such that $h \in E_n$ for many $h \in H$ (i.e., E_n is a dense subset of H).*
- Zero density. *For almost all $h \in H$, one has $h \notin E_n$ for all n. (In other words, the (external) set $\bigcup_{n \in \mathbf{N}} E_n$ is contained in a sparse subset of H.)*

This lemma is sufficient to resolve all the measurability issues raised in the previous proof. It is analogous to the trivial statement in measure theory that given a countable collection of measurable subsets of a space of positive measure, either one of the measurable sets has positive measure, or else their union has measure zero (i.e., the sets fail to cover almost all of the space).

Proof. If any of the E_n are dense, we are done. So suppose this is not the case. Since E_n is a definable subset of H which is not dense, it is sparse, thus $|E_n| = o(|H|)$. Now it is convenient to undo the ultralimit and work in the finite sets H_K that H is the ultralimit of. Note that each E_n, being internal, is also an ultralimit of some finite subsets $E_{n,K}$ of H_K.

For each standard integer $M > 0$, the set $E_1 \cup \cdots \cup E_M$ is sparse in H, and in particular has density less than $1/M$. Thus, one can find a p-large set $S_M \subset \mathbf{N}$ such that

$$|E_{1,K} \cup \cdots \cup E_{M,K}| \leq |H_K|/M$$

for all $K \in S_M$. One can arrange matters so that the S_M are decreasing in M. One then sets the set E_K to equal $E_{1,K} \cup \cdots \cup E_{M,K}$, where M is the smallest integer for which $K \in S_M$ (or E_K is empty if K lies in all the S_M, or in none), and let E be the ultralimit of the E_K. Then we see that $|E| \leq |H|/M$ for every standard M, and so E is a sparse subset of H. Furthermore, E contains E_M for every standard M, and so we are in the zero density conclusion of the argument. □

Remark 2.11.9. Curiously, I do not see how to prove this lemma without unpacking the limit; it does not seem to follow just from, say, the *overspill principle*. Instead, it seems to be exploiting the weak countable saturation property I mentioned in Section 2.10. But perhaps I missed a simple argument.

2.11.4. Summary.

Let me summarise with a brief list of pros and cons of switching to a nonstandard framework. First, the pros:

- Many "first-order" parameters such as ε or N disappear from view, as do various "negligible" errors. More importantly, "second-order" parameters, such as the function F appearing in Theorem 2.11.2, also disappear from view. (In principle, third-order and higher parameters would also disappear, though I do not yet know of an actual finitary argument in my fields of study which would have used such parameters (with the exception of Ramsey theory, where such parameters must come into play in order to generate such enormous quantities as *Graham's number*).) As such, a lot of tedious "epsilon management" disappears.
- Iterative (and often parameter-heavy) arguments can often be replaced by minimisation (or more generally, extremisation) arguments, taking advantage of such properties as the *well-ordering principle*, the *least upper bound axiom*, or compactness.

- The *transfer principle* lets one use "for free" any (first-order) statement about standard mathematics in the nonstandard setting (provided that all objects involved are *internal*; see below).
- Mature and powerful theories from infinitary mathematics (e.g., linear algebra, real analysis, representation theory, topology, functional analysis, measure theory, Lie theory, ergodic theory, model theory, etc.) can be used rigorously in a nonstandard setting (as long as one is aware of the usual infinitary pitfalls, of course; see below).
- One can formally define terms that correspond to what would otherwise only be heuristic (or heavily parameterised and quantified) concepts such as "small", "large", "low rank", "independent", "uniformly distributed", etc.
- The conversion from a standard result to its nonstandard counterpart, or vice versa, is fairly quick (but see below), and generally only needs to be done only once or twice per paper.

Next, the cons:

- Nonstandard analysis often requires the axiom of choice, as well as a certain amount of set theory. (There are however weakened versions of nonstandard analysis that can avoid choice that are still suitable for many applications.)
- One needs the machinery of ultralimits and ultraproducts to set up the conversion from standard to nonstandard structures.
- The conversion usually proceeds by a proof by contradiction, which (in conjunction with the use of ultralimits) may not be particularly intuitive.
- One cannot efficiently discern what quantitative bounds emerge from a nonstandard argument (other than by painstakingly converting it back to a standard one or by applying the tools of *proof mining*). (On the other hand, in particularly convoluted standard arguments, the quantitative bounds are already so poor, e.g., of iterated tower-exponential type, that letting go of these bounds is no great loss.)
- One has to take some care to distinguish between standard and nonstandard objects (and also between internal and external sets and functions, which are concepts somewhat analogous to measurable and nonmeasurable sets and functions in measurable theory). More generally, all the usual pitfalls of infinitary analysis (e.g.,

interchanging limits, or the need to ensure measurability or continuity) emerge in this setting, in contrast to the finitary setting where they are usually completely trivial.

- It can be difficult at first to conceptually visualise what nonstandard objects look like (although this becomes easier once one maps nonstandard analysis concepts to heuristic concepts such as "small" and "large" as mentioned earlier, thus for instance one can think of an unbounded nonstandard natural number as being like an incredibly large standard natural number).
- It is inefficient for both nonstandard and standard arguments to coexist within a paper; this makes things a little awkward if one for instance has to cite a result from a standard mathematics paper in a nonstandard mathematics one.
- There are philosophical objections to using mathematical structures that only exist abstractly, rather than corresponding to the "real world". (Note though that similar objections were also raised in the past with regard to the use of, say, complex numbers, non-Euclidean geometries, or even negative numbers.)
- Formally, there is no increase in logical power gained by using nonstandard analysis (at least if one accepts the axiom of choice); anything which can be proven by nonstandard methods can also be proven by standard ones. In practice, though, the length and clarity of the nonstandard proof may be substantially better than the standard one.

In view of the pros and cons, I would not say that nonstandard analysis is suitable in all situations, nor is it unsuitable in all situations, but one needs to carefully evaluate the costs and benefits in a given setting. Also, in some cases having both a finitary and infinitary proof side by side for the same result may be more valuable than just having one of the two proofs. My rule of thumb is that if a finitary argument is already spitting out iterated tower-exponential type bounds or worse in an argument, this is a sign that the argument "wants" to be infinitary, and it may be simpler to move over to an infinitary setting (such as the nonstandard setting).

Notes. This article first appeared at

`terrytao.wordpress.com/2009/12/13`.

2.12. The double Duhamel trick and the in/out decomposition

This is a technical post inspired by separate conversations with Jim Colliander and with Soonsik Kwon on the relationship between two techniques used to control nonradiating solutions to dispersive nonlinear equations, namely the "double Duhamel trick" and the "in/out decomposition". See for instance [**KiVi2009**] for a survey of these two techniques and other related methods in the subject. (I should caution that this article is likely to be unintelligible to anyone not already working in this area.)

For sake of discussion we shall focus on solutions to a nonlinear Schrödinger equation

$$iu_t + \Delta u = F(u),$$

and we will not concern ourselves with the specific regularity of the solution u or the specific properties of the nonlinearity F here. We will also not address the issue of how to justify the formal computations being performed here.

Solutions to this equation enjoy the *forward Duhamel formula*

$$u(t) = e^{i(t-t_0)\Delta}u(t_0) - i\int_{t_0}^{t} e^{i(t-t')\Delta}F(u(t'))\,dt'$$

for times t to the future of t_0 in the lifespan of the solution, as well as the *backward Duhamel formula*

$$u(t) = e^{i(t-t_1)\Delta}u(t_1) + i\int_{t}^{t_1} e^{i(t-t')\Delta}F(u(t'))\,dt'$$

for all times t to the past of t_1 in the lifespan of the solution. The first formula asserts that the solution at a given time is determined by the initial state and by the immediate past, while the second formula is the time reversal of the first, asserting that the solution at a given time is determined by the final state and the immediate future. These basic causal formulae are the foundation of the local theory of these equations, and in particular play an instrumental role in establishing local well-posedness for these equations. In this local theory, the main philosophy is to treat the homogeneous (or *linear*) term $e^{i(t-t_0)\Delta}u(t_0)$ or $e^{i(t-t_1)\Delta}u(t_1)$ as the main term, and the inhomogeneous (or *nonlinear*, or *forcing*) integral term as an error term.

The situation is reversed when one turns to the global theory, and looks at the asymptotic behaviour of a solution as one approaches a limiting time T (which can be infinite if one has global existence, or finite if one has finite time blowup). After a suitable rescaling, the linear portion of the solution often disappears from view, leaving one with an *asymptotic blowup profile* solution which is *nonradiating* in the sense that the linear components of the Duhamel formulae vanish, thus

$$u(t) = -i \int_{t_0}^{t} e^{i(t-t')\Delta} F(u(t'))\, dt' \tag{2.42}$$

and

$$u(t) = i \int_{t}^{t_1} e^{i(t-t')\Delta} F(u(t'))\, dt', \tag{2.43}$$

where t_0, t_1 are the endpoint times of existence. (This type of situation comes up for instance in the Kenig-Merle approach to critical regularity problems, by reducing to a minimal blowup solution which is almost periodic modulo symmetries, and hence nonradiating.) These types of nonradiating solutions are propelled solely by their own nonlinear self-interactions from the immediate past or immediate future; they are generalisations of "nonlinear bound states" such as solitons.

A key task is then to somehow combine the forward representation (2.42) and the backward representation (2.43) to obtain new information on $u(t)$ itself, that cannot be obtained from either representation alone; it seems that the immediate past and immediate future can collectively exert more control on the present than they each do separately. This type of problem can be abstracted as follows. Let $\|u(t)\|_{Y_+}$ be the infimal value of $\|F_+\|_N$ over all forward representations of $u(t)$ of the form

$$u(t) = \int_{t_0}^{t} e^{i(t-t')\Delta} F_+(t')\, dt', \tag{2.44}$$

where N is some suitable spacetime norm (e.g., a Strichartz-type norm), and similarly let $\|u(t)\|_{Y_-}$ be the infimal value of $\|F_-\|_N$ over all backward representations of $u(t)$ of the form

$$u(t) = \int_{t}^{t_1} e^{i(t-t')\Delta} F_-(t')\, dt'. \tag{2.45}$$

Typically, one already has (or is willing to assume as a bootstrap hypothesis) control on $F(u)$ in the norm N, which gives control of $u(t)$ in the norms Y_+, Y_-. The task is then to use the control of both the Y_+ and Y_- norm of $u(t)$ to gain control of $u(t)$ in a more conventional Hilbert space norm X, which is typically a Sobolev space such as H^s or L^2.

One can use some classical functional analysis to clarify this situation. By the closed graph theorem, the above task is (morally, at least) equivalent to establishing an a priori bound of the form

$$\|u\|_X \lesssim \|u\|_{Y_+} + \|u\|_{Y_-} \tag{2.46}$$

for all reasonable u (e.g., test functions). The *double Duhamel trick* accomplishes this by establishing the stronger estimate

$$|\langle u, v\rangle_X| \lesssim \|u\|_{Y_+} \|v\|_{Y_-} \tag{2.47}$$

for all reasonable u, v; note that setting $u = v$ and applying the arithmetic-geometric inequality then gives (2.46). The point is that if u has a forward representation (2.44) and v has a backward representation (2.45), then the inner product $\langle u, v\rangle_X$ can (formally, at least) be expanded as a double integral

$$\int_{t_0}^{t} \int_{t}^{t_1} \langle e^{i(t''-t')\Delta} F_+(t'), e^{i(t''-t')\Delta} F_-(t')\rangle_X \, dt'' dt'.$$

The dispersive nature of the linear Schrödinger equation often causes $\langle e^{i(t''-t')\Delta} F_+(t'), e^{i(t''-t')\Delta} F_-(t')\rangle_X$ to decay, especially in high dimensions. In high enough dimension (typically one needs five or higher dimensions, unless one already has some space-time control on the solution), the decay is stronger than $1/|t' - t''|^2$, so that the integrand becomes absolutely integrable and one recovers (2.47).

Unfortunately it appears that estimates of the form (2.47) fail in low dimensions (for the type of norms N that actually show up in applications); there is just too much interaction between past and future to hope for any reasonable control of this inner product. But one can try to obtain (2.46) by other means. By the Hahn-Banach theorem (and ignoring various issues related to reflexivity), (2.46) is equivalent to the assertion that every $u \in X$ can be decomposed as $u = u_+ + u_-$, where $\|u_+\|_{Y_+^*} \lesssim \|u\|_X$ and $\|u_-\|_{Y_-^*} \lesssim \|v\|_X$. Indeed once one has such a decomposition, one obtains (2.46) by computing the inner product of u with $u = u_+ + u_-$ in X in two different ways. One can also (morally at least) write $\|u_+\|_{Y_+^*}$ as $\|e^{i(\cdot-t)\Delta} u_+\|_{N^*([t_0,t])}$ and similarly write $\|u_-\|_{Y_-^*}$ as $\|e^{i(\cdot-t)\Delta} u_-\|_{N^*([t,t_1])}$

So one can dualise the task of proving (2.46) as that of obtaining a decomposition of an arbitrary initial state u into two components u_+ and u_-, where the former disperses into the past and the latter disperses into the future under the linear evolution. We do not know how to achieve this type of task efficiently in general—and doing so would likely lead to a significant advance in the subject (perhaps one of the main areas in this topic where serious harmonic analysis is likely to play a major role). But in the model case of spherically symmetric data u, one can perform such a decomposition quite easily: one uses microlocal projections to set u_+ to be the "inward" pointing component of u, which propagates towards the origin in the future and away from the origin in the past, and u_- to simimlarly be the "outward" component of u. As spherical symmetry significantly dilutes the amplitude of the solution (and hence the strength of the nonlinearity) away from the origin, this decomposition tends to work quite well for applications, and is one of the main reasons (though not the only one) why we have a global theory for low-dimensional nonlinear Schrödinger equations in the radial case, but not in general.

The in/out decomposition is a linear one, but the Hahn-Banach argument gives no reason why the decomposition needs to be linear. (Note that other well-known decompositions in analysis, such as the Fefferman-Stein decomposition of BMO, are necessarily nonlinear, a fact which is ultimately equivalent to the noncomplemented nature of a certain subspace of a Banach space; see Section 1.7 of *Volume I*.) So one could imagine a sophisticated nonlinear decomposition as a general substitute for the in/out decomposition. See for instance [**BoBr2003**] for some of the subtleties of decomposition even in very classical function spaces such as $H^{1/2}(R)$. Alternatively, there may well be a third way to obtain estimates of the form (2.46) that do not require either decomposition or the double Duhamel trick; such a method may well clarify the relative relationship between past, present, and future for critical nonlinear dispersive equations, which seems to be a key aspect of the theory that is still only partially understood. (In particular, it seems that one needs a fairly strong decoupling of the present from both the past and the future to get the sort of elliptic-like regularity results that allow us to make further progress with such equations.)

Notes. This article first appeared at

`terrytao.wordpress.com/2009/12/17.`

Thanks to Kareem Carr, hezhigang, and anonymous commenters for corrections.

2.13. The free nilpotent group

In a multiplicative group G, the *commutator* of two group elements g, h is defined as $[g, h] := g^{-1}h^{-1}gh$ (other conventions are also in use, though they are largely equivalent for the purposes of this discussion). A group is said to be *nilpotent of step s* (or more precisely, step $\leq s$), if all iterated commutators of order $s+1$ or higher necessarily vanish. For instance, a group is nilpotent of order 1 if and only if it is abelian, and it is nilpotent of order 2 if and only if $[[g_1, g_2], g_3] = id$ for all g_1, g_2, g_3 (i.e., all commutator elements $[g_1, g_2]$ are *central*), and so forth. A good example of an s-step nilpotent group is the group of $s+1 \times s+1$ upper-triangular *unipotent* matrices (i.e., matrices with 1's on the diagonal and zero below the diagonal), and taking values in some ring (e.g., reals, integers, complex numbers, etc.).

Another important example of nilpotent groups arise from operations on polynomials. For instance, if $V_{\leq s}$ is the vector space of real polynomials of one variable of degree at most s, then there are two natural affine actions on $V_{\leq s}$. Firstly, every polynomial Q in $V_{\leq s}$ gives rise to a "vertical" shift $P \mapsto P + Q$. Secondly, every $h \in \mathbf{R}$ gives rise to a "horizontal" shift $P \mapsto P(\cdot + h)$. The group generated by these two shifts is a nilpotent group of step $\leq s$; this reflects the well-known fact that a polynomial of degree $\leq s$ vanishes once one differentiates more than s times. Because of this link between nilpotentcy and polynomials, one can view nilpotent algebra as a generalisation of polynomial algebra.

Suppose one has a finite number $g_1, \ldots, g_n$ of generators. Using abstract algebra, one can then construct the *free nilpotent group* $\mathcal{F}_{\leq s}(g_1, \ldots, g_n)$ of step $\leq s$, defined as the group generated by the $g_1, \ldots, g_n$ subject to the relations that all commutators of order $s+1$ involving the generators are trivial. This is the *universal object* in the category of nilpotent groups of step $\leq s$ with n marked elements $g_1, \ldots, g_n$. In other words, given any other $\leq s$-step nilpotent group G' with n marked elements $g'_1, \ldots, g'_n$, there is a unique homomorphism from the free nilpotent group to G' that maps each g_j to g'_j for $1 \leq j \leq n$. In particular, the free nilpotent group is well defined up to isomorphism in this category.

In many applications, one wants to have a more concrete description of the free nilpotent group, so that one can perform computations more easily (and in particular, be able to tell when two words in the group are equal or not). This is easy for small values of s. For instance, when $s = 1$, $\mathcal{F}_{\leq 1}(g_1, \ldots, g_n)$ is simply the *free abelian group* generated by $g_1, \ldots, g_n$, and so every element g of $\mathcal{F}_{\leq 1}(g_1, \ldots, g_n)$ can be described uniquely as

$$g = \prod_{j=1}^{n} g_j^{m_j} := g_1^{m_1} \cdots g_n^{m_n} \tag{2.48}$$

for some integers $m_1, \ldots, m_n$, with the obvious group law. Indeed, to obtain existence of this representation, one starts with any representation of g in terms of the generators $g_1, \ldots, g_n$, and then uses the abelian property to push the g_1 factors to the far left, followed by the g_2 factors, and so forth. To show uniqueness, we observe that the group G of formal abelian products $\{g_1^{m_1} \cdots g_n^{m_n} : m_1, \ldots, m_n \in \mathbf{Z}\} \equiv \mathbf{Z}^k$ is already a ≤ 1-step nilpotent group with marked elements $g_1, \ldots, g_n$, and so there must be a homomorphism from the free group to G. Since G distinguishes all the products $g_1^{m_1} \cdots g_n^{m_n}$ from each other, the free group must also.

It is only slightly more tricky to describe the free nilpotent group $\mathcal{F}_{\leq 2}(g_1, \ldots, g_n)$ of step ≤ 2. Using the identities

$$\begin{aligned} gh &= hg[g,h], \\ gh^{-1} &= ([g,h]^{-1})^{g^{-1}} h^{-1} g, \\ g^{-1}h &= h[g,h]^{-1}g^{-1}, \\ g^{-1}h^{-1} &:= [g,h]g^{-1}h^{-1}, \end{aligned}$$

where $g^h := h^{-1}gh$ is the conjugate of g by h, we see that whenever $1 \leq i < j \leq n$, one can push a positive or negative power of g_i past a positive or negative power of g_j, at the cost of creating a positive or negative power of $[g_i, g_j]$ or one of its conjugates. Meanwhile, in a $\leq$ 2-step nilpotent group, all the commutators are central, and one can pull all the commutators out of a word and collect them as in the abelian case. Doing all this, we see that every element g of $\mathcal{F}_{\leq 2}(g_1, \ldots, g_n)$ has a representation of the form

$$g = (\prod_{j=1}^{n} g_j^{m_j})(\prod_{1 \leq i < j \leq n} [g_i, g_j]^{m_{[i,j]}}) \tag{2.49}$$

for some integers m_j for $1 \leq j \leq n$ and $m_{[i,j]}$ for $1 \leq i < j \leq n$. Note that we do not need to consider commutators $[g_i, g_j]$ for $i \geq j$, since

$$[g_i, g_i] = id$$

and

$$[g_i, g_j] = [g_j, g_i]^{-1}.$$

It is possible to show also that this representation is unique, by repeating the previous argument, i.e., by showing that the set of formal products

$$G := \{(\prod_{j=1}^{k} g_j^{m_j})(\prod_{1 \leq i < j \leq n} [g_i, g_j]^{m_{[i,j]}}) : m_j, m_{[i,j]} \in \mathbf{Z}\}$$

forms a $\leq$ 2-step nilpotent group, after using the above rules to define the group operations. This can be done, but verifying the group axioms (particularly the associative law) for G is unpleasantly tedious.

Once one sees this, one rapidly loses an appetite for trying to obtain a similar explicit description for free nilpotent groups for higher step, especially once one starts seeing that higher commutators obey some nonobvious identities such as the *Hall-Witt identity*

$$[[g, h^{-1}], k]^h \cdot [[h, k^{-1}], g]^k \cdot [[k, g^{-1}], h]^g = 1 \tag{2.50}$$

(a nonlinear version of the *Jacobi identity* in the theory of Lie algebras), which make one less certain as to the existence or uniqueness of various proposed generalisations of the representations (2.48) or (2.49). For instance, in the free $\leq$ 3-step nilpotent group, it turns out that for representations of the form

$$g = (\prod_{j=1}^{n} g_j^{m_j})(\prod_{1\leq i<j\leq n} [g_i, g_j]^{m_{[i,j]}})(\prod_{1\leq i<j<k\leq n} [[g_i, g_j], g_k]^{n_{[[i,j],k]}}),$$

one has uniqueness but not existence (e.g., even in the simplest case $n = 3$, there is no place in this representation for, say, $[[g_1, g_3], g_2]$ or $[[g_1, g_2], g_2]$). But if one tries to insert more triple commutators into the representation to make up for this, one has to be careful not to lose uniqueness due to identities such as (2.50). One can paste these in by ad hoc means in the $s = 3$ case, but the $s = 4$ case looks more fearsome still, especially now that the quadruple commutators split into several distinct-looking species such as $[[g_i, g_j], [g_k, g_l]]$ and $[[[g_i, g_j], g_k], g_l]$ which are nevertheless still related to each other by identities such as (2.50). While one can eventually disentangle this mess for any fixed n and s by a finite amount of combinatorial computation, it is not immediately obvious how to give an explicit description of $\mathcal{F}_{\leq s}(g_1, \ldots, g_n)$ uniformly in n and s.

Nevertheless, it turns out that one can give a reasonably tractable description of this group if one takes a *polycyclic perspective* rather than a nilpotent one; i.e., one views the free nilpotent group as a tower of group extensions of the trivial group by the cyclic group $\mathbf{Z}$. This seems to be a fairly standard observation in group theory—I found it in [**MaKaSo2004**] and [**Le2009**]—but seems not to be so widely known outside of that field, so I wanted to record it here.

2.13.1. Generalisation. The first step is to generalise the concept of a free nilpotent group to one where the generators have different "degrees". Define a *graded sequence* to be a finite ordered sequence $(g_\alpha)_{\alpha\in A}$ of formal group elements g_α, indexed by a finite, totally ordered set A, where each g_α is assigned a positive integer $\deg(g_\alpha)$, which we call the *degree* of g_α. We then define the degree of any formal iterated commutator of the g_α by declaring the degree of $[g, h]$ to be the sum of the degrees of g and h. Thus for instance $[[g_{\alpha_1}, g_{\alpha_2}], g_{\alpha_3}]$ has degree $\deg(g_{\alpha_1}) + \deg(g_{\alpha_2}) + \deg(g_{\alpha_3})$. (The ordering on A is not presently important, but will become useful for the polycyclic representation; note that such ordering has already appeared implicitly in (2.48) and (2.49).)

Define the *free $\leq$ s-step nilpotent group* $\mathcal{F}_{\leq s}((g_\alpha)_{\alpha\in A})$ generated by a graded sequence $(g_\alpha)_{\alpha\in A}$ to be the group generated by the g_α, subject to the constraint that any iterated commutator of the g_α of degree greater than s is trivial. Thus the free group $\mathcal{F}_{\leq s}(g_1, \ldots, g_k)$ corresponds to the case when all the g_i are assigned a degree of 1.

Note that any element of a graded sequence of degree greater than s is automatically trivial (we view it as a 0-fold commutator of itself) and so can be automatically discarded from that sequence.

We will recursively define the free $\leq s$-step nilpotent group of some graded sequence $(g_\alpha)_{\alpha\in A}$ in terms of simpler sequences, which have fewer low-degree terms at the expense of introducing higher-degree terms, though as mentioned earlier there is no need to introduce terms of degree larger than s. Eventually this process exhausts the sequence, and at that point the free nilpotent group will be completely described.

2.13.2. Shift. It is convenient to introduce the iterated commutators $[g, mh]$ for $m = 0, 1, 2, \ldots$ by declaring $[g, 0h] := g$ and $[g, (m+1)h] := [[g, mh], h]$, thus for instance $[g, 3h] = [[[g, h], h], h]$.

Definition 2.13.1 (Shift). Let $s \geq 1$ be an integer, let $(g_\alpha)_{\alpha\in A}$ be a nonempty graded sequence, and let α_0 be the minimal element of A. We define the (degree $\leq s$) *shift* $(g_\alpha)_{\alpha\in A'}$ of $(g_\alpha)_{\alpha\in A}$ by defining A' to be formed from A by removing α_0, and then adding at the end of A all commutators $[\beta, m\alpha_0]$ of degree at most s, where $\beta \in A\backslash\{\alpha_0\}$ and $m \geq 1$. For sake of concreteness we order these commutators lexicographically, so that $[\beta, m\alpha_0] \geq [\beta', m'\alpha_0]$ if $\beta > \beta'$, or if $\beta = \beta'$ and $m > m'$. (These commutators are also considered to be larger than any element of $A\backslash\{\alpha_0\}$). We give each $[\beta, m\alpha_0]$ a degree of $\deg(\beta) + m\deg(\alpha_0)$, and define the group element $g_{[\beta,m\alpha_0]}$ to be $[g_\beta, mg_{\alpha_0}]$.

Example 2.13.2. If $s \leq 3$, and the graded sequence g_a, g_b, g_c consists entirely of elements of degree 1, then the shift of this sequence is given by

$$g_b, g_c, g_{[b,a]}, g_{[b,2a]}, g_{[c,a]}, g_{[c,2a]},$$

where $[b, a], [c, a]$ have degree 2, and $[b, 2a]$, $[c, 2a]$ have degree 3, and $g_{[b,a]} = [g_b, g_a]$, $g_{[b,2a]} = [g_b, 2g_a]$, etc.

The key lemma is then

Lemma 2.13.3 (Recursive description of free group). *Let $s \geq 1$ be an integer, let $(g_\alpha)_{\alpha\in A}$ be a nonempty graded sequence, and let α_0 be the minimal element of A. Let $(g_\alpha)_{\alpha\in A'}$ be the shift of $(g_\alpha)_{\alpha\in A}$. Then $\mathcal{F}_{\leq s}((g_\alpha)_{\alpha\in A})$ is generated by g_{α_0} and $\mathcal{F}_{\leq s}((g_\alpha)_{\alpha\in A'})$, and furthermore the latter group is a normal subgroup of $\mathcal{F}_{\leq s}((g_\alpha)_{\alpha\in A})$ that does not contain g_{α_0}. In other words, we have a semidirect product representation*

$$\mathcal{F}_{\leq s}((g_\alpha)_{\alpha\in A}) = \mathbf{Z} \ltimes \mathcal{F}_{\leq s}((g_\alpha)_{\alpha\in A'})$$

with g_{α_0} being identified with $(1, id)$ and the action of $\mathbf{Z}$ being given by the conjugation action of g_{α_0}. In particular, every element g in $\mathcal{F}_{\leq s}((g_\alpha)_{\alpha\in A})$ can be uniquely expressed as $g = g_\alpha^{n_\alpha} g'$, where $g' \in \mathcal{F}_{\leq s}((g_\alpha)_{\alpha\in A'})$.

Proof. It is clear that $\mathcal{F}_{\leq s}((g_\alpha)_{\alpha\in A'})$ is a subgroup of $\mathcal{F}_{\leq s}((g_\alpha)_{\alpha\in A})$, and that it together with g_{α_0} generates $\mathcal{F}_{\leq s}((g_\alpha)_{\alpha\in A})$. To show that this subgroup is normal, it thus suffices to show that the conjugation action of g_{α_0} and $g_{\alpha_0}^{-1}$ preserve $\mathcal{F}_{\leq s}((g_\alpha)_{\alpha\in A'})$. It suffices to check this on generators. But this is clear from the identity

$$g_{\alpha_0}^{-1}[g_\beta, mg_{\alpha_0}]g_{\alpha_0} = [g_\beta, mg_{\alpha_0}][g_\beta,(m+1)g_{\alpha_0}]$$

and its inverse

$$g_{\alpha_0}[g_\beta, mg_{\alpha_0}]g_{\alpha_0}^{-1} = [g_\beta, mg_{\alpha_0}][g_\beta,(m+1)g_{\alpha_0}]^{-1}[g_\beta,(m+2)g_{\alpha_0}]\cdots$$

(note that the product terminates in finite time due to nilpotency).

Finally, we need to show that g_{α_0} is not contained in $\mathcal{F}_{\leq s}((g_\alpha)_{\alpha\in A'})$. But because the conjugation action of g_{α_0} preserves the latter group, we can form the semidirect product $G := \mathbf{Z} \ltimes \mathcal{F}_{\leq s}((g_\alpha)_{\alpha\in A'})$. By the universal nature of the free group, there must thus be a homomorphism from $\mathcal{F}_{\leq s}((g_\alpha)_{\alpha\in A})$ to G which maps g_{α_0} to $(1, id)$ and maps $\mathcal{F}_{\leq s}((g_\alpha)_{\alpha\in A'})$ to $0 \times \mathcal{F}_{\leq s}((g_\alpha)_{\alpha\in A'})$. This implies that g_{α_0} cannot lie in $\mathcal{F}_{\leq s}((g_\alpha)_{\alpha\in A'})$, and the claim follows. □

We can now iterate this. Observe that every time one shifts a nonempty graded sequence, one removes one element (the minimal element g_{α_0}) but replaces it with zero or more elements of higher degree. Iterating this process, we eventually run out of elements of degree one, then degree two, and so forth, until the sequence becomes completely empty. We glue together all the elements encountered this way and refer to the full sequence as the *completion* $(g_\alpha)_{\alpha\in\overline{A}}$ of the original sequence $(g_\alpha)_{\alpha\in A}$. As a corollary of the above lemma we thus have

Corollary 2.13.4 (Explicit description of free nilpotent group). *Let $s \geq 1$ be an integer, and let $(g_\alpha)_{\alpha\in A}$ be a graded sequence. Then every element g of $\mathcal{F}_{\leq s}((g_\alpha)_{\alpha\in A})$ can be represented uniquely as*

$$\prod_{\alpha\in\overline{A}} g_\alpha^{n_\alpha},$$

where n_α is an integer and $\overline{A}$ is the completion of A.

Example 2.13.5. We continue with the sequence g_a, g_b, g_c from Example 2.13.2, with $s = 3$. We already saw that shifting once yielded the sequence

$$g_b, g_c, g_{[b,a]}, g_{[b,2a]}, g_{[c,a]}, g_{[c,2a]}.$$

Another shift gives

$$g_c, g_{[b,a]}, g_{[b,2a]}, g_{[c,a]}, g_{[c,2a]}, g_{[c,b]}, g_{[c,2b]}, g_{[[b,a],b]}, g_{[[c,a],b]},$$

and shifting again gives

$$g_{[b,a]}, g_{[b,2a]}, g_{[c,a]}, g_{[c,2a]}, g_{[c,b]}, g_{[c,2b]}, g_{[[b,a],b]}, g_{[[c,a],b]}, g_{[[b,a],c]}, g_{[[c,a],c]}.$$

At this point, all remaining terms in the sequence have degree at least two, and further shifting simply removes the first element without adding any new elements. Thus the completion is

$$g_a, g_b, g_c, g_{[b,a]}, g_{[b,2a]}, g_{[c,a]}, g_{[c,2a]},$$

$$g_{[c,b]}, g_{[c,2b]}, g_{[[b,a],b]}, g_{[[c,a],b]}, g_{[[b,a],c]}, g_{[[c,a],c]}$$

and every element of $\mathcal{F}_{\leq 3}(g_a, g_b, g_c)$ can be uniquely expressed as

$$\begin{aligned} & g_a^{n_a} g_b^{n_b} g_c^{b_c} [g_b, g_a]^{n_{[b,a]}} [g_b, 2g_a]^{n_{[b,2a]}} [g_c, g_a]^{n_{[c,a]}} \\ & \quad [g_c, 2g_a]^{n_{[c,2a]}} [g_c, g_b]^{n_{[c,b]}} [g_c, 2g_b]^{n_{[c,2b]}} [[g_b, g_a], g_b]^{n_{[[b,a],b]}} \\ & \quad\quad [[g_c, g_a], g_b]^{n_{[[c,a],b]}} [[g_b, g_a], g_c]^{n_{[[b,a],c]}} [[g_c, g_a], g_c]^{n_{[[c,a],c]}}. \end{aligned}$$

In [**Le2009**], a related argument was used to expand bracket polynomials (a generalisation of ordinary polynomials in which the integer part operation $x \mapsto \lfloor x \rfloor$ is introduced) of degree $\leq s$ in several variables $(x_\alpha)_{\alpha \in A}$ into a canonical basis $(x_\alpha)_{\alpha \in \overline{A}}$, where $\overline{A}$ is the same completion of A that was encountered here. This was used to show a close connection between such bracket polynomials and nilpotent groups (or more precisely, nilsequences).

Notes. This article first appeared at

`terrytao.wordpress.com/2009/12/21.`

Thanks to Dylan Thurston for corrections.

Bibliography

[AgKaSa2004] M. Agrawal, N. Kayal, N. Saxena, *PRIMES is in P*, Annals of Mathematics **160** (2004), no. 2, pp. 781–793.

[AjSz1974] M. Ajtai, E. Szemerédi, *Sets of lattice points that form no squares*, Stud. Sci. Math. Hungar. **9** (1974), 9–11 (1975).

[AlDuLeRoYu1994] N. Alon, R. Duke, H. Lefmann, Y. Rödl, R. Yuster, *The algorithmic aspects of the regularity lemma*, J. Algorithms **16** (1994), no. 1, 80–109.

[AlSh2008] N. Alon, A. Shapira, *Every monotone graph property is testable*, SIAM J. Comput. **38** (2008), no. 2, 505–522.

[AlSp2008] N. Alon, J. Spencer, The probabilistic method. Third edition. With an appendix on the life and work of Paul Erdős. Wiley-Interscience Series in Discrete Mathematics and Optimization. John Wiley & Sons, Inc., Hoboken, NJ, 2008.

[Au2008] T. Austin, *On exchangeable random variables and the statistics of large graphs and hypergraphs*, Probab. Surv. **5** (2008), 80–145.

[Au2009] T. Austin, *Deducing the multidimensional Szemerédi Theorem from an infinitary removal lemma*, preprint.

[Au2009b] T. Austin, *Deducing the Density Hales-Jewett Theorem from an infinitary removal lemma*, preprint.

[AuTa2010] T. Austin, T. Tao, *On the testability and repair of hereditary hypergraph properties*, preprint.

[Ax1968] J. Ax, *The elementary theory of finite fields*, Ann. of Math. **88** (1968) 239–271.

[BaGiSo1975] T. Baker, J. Gill, R. Solovay, *Relativizations of the $\mathcal{P} =?\mathcal{NP}$ question*, SIAM J. Comput. **4** (1975), no. 4, 431–442.

[Be1975] W. Beckner, *Inequalities in Fourier analysis*, Ann. of Math. **102** (1975), no. 1, 159–182.

[BeTaZi2009] V. Bergelson, T. Tao, T. Ziegler, *An inverse theorem for the uniformity seminorms associated with the action of F^{ω}*, preprint.

[BeLo1976] J. Bergh, J. Löfström, Interpolation spaces. An introduction. Grundlehren der Mathematischen Wissenschaften, No. 223. Springer-Verlag, Berlin-New York, 1976.

[BiRo1962] A. Białynicki-Birula, M. Rosenlicht, *Injective morphisms of real algebraic varieties*, Proc. Amer. Math. Soc. **13** (1962) 200–203.

[BoKe1996] E. Bogomolny, J. Keating, *Random matrix theory and the Riemann zeros. II. n-point correlations*, Nonlinearity **9** (1996), no. 4, 911–935.

[Bo1969] A. Borel, *Injective endomorphisms of algebraic varieties*, Arch. Math. (Basel) **20** (1969), 531–537.

[Bo1999] J. Bourgain, *On the dimension of Kakeya sets and related maximal inequalities*, Geom. Funct. Anal. **9** (1999), no. 2, 256–282.

[BoBr2003] J. Bourgain, H. Brezis, *On the equation* $\operatorname{div} Y = f$ *and application to control of phases*, J. Amer. Math. Soc. **16** (2003), no. 2, 393–426.

[BudePvaR2008] G. Buskes, B. de Pagter, A. van Rooij, *The Loomis-Sikorski theorem revisited*, Algebra Universalis **58** (2008), 413–426.

[ChPa2009] T. Chen, N. Pavlovic, *The quintic NLS as the mean field limit of a boson gas with three-body interactions*, preprint.

[ClEdGuShWe1990] K. Clarkson, H. Edelsbrunner, L. Guibas, M. Sharir, E. Welzl, *Combinatorial complexity bounds for arrangements of curves and spheres*, Discrete Comput. Geom. **5** (1990), no. 2, 99–160.

[Co1989] J. B. Conrey, *More than two fifths of the zeros of the Riemann zeta function are on the critical line*, J. Reine Angew. Math. **399** (1989), 1–26.

[Dy1970] F. Dyson, *Correlations between eigenvalues of a random matrix*, Comm. Math. Phys. **19** (1970), 235–250.

[ElSz2008] G. Elek, B. Szegedy, *A measure-theoretic approach to the theory of dense hypergraphs*, preprint.

[ElObTa2009] J. Ellenberg, R. Oberlin, T. Tao, *The Kakeya set and maximal conjectures for algebraic varieties over finite fields*, preprint.

[ElVeWe2009] J. Ellenberg, A. Venkatesh, C. Westerland, *Homological stability for Hurwitz spaces and the Cohen-Lenstra conjecture over function fields*, preprint.

[ErKa1940] P. Erdős, M. Kac, *The Gaussian Law of Errors in the Theory of Additive Number Theoretic Functions*, Amer. J. Math., **62** (1940), no. 1/4, pages 738–742.

[EsKePoVe2008] L. Escauriaza, C. E. Kenig, G. Ponce, L. Vega, *Hardy's uncertainty principle, convexity and Schrödinger evolutions*, J. Eur. Math. Soc. (JEMS) **10** (2008), no. 4, 883–907.

[Fa2003] K. Falconer, *Fractal geometry*, Mathematical Foundations and Applications. Second edition. John Wiley & Sons, Inc., Hoboken, NJ, 2003.

[FeSt1972] C. Fefferman, E. M. Stein, H^p *spaces of several variables*, Acta Math. **129** (1972), no. 3-4, 137–193.

[FiMaSh2007] E. Fischer, A. Matsliach, A. Shapira, *Approximate Hypergraph Partitioning and Applications*, Proc. of FOCS 2007, 579–589.

[Fo2000] G. Folland, *Real analysis, modern techniques and their applications*. Second edition. Pure and Applied Mathematics (New York). A Wiley-Interscience Publication. John Wiley & Sons, Inc., New York, 1999.

[Fo1955] E. Følner, *On groups with full Banach mean value*, Math. Scand. **3** (1955), 243–254.

[Fo1974] J. Fournier, *Majorants and* L^p *norms*, Israel J. Math. **18** (1974), 157–166.

[Fr1973] G. Freiman, *Groups and the inverse problems of additive number theory*, Number-theoretic studies in the Markov spectrum and in the structural theory of set addition, pp. 175-183. Kalinin. Gos. Univ., Moscow, 1973.

[Fu1977] H. Furstenberg, *Ergodic behavior of diagonal measures and a theorem of Szemerédi on arithmetic progressions*, J. Analyse Math. **31** (1977), 204–256.

[FuKa1989] H. Furstenberg, Y. Katznelson, *A density version of the Hales-Jewett theorem for $k = 3$*, Graph theory and combinatorics (Cambridge, 1988). Discrete Math. **75** (1989), no. 1–3, 227–241.

[FuKa1991] H. Furstenberg, Y. Katznelson, *A density version of the Hales-Jewett theorem*, J. Anal. Math. **57** (1991), 64–119.

[GiTr1998] D. Gilbarg, N. Trudinger, Elliptic partial differential equations of second order. Reprint of the 1998 edition. Classics in Mathematics. Springer-Verlag, Berlin, 2001.

[GoMo1987] D. Goldston, H. Montgomery, *Pair correlation of zeros and primes in short intervals*, Analytic number theory and Diophantine problems (Stillwater, OK, 1984), 183–203, Progr. Math., 70, Birkhäuser Boston, Boston, MA, 1987.

[Go1993] W. T. Gowers, B. Maurey, *The unconditional basic sequence problem*, J. Amer. Math. Soc. **6** (1993), no. 4, 851–874.

[Gr1992] A. Granville, *On elementary proofs of the prime number theorem for arithmetic progressions, without characters*, Proceedings of the Amalfi Conference on Analytic Number Theory (Maiori, 1989), 157–194, Univ. Salerno, Salerno, 1992.

[Gr2005] A. Granville, *It is easy to determine whether a given integer is prime*, Bull. Amer. Math. Soc. (N.S.) **42** (2005), no. 1, 3–38.

[GrSo2007] A. Granville, K. Soundararajan, *Large character sums: pretentious characters and the Pólya-Vinogradov theorem*, J. Amer. Math. Soc. **20** (2007), no. 2, 357–384.

[GrTa2007] B. Green, T. Tao, *The distribution of polynomials over finite fields, with applications to the Gowers norms*, preprint.

[GrTaZi2009] B. Green, T. Tao, T. Ziegler, *An inverse theorem for the Gowers U^4 norm*, preprint.

[Gr1999] M. Gromov, *Endomorphisms of symbolic algebraic varieties*, J. Eur. Math. Soc. (JEMS) **1** (1999), no. 2, 109–197.

[Gr1966] A. Grothendieck, *Éléments de géométrie algébrique. IV. Étude locale des schémas et des morphismes de schémas. III.*, Inst. Hautes Études Sci. Publ. Math. No. 28, 1966, 255 pp.

[GyMaRu2008] K. Gyarmati, M. Matolcsi, I. Ruzsa, *Plünnecke's inequality for different summands*, Building Bridges, 309–320, Bolyai Soc. Math. Stud., 19, Springer, Berlin, 2008.

[Ho1990] L. Hörmander, The analysis of linear partial differential operators. I.-IV. Reprint of the second (1990) edition. Classics in Mathematics. Springer-Verlag, Berlin, 2003.

[HoKrPeVi2006] J. Hough, M. Krishnapur, Y. Peres, B. Virág, Determinantal processes and independence, Probab. Surv. **3** (2006), 206–229.

[Hr2009] E. Hrushovski, *Stable group theory and approximate subgroups*, preprint.

[Hu1968] R. Hunt, *On the convergence of Fourier series*, 1968 Orthogonal Expansions and their Continuous Analogues (Proc. Conf., Edwardsville, Ill., 1967) pp. 235–255 Southern Illinois Univ. Press, Carbondale, Ill.

[Is2006] Y. Ishigami, *A Simple Regularization of Hypergraphs*, preprint.

[IwKo2004] H. Iwaniec, E. Kowalski, *Analytic number theory*, American Mathematical Society Colloquium Publications, 53. American Mathematical Society, Providence, RI, 2004.

[Jo1986] D. Joyner, *Distribution theorems of L-functions*, Pitman Research Notes in Mathematics Series, 142. Longman Scientific & Technical, Harlow; John Wiley & Sons, Inc., New York, 1986.

[KaVe1983] V. Kaimanovich, A. Vershik, *Random walks on discrete groups: boundary and entropy*, Ann. Probab. **11** (1983), no. 3, 457–490.

[KiVi2009] R. Killip, M. Visan, *Nonlinear Schrödinger Equations at critical regularity*, preprint.

[KiScSt2008] K. Kirkpatrick, B. Schlein, G. Staffilani, *Derivation of the two dimensional nonlinear Schrodinger equation from many body quantum dynamics*, preprint.

[KlMa2008] S. Klainerman, M. Machedon, *On the uniqueness of solutions to the Gross-Pitaevskii hierarchy*, Comm. Math. Phys. **279** (2008), no. 1, 169–185.

[Ku1999] K. Kurdyka, *Injective endomorphisms of real algebraic sets are surjective*, Math. Ann. **313** (1999), no. 1, 69–82.

[La2001] I. Łaba, *Fuglede's conjecture for a union of two intervals*, Proc. Amer. Math. Soc. **129** (2001), no. 10, 2965–2972.

[LaTa2001] I. Łaba, T. Tao, *An x-ray transform estimate in* $\mathbf{R}^n$, Rev. Mat. Iberoamericana **17** (2001), no. 2, 375–407.

[La1996] A. Laurincikas, Limit theorems for the Riemann zeta-function. Mathematics and Its Applications, 352. Kluwer Academic Publishers Group, Dordrecht, 1996.

[Le2009] A. Leibman, *A canonical form and the distribution of values of generalised polynomials,* preprint.

[Le2000] V. Lev, *Restricted Set Addition in Groups I: The Classical Setting*, J. London Math. Soc. **62** (2000), no. 1, 27–40.

[LiLo2000] E. Lieb, E. Loss, *Analysis.* Second edition. Graduate Studies in Mathematics, 14. American Mathematical Society, Providence, RI, 2001.

[Li1853] J. Liouville, *Sur l'equation aux differences partielles*, J. Math. Pure et Appl. **18** (1853), 71–74.

[LiTz1971] J. Lindenstrauss, L. Tzafriri, *On the complemented subspaces problem*, Israel J. Math. **9** (1971) 263–269.

[Lo1946] L. H. Loomis, *On the representation of* σ*-complete Boolean algebras*, Bull. Amer. Math Soc. **53** (1947), 757–760.

[LoSz2007] L. Lovász, B. Szegedy, *Szemerédi's lemma for the analyst*, Geom. Funct. Anal. **17** (2007), no. 1, 252–270.

[Ly2003] R. Lyons, *Determinantal probability measures*, Publ. Math. Inst. Hautes Études Sci. No. **98** (2003), 167–212.

[Ma2008] M. Madiman, *On the entropy of sums*, preprint.

[MaKaSo2004] W. Magnus, A. Karras, and D. Solitar, *Presentations of Groups in Terms of Generators and Relations*, Dover Publications, 2004.

[Ma1999] R. Matthews, *The power of one*, New Scientist, 10 July 1999, p. 26.

[Ma1995] P. Mattila, *Geometry of sets and measures in Euclidean spaces. Fractals and rectifiability.* Cambridge Studies in Advanced Mathematics, 44. Cambridge University Press, Cambridge, 1995.

[Ma1959] B. Mazur, *On embeddings of spheres*, Bull. Amer. Math. Soc. **65** (1959), 59–65.

[Mo2009] R. Moser, *A constructive proof of the Lovász local lemma*, Proceedings of the 41st Annual ACM Symposium on Theory of Computing, 2009, 343–350.

[Na1964] I. Namioka, *Følner's conditions for amenable semi-groups*, Math. Scand. **15** (1964), 18–28.

[ON1963] R. O'Neil, *Convolution operators and* $L(p, q)$ *spaces*, Duke Math. J. **30** (1963), 129–142.

[Po2009] D.H.J. Polymath, A new proof of the density Hales-Jewett theorem, preprint.

[PCM] *The Princeton Companion to Mathematics.* Edited by Timothy Gowers, June Barrow-Green and Imre Leader. Princeton University Press, Princeton, NJ, 2008.

[Ra1959] H. Rademacher, *On the Phragmén-Lindelöf theorem and some applications*, Math. Z. **72** (1959/1960), 192–204.

[RaRu1997] A. Razborov, S. Rudich, *Natural proofs*, 26th Annual ACM Symposium on the Theory of Computing (STOC '94) (Montreal, PQ, 1994). J. Comput. System Sci. **55** (1997), no. 1, part 1, 24–35.

[Ro1982] J.-P. Rosay, *Injective holomorphic mappings*, Amer. Math. Monthly **89** (1982), no. 8, 587–588.

[Ro1953] K. Roth, *On certain sets of integers, I*, J. London Math. Soc. **28** (1953), 104–109.

[Ru1962] W. Rudin, *Fourier Analysis on Groups.* Reprint of the 1962 original. Wiley Classics Library. A Wiley-Interscience Publication. John Wiley & Sons, Inc., New York, 1990.

[Ru1995] W. Rudin, *Injective polynomial maps are automorphisms*, Amer. Math. Monthly **102** (1995), no. 6, 540–543.

[Ru1989] I. Ruzsa, *An application of graph theory to additive number theory*, Sci. Ser. A Math. Sci. (N.S.) **3** (1989), 97–109.

[RuSz1978] I. Ruzsa, E. Szemerédi, *Triple systems with no six points carrying three triangles*, Colloq. Math. Soc. J. Bolyai, **18** (1978), 939–945.

[Sc2006] B. Schlein, *Dynamics of Bose-Einstein Condensates*, preprint.

[Se2009] J. P. Serre, *How to use finite fields for problems concerning infinite fields*, preprint.

[So2000] A. Soshnikov, *Determinantal random point fields*, Uspekhi Mat. Nauk **55** (2000), no. 5 (335), 107–160; translation in Russian Math. Surveys **55** (2000), no. 5, 923–975.

[St1961] E. M. Stein, *On limits of seqences of operators*, Ann. of Math. **74** (1961) 140–170.

[St1970] E. M. Stein, *Singular integrals and differentiability properties of functions.* Princeton Mathematical Series, No. 30 Princeton University Press, Princeton, N.J. 1970

[St1993] E. M. Stein, *Harmonic analysis: real-variable methods, orthogonality, and oscillatory integrals.* With the assistance of Timothy S. Murphy. Princeton Mathematical Series, 43. Monographs in Harmonic Analysis, III. Princeton University Press, Princeton, NJ, 1993.

[St1969] S. A. Stepanov, *The number of points of a hyperelliptic curve over a finite prime field*, Izv. Akad. Nauk SSSR Ser. Mat. **33** (1969) 1171–1181.

[St1948] A. H. Stone, *Paracompactness and product spaces*, Bull. Amer. Math. Soc. **54**, (1948), 977–982.

[Sz2009] B. Szegedy, *Higher order Fourier analysis as an algebraic theory I*, preprint.

[SzTr1983] E. Szemerédi, W. Trotter, *Extremal problems in discrete geometry*, Combinatorica **3** (1983), no. 3–4, 381–392.

[Ta1996] M. Talagrand, *Concentration of measure and isoperimetric inequalities in product spaces*, Inst. Hautes Études Sci. Publ. Math. No. **81** (1995), 73–205

[Ta2005] M. Talagrand, *The Generic Chaining. Upper and Lower Bounds of Stochastic Processes.* Springer Monographs in Mathematics. Springer-Verlag, Berlin, 2005.

[Ta1951] A. Tarski, *A Decision Method for Elementary Algebra and Geometry*, 2nd ed. University of California Press, Berkeley and Los Angeles, Calif., 1951.

[Ta] T. Tao, *Summability of functions*, unpublished preprint.

[Ta2006] T. Tao, *Nonlinear Dispersive Equations. Local and Global Analysis.* CBMS Regional Conference Series in Mathematics, 106. Published for the Conference Board of the Mathematical Sciences, Washington, DC; by the American Mathematical Society, Providence, RI, 2006.

[Ta2006b] T. Tao, *A quantitative ergodic theory proof of Szemerédi's theorem*, Electron. J. Combin. **13** (2006), no. 1.

[Ta2006c] T. Tao, *Szemerédi's regularity lemma revisited*, Contrib. Discrete Math. **1** (2006), no. 1, 8–28.

[Ta2007] T. Tao, *A correspondence principle between (hyper)graph theory and probability theory, and the (hyper)graph removal lemma*, J. Anal. Math. **103** (2007), 1–45.

[Ta2007b] T. Tao, *Structure and randomness in combinatorics*, Proceedings of the 48th annual symposium on Foundations of Computer Science (FOCS) 2007, 3–18.

[Ta2008] T. Tao, *Structure and Randomness: pages from year one of a mathematical blog*, American Mathematical Society, Providence RI, 2008.

[Ta2009] T. Tao, *Poincaré's Legacies: pages from year two of a mathematical blog*, Vols. I, II, American Mathematical Society, Providence RI, 2009.

[Ta2010] T. Tao, *The high exponent limit $p \to \infty$ for the one-dimensional nonlinear wave equation*, preprint.

[Ta2010b] T. Tao, *A remark on partial sums involving the Möbius function*, preprint.

[Ta2010c] T. Tao, *Sumset and inverse sumset theorems for Shannon entropy*, preprint.

[TaVu2006] T. Tao, V. Vu, *On random ± 1 matrices: singularity and determinant*, Random Structures Algorithms **28** (2006), no. 1, 1–23

[TaVu2006b] T. Tao, V. Vu, *Additive Combinatorics.* Cambridge Studies in Advanced Mathematics, 105. Cambridge University Press, Cambridge, 2006.

[TaVu2007] T. Tao, V. Vu, *On the singularity probability of random Bernoulli matrices*, J. Amer. Math. Soc. **20** (2007), 603–628.

[TaWr2003] T. Tao, J. Wright, *L^p improving bounds for averages along curves*, J. Amer. Math. Soc. **16** (2003), no. 3, 605–638.

[Th1994] W. Thurston, *On proof and progress in mathematics*, Bull. Amer. Math. Soc. (N.S.) **30** (1994), no. 2, 161–177.

[To2005] C. Toth, *The Szemerédi-Trotter Theorem in the Complex Plane*, preprint.

[Uc1982] A. Uchiyama, *A constructive proof of the Fefferman-Stein decomposition of BMO (R^n)*, Acta Math. **148** (1982), 215–241.

[VuWoWo2010] V. Vu, M. Wood, P. Wood, *Mapping incidences*, preprint.

[Wo1995] T. Wolff, *An improved bound for Kakeya type maximal functions*, Rev. Mat. Iberoamericana **11** (1995), no. 3, 651–674.

[Wo1998] T. Wolff, *A mixed norm estimate for the X-ray transform*, Rev. Mat. Iberoamericana **14** (1998), no. 3, 561–600.

[Wo2003] T. Wolff, *Lectures on Harmonic Analysis.* With a foreword by Charles Fefferman and preface by Izabella Laba. Edited by Laba and Carol Shubin. University Lecture Series, 29. American Mathematical Society, Providence, RI, 2003.

Index